AF753316

GENETIC TRANSLOCATIONS AND OTHER CHROMOSOME ABERRATIONS

GENETIC TRANSLOCATIONS AND OTHER CHROMOSOME ABERRATIONS

GUY T. LEYDEN
EDITOR

Nova Biomedical Books
New York

Copyright © 2008 by Nova Science Publishers, Inc.

All rights reserved. No part of this book may be reproduced, stored in a retrieval system or transmitted in any form or by any means: electronic, electrostatic, magnetic, tape, mechanical photocopying, recording or otherwise without the written permission of the Publisher.

For permission to use material from this book please contact us:
Telephone 631-231-7269; Fax 631-231-8175
Web Site: http://www.novapublishers.com

NOTICE TO THE READER

The Publisher has taken reasonable care in the preparation of this book, but makes no expressed or implied warranty of any kind and assumes no responsibility for any errors or omissions. No liability is assumed for incidental or consequential damages in connection with or arising out of information contained in this book. The Publisher shall not be liable for any special, consequential, or exemplary damages resulting, in whole or in part, from the readers' use of, or reliance upon, this material.

Independent verification should be sought for any data, advice or recommendations contained in this book. In addition, no responsibility is assumed by the publisher for any injury and/or damage to persons or property arising from any methods, products, instructions, ideas or otherwise contained in this publication.

This publication is designed to provide accurate and authoritative information with regard to the subject matter cover herein. It is sold with the clear understanding that the Publisher is not engaged in rendering legal or any other professional services. If legal, medical or any other expert assistance is required, the services of a competent person should be sought. FROM A DECLARATION OF PARTICIPANTS JOINTLY ADOPTED BY A COMMITTEE OF THE AMERICAN BAR ASSOCIATION AND A COMMITTEE OF PUBLISHERS.

Library of Congress Cataloging-in-Publication Data

Genetic translocations and other chromosome aberrations / Guy T. Leyden, editor.
p. ; cm.
Includes bibliographical references and index.
ISBN 978-1-60456-344-3 (hardcover)
1. Translocation (Genetics) I. Leyden, Guy T.
[DNLM: 1. Translocation, Genetic. 2. Chromosome Aberrations. 3. Chromosome Disorders. 4. Neoplasms--genetics. QS 677 G3278 2008]
QH462.T7G46 2008
572.8'77--dc22

2007050811

Published by Nova Science Publishers, Inc. ✣ New York

Contents

Preface

Chromosomal aberrations are disruptions in the normal chromosomal content of a cell, and are a major cause of genetic conditions in humans, such as Down syndrome. Some chromosome abnormalities do not cause disease in carriers, such as translocations, or chromosomal inversions, although they may lead to a higher chance of having a child with a chromosome disorder. Abnormal numbers of chromosomes or chromosome sets, aneuploidy, may be lethal or give rise to genetic disorders. Genetic counseling is offered for families that may carry a chromosome rearrangement. This new book offers leading-edge research from around the world.

As presented in Chapter 1, recurrent gene translocations occur in 50%-70% of AML cases. Of these cyogenetic abnormalities, the most frequent are the t(15;17), t(8;21), inv(16)/t(16;16) and 11q23 aberrations. More importantly, the genes affected by these translocations encode transcription factors or cofactors that govern various aspects of normal hematopoiesis, including self-renewal, proliferation, differentiation and apoptosis. The most frequently implicated transcription factors include RARα, AML-1, CBFβ and MLL. The resulting chimeric proteins acquire new properties in addition to the alteration of the transcriptional activity of the molecules involved. Interestingly, although these fusion proteins play major roles in inducing the leukemic phenotype by disrupting the normal hematopoietic differentiation process and giving rise to an immortalized leukemic stem cells or progenitors, they are not sufficient alone for full leukemic transformation. In fact, additional molecular dysregulations that control proliferation and/or apoptosis, such as constitutive activation of the tyrosine kinase Flt3, increased activity of IL-3 pathway, overexpression of the antiapoptotic gene BCL2 or upregulation of the transcription factor plagl2, seem to be necessary mechanisms for overt leukemic transformation.

Soft tissue tumors are rare and a heterogeneous group of neoplastic lesions that mostly display variable morphologic features of non-epithelial or mesenchymal tissues. Because of the rarity and morphologic diversity, the diagnoses of soft tissue tumors are often challenging and difficult to make. Besides, pathogenetic mechanisms underlying the development of soft tissue tumors still remain largely unknown. In the past decades, great strides in our understanding of molecular or genetic aberrations of various neoplastic diseases have been made. One of the most profound impacts in this field is an identification of tumor-associated chromosomal translocations that result in formation of unique fusion (or chimeric) genes. It has been also shown that such aberrations are specifically found in certain morphologic types of neoplasms, many of which are belonging to soft tissue tumors with malignant biological potential, namely sarcomas. Currently more than ten types of soft tissue sarcomas are known to be translocation-associated. Many tumor-associated chromosomal translocations or fusion genes have been repeatedly and increasingly detected by standard karyotyping in combination with recent molecular techniques such as fluorescence *in situ* hybridization and reverse transcription-polymerase chain reaction, which can be also applied to archival formalin-fixed, paraffin-embedded tumor tissues. In addition to an insight into their oncogenic roles, molecular detection of such cytogenetic aberrations can provide valuable information with respect to tumor diagnosis, because such alterations are most often tumor type-specific. Notably, most translocation-associated tumors have some variations in the chromosomal regions or genes involved by the translocations or in resultant fusion gene transcripts. This phenomenon implies that morphologic features of the tumors and their biological behavior or therapeutic susceptibility may also vary according to such cytogenetic or molecular variations. In Chapter 2, summarized and updated molecular genetic and clinicopathologic features of translocation-associated soft tissue sarcomas are described.

As explained in Chapter 3, the sex chromosomes of the silkworm, *Bombyx mori*, are designated ZW (XY) for females and ZZ (XX) for males. The female mode of development is determined by the presence of a single W chromosome. Therefore, it is presumed that the female-determining gene (*Fem*) is present on the W chromosome. This chromosome does not recombine with either the Z chromosome or autosomes and as yet, no genes for morphological traits have been mapped to the W chromosome. The silkworm has been reared in Japan for more than 1,700 years. As heterosis helps to ensure stable cocoon production, almost all silkworm strains reared at sericultural farms are F_1 hybrids, a practice that has been in place since about 1910. Therefore, discriminating the sexes of would-be silkworm parents for the purpose of controlled hybridization is very important. This was traditionally accomplished by inspecting the post-ventral surface of final instar larvae. However, the method was problematic and inconvenient. For this reason, identification of a translocation to the W chromosome that carries one or more dominant markers would be helpful. Several researchers undertook the challenge of inducing translocation of an autosomal fragment containing dominant alleles of various genes (including genes that control egg color, larval marking, and cocoon color) to the W chromosome using X-ray or γ -ray irradiation, and were successful. One of the W-translocation chromosomes, the sex-limited Zebra-W strain or T(W;3)*Ze*, was thought to be the product of reciprocal translocation between W and chromosome 3. Thus, the W chromosome in this strain should contain deletions; however this cannot be confirmed using conventional genetic methods. To further test for the presence of a deletion, the authors compared the T(W;3)*Ze* chromosome to the parental W chromosome using 12 W-specific random amplified polymorphic DNA (RAPD) markers. The T(W;3)*Ze* chromosome lacked two W-specific RAPD markers. These results strongly indicate that the region containing these two W-specific RAPD markers was deleted. Additionally, the authors investigated T(W;10)$+^{w-2}$ (the sex-limited Black-egg-W strain) and T(W;2)*Y* (the sex-limited Yellow-cocoon-W strain). The T(W;10)$+^{w-2}$ chromosome lacks one W-specific RAPD marker and the T(W;2)*Y* chromosome lacks 11 W-specific RAPD markers. These results indicate that the W chromosomes of sex-limited strains are not the results of simple fusion of an autosomal fragment to the end of an unaltered W chromosome but rather, they are the products of reciprocal translocation accompanied by deletion. Furthermore, the authors investigated a female-killing chromosome derived from the translocation-carrying W chromosome and found that this chromosome was generated by translocation between the W and Z chromosomes. Our analysis of W chromosome variants helped to elucidate the relative positions of molecular markers and the putative *Fem* gene(s).

Apoptosis is a programmed cell death that plays an important role in the physiological process of development in animals and plants, and has also been reported to be involved in the pathogenesis and pathophysiology of various diseases. It is expected that control of apoptosis will be an important goal of new therapies for apoptosis related diseases. Glaucoma is a progressive optic neuropathy related to apoptosis that finally leads to blindness. Clinical features of the disease include visual field loss and a large cupping of the optic nerve head due to a glaucoma-specific deterioration of the retinal nerve fibers, which consist of ganglion cell axons, in association with elevated intraocular pressure (IOP). Although abnormally elevated IOP is a risk factor, apoptosis of retinal ganglion cells is a cause of the disease. The relationships between neuro-retinal tissue damage and apoptosis have been investigated and clarified in several experimental models, such as N-methyl-D-aspartate (NMDA)-induced retinal damage and secondary cell death caused by transection of the optic nerve, as well as in ocular hypertensive rats. Thus, the prevention of apoptosis of retinal ganglion cells is considered to be a novel strategy for treatment of glaucoma. Nipradilol [3,4-dihydro-8-(2-hydroxy-3-isopropyl-amino)-propoxy-3-nitroxy-2H-1-beozopyran]is a nonselective β- and selective α_1-adrenoceptor antagonist that has been used as an anti-glaucomatous agent due to its ability to lower IOP. Recently, it was reported that nipradilol also had a cytoprotective effect toward neuronal cells by nitric oxide (NO) donative action. NO has been shown to regulate apoptosis in both positive and negative directions in several experimental models, and many reports have shown that NO inhibits apoptosis by increasing caspase-3-like activity and cGMP-mediated mechanisms, as well as by inhibiting Bcl-2 cleavage and cytochrome *c* release *in vitro*. The gene expressions of some apoptosis associated factors are also regulated by the activity of the transcription factor nuclear factor-κB (NF-κB). However, whether nipradilol regulates the expression of apoptosis associated genes and/or NF-κB activity remains unclear. In Chapter 4, the authors found that nipradilol down-regulated Bax, Caspase-9, and Smac/DIABLO gene expression using a real-time polymerase chain reaction method, and increased levels of translocation of NF-κB into nuclei were shown using an electrophoresis mobility shift assay of PC12 cells during apoptosis induced by serum deprivation. Activation of NF-κB was seen prior to the alterations of gene expression, therefore, the down-regulation of some apoptosis related genes in PC12 cells by nipradilol may be the result of NF-κB activation. Our results suggest that nipradilol regulates apoptosis associated gene expression and the translocation of NF-κB into nuclei as a means of neuro-protection against apoptosis via NO donative activity.

By *PRINS,* unlabeled oligonucleotide primers are annealed to denatured complementary sequences on chromosomes *in situ,* and extended in the presence of labeled nucleotides and Taq DNA polymerase, as explained in Chapter 5. Chromosomal DNA acts as template for the extension, and the labeled nucleotides act as substrate for the DNA polymerase. The labeled extended target is visualized by fluorescence microscopy. Thus PRINS can be used to detect subtle chromosome rearrangements and microdeletions that are not amenable to analysis by standard analytical methods such as FISH. The PRINS method can also be used to identify single copy sequences and individual genes.

A self-organization mechanism of establishment of interphase chromosome organization formerly proposed for large-scale interphase chromatin/chromosome self-organization is investigated in Chapter 6 by computer simulation. Large-scale structure of interphase chromosome is suggested to be determined by volumetric interactions between chromosome subunits resulting in folding of subunit chain in a globular state. Structure units of interphase chromosome are studied as a spherical nonoverlapped structures consisting of a cluster of compact chromatin domains. These structure units/domain clusters were termed introduced as a superdomains (SD) to emphasize their supradomain organization as well as domain cooperative behaviour within SDs. Unlike Gauss chain model for interphase chromosomes it is proposed the existence of globular state for interphase chromosomes. Globular state of interphase chromosomes is studied by dynamic Monte-Carlo approach. superdomain chain (SD chain) with excluded volume as well as together with phase transitions between condensed, globular and decondensed, coiled states. Globular model/structural state is supported in agreement with by confocal microscopy data. It extends current view on interphase chromosome as a random ideal Gauss polymer chain. Radiation-induced intrachromosome exchange aberrations (intrachanges) are modeled on the basis of the contact hypothesis. It implies that Intrachanges are resulted with high frequency by misrejoining of damaged chromatin subunits (SDs) being in contacts. Calculated intrachange frequencies for X-rays agree well with experimental data. Results of calculations for globule provide biophysical modeling evidence that high frequency of localized multiple damaged sites per chromosome exists leading to enhanced interactions within the chromosome may be responsible for observed high frequency of interstitial deletions visible as a fragments at low doses of alpha particles.

As explained in Chapter 7, structural chromosomal aberrations such as deletions, duplications, isochromosomes and translocations are common in epithelial tumors. However, the mechanisms leading to tumor type- and stage-specific chromosomal aberrations, as well as their biological role in tumor development and progression, are not well understood.

The authors discovered that in squamous cell carcinoma cell lines of the head and neck up to 40% of all chromosomal breakpoints occur in centromeric regions leading to whole arm translocations. When the authors studied epithelial tumors of other histological type and origin, such as colorectal adenocarcinomas, they found only 10% of all rearrangements to be whole arm translocations. More importantly, the authors demonstrated tissue-specific differences in the formation of these centromeric translocations. In head and neck squamous cell carcinoma, whole arm translocation chromosomes contained centromeric sequences of both contributing chromosomes, whereas in colorectal adenocarcinoma centromeric material of only one chromosome was present. This suggests that the exact location of DNA breakage and refusion resulting in whole arm translocation chromosomes differs between squamous cell and adenocarcinomas. It occurs within centromeric repetitive sequences in head and neck squamous cell carcinomas and, in contrast, within pericentromeric sequences in colorectal adenocarcinoma.

The mechanisms leading to these types of chromosomal abnormalities are unknown, but clues can be found in two genetic instability syndromes. First, Fanconi Anemia (FA) predisposes to acute myeloid leukemia (AML) and head and neck squamous cell carcinoma. Centromeric translocations in head and neck squamous cell carcinoma from FA patients are as frequent as in sporadic head and neck squamous cell carcinoma and mutations in FA genes have been found in sporadic head and neck squamous cell carcinoma. FA cells are typically hypersensitive to DNA crosslinking agents, which is related to the role of FA genes in homologous repair. Second, centromeric instability of chromosomes 1, 9 and 16 is a characteristic of ICF syndrome (immunodeficiency, centromere instability, facial abnormalities) and is caused by hypomethylation of heterochromatic classical satellite DNA, due to mutations in a DNA methyltransferase protein.

Therefore it may be speculated that sporadic head and neck squamous cell carcinoma, or at least a subgroup of these tumors, have a defect in homologous repair leading to centromeric translocations upon exposure to crosslinking agents present in tobacco (a strong etiological factor in head and neck squamous cell carcinoma). Alternatively, sporadic head and neck squamous cell carcinoma may have hypomethylation of centromeric repeat sequences, making them vulnerable for DNA damaging compounds, or they may have a combination of centromeric hypomethylation and defective homologous repair which together lead to frequent centromeric breakage and illegitimate recombination.

Fragile sites on human chromosomes were first described in 1965 by Dekaban et al. In the early years of human cytogenetics these were rather enigmatic phenomena that were seen by some laboratories and not by others. It was not until 1977 that it was demonstrated that the type of culture medium in which lymphocytes were cultured to prepare chromosomes was a critical factor in determining whether fragile sites were seen (Sutherland 1977). It was subsequently shown that folic acid and thymidine needed to be absent from the culture medium for fragile sites to manifest themselves (Sutherland 1979). Some rare fragile sites were found not to be sensitive to folic acid but inducible by a range of other compounds.

In this chapter the classification of fragile sites will be given, the cytogenetics presented, the clinical significance of the rare fragile sites discussed and the inheritance and epidemiology of fragile sites presented.

The identification of the non-random genetic alterations associated with cancer has great importance not only to basic research but also to disease management as diagnostic and prognostic markers. In contrast to other types of lymphomas, translocations are rare events in chronic lymphocytic leukemia/small lymphocytic lymphoma (CLL/SLL) and not any CLL/SLL-specific translocation has been found yet. Moreover, translocations have been reported to be a prognostic marker for an unfavorable clinical outcome. The purpose of our study was to contribute to the understanding of the chromosome alterations associated with this group of lymphoid malignancies. The combination of conventional cytogenetics with FISH (fluorescence *in situ* hybridization) analysis allowed us a more accurate definition of the genomic aberrations involved in CLL/SLL. Novel and uncommon chromosome rearrangements were found in our series. In the Chapter 9, the authors describe the cytogenetic and FISH analyses of 12 CLL/SLL patients out of 150 cases with successful cytogenetic study, that interestingly showed in their karyotypes new or infrequent clonal chromosome rearrangements, all of them associated to adverse clinical course with partial or null response to different chemotherapeutic agents. The authors identified 16 different chromosome rearrangements and translocations were the most frequent aberration (7/16; 43,8%). Some of these findings were recently reported by us as a new alterations in SLL/CLL patients: t(2;17)(p21;q23); der(17)t(12;17)(q13;q25); der(14)(t(11;14)(q13;q24), t(5;12;19)(q15;p11;q 13); psu dic (17;2)(p11.2;p21), psu dic (12;21)(q24;q10), t(10;14)(q22;q32), del(3)(p11) and del(5)(q15). Translocation t(11;13)(p15;q22) and del(18)(p11) were not previously described, meanwhile, dic(17;18)(p11.2;p11.2), dup(12)(q13q24), del(7)(q34), t(2;14)(p11;q32) and del(7)(q11) were infrequently reported in the literature. Translocation t(2;17)(p21;q23) is recurrent in our series and t(2;14)(p11;q32) became recurrent from our data. Chromosomes 17, 2, 12 and 14 were the most frequently involved. Twenty one different breakpoints were found and 38.1% (8/21) of them, such as 2p21, 5q15, 17p11.2, 17q23, 12q13, 12q24, 14q32 and 18p11 were recurrent in our series. All but one patient of our series showed an unfavorable clinical outcome. A detailed analysis and a review of the cytogenetic and molecular cytogenetics findings in CLL/SLL were performed. Our findings may facilitate the understanding of genetic events associated with development and progression of CLL/SLL and may have implications in the clinical management of patients with this disease.

As explained in Chapter 10, the authors have reviewed 4884 patients with different hematologic disorders, cytogenetically studied in our laboratory during the last 27 years, in order to analyze the frequency of constitutional chromosome aberrations (CCA) and to evaluate their association with hematologic malignancies. Our population included 1235 pediatric patients and 3649 adults. Twenty-six (0.53%) cases showed CCA. They included 4 patients with Robertsonian translocations, one patient with a balanced translocation, 2 patients with sex chromosome abnormalities and 19 cases with Down´s Syndrome (DS). Non significant differences between the frequency of patients with CCA from our hematologic series and those observed in the two largest combined published surveys of liveborns (0.65%-0.78%) were found. The incidence of DS patients in our population (0.38%) was about three times higher than that seen at birth (0.12%-0.14%) ($p<0.001$). The total incidence of constitutional chromosome abnormalities in the non DS hematologic patients was 0.14% (7/4884) lower than that observed in the newborn population (0.53%-0.64%) ($p<0.001$), supporting the idea that DS patients are at greater risk of developing acute leukemia than normal subjects of the same age. Non-significant differences were found when the incidences of structural aberrations and sex chromosome anomalies were individually compared with the data of the overall population. Moreover, our hematologic population was stratified in adults (over 17 years of age), children (1-17 years of age) and infants (0-1 year of age). Higher frequencies of chromosome aberrations in children and infants compared to adults ($p<0.0001$) and in infants respect to children ($p<0.05$) were found. These differences were related to the DS patients that showed a clear different distribution by age: infants had the highest incidence compared to adults ($p<0.0001$) and children ($p<0.03$). In conclusion, our results did not show significant differences in the incidence of individuals with CCA compared to the overall population, suggesting that the presence of a CCA, other than DS, would not predispose hematologic malignancies.

In: Genetic Translocations and Other Chromosome Aberrations
Editor: Guy T. Leyden, pp. 1-63
ISBN: 978-1-60456-344-3
© 2008 Nova Science Publishers, Inc.

Chapter 1

The Role of Gene Translocations in the Molecular Pathogenesis of Acute Myeloid Leukemia

Haytham Khoury
Department of Stem Cell and Developmental Biology
Princess Margaret Hospital, Room 9-111
610 University Ave.
Toronto, ON, M5G 2M9 Canada

Abstract

Recurrent gene translocations occur in 50%-70% of AML cases. Of these cyogenetic abnormalities, the most frequent are the t(15;17), t(8;21), inv(16)/t(16;16) and 11q23 aberrations. More importantly, the genes affected by these translocations encode transcription factors or cofactors that govern various aspects of normal hematopoiesis, including self-renewal, proliferation, differentiation and apoptosis. The most frequently implicated transcription factors include RARα, AML-1, CBFβ and MLL. The resulting chimeric proteins acquire new properties in addition to the alteration of the transcriptional activity of the molecules involved. Interestingly, although these fusion proteins play major roles in inducing the leukemic phenotype by disrupting the normal hematopoietic differentiation process and giving rise to an immortalized leukemic stem cells or progenitors, they are not sufficient alone for full leukemic transformation. In fact, additional molecular dysregulations that control proliferation and/or apoptosis, such as constitutive activation of the tyrosine kinase Flt3, increased activity of IL-3 pathway, overexpression of the antiapoptotic gene BCL2 or upregulation of the transcription factor plagl2, seem to be necessary mechanisms for overt leukemic transformation.

Introduction

Although AML rarely arises from recognizable heritable genetic alterations, it should be considered a disease of genetic origin [1,2]. Somatic aberrations in form of chromosomal abnormalities and/or nucleotide mutation could, virtually, be found in all cases of AML [3,4,5,6]. An archetype of these alterations is gene rearrangements that result from chromosomal translocations, inversions or submicroscopic insertions. Up to date, more than 100 translocations/inversions have been identified in AML. Of these, the most commonly found are the balanced translocations t(15;17), t(8;21), inv(16)/t(16;16) and various aberrations involving 11q23 [7]. These gene rearrangements account for more than 80% of all balanced translocations encountered in AML [7].

It has been recognized that these specific cytogenetic abnormalities define distinct clinico-pathological entities associated with specific morphological features, immunophentypic characteristics and clinical behavior, particularly response to therapy and survival. Thus, this assiciation has given evidence for a correlation between phenotype and genotype, with a possible causal relationship, and also provided ground for a clinically meaningful AML classification [8].

Moreover, the molecular cloning of these rearrangements has permitted to identify genes that play important roles in different cellular processes that govern the self-renewal, growth, differentiation, and survival of normal embryonic and adult hematopoietic cells [9,10,11,12]. In these rearrangements, one gene encodes a transcription factor, whereas the other gene is most likely to code for a different transcription factor or a molecule with a transcription coregulatory function [13]. The resulting hybrid proteins lead to functional disruption of the transcription factors involved through recruitment of corepressors, formation of dimers or multimers with potent transactivation efficiency or aberrant subnuclear localization leading to diversion of these transcription factors from their normal targets. These mechanisms have been extensively investigated in the recent years, and there is large amount of evidence supporting their relevance for the pathogenesis of AML. Interestingly, multiple *in vivo* studies using transgenic mice have shown that although these chimeric genes are crucial for the pathogeneis of AML, they are not sufficient for full leukemic transformation. Indeed, they act in concert with other oncogenic events in multistep processes to culminate in full leukemic transformation [14].

Indeed, our current understanding to the molecular pathogenesis of AML, however partial, has provided insights into the mechanism of action of some successfully employed targeted-therapy, namely ATRA [15,16], and fueled research efforts to develop more agents aimed to alter the leukemic molecular pathways [17], offering promise to design new treatment regimens that are more effective and less toxic than the currently available therapies [17,18].

The t(15;17)(q22;q21)

The t(15;17) Occurs in 10-25% of AML. It is invariably associated with a distinct AML subtype, the acute promyelocytic leukemia (APL). APL has specific clinical, morphological and immunophenotypic characteristics. Clinically, it is characterized by low white blood count, increased incidence of potentially fatal coagulopathy and remarkable response to all-trans-retinoic acid (ATRA) and arsenic trioxide (AS_2O_3) [19]. The cytological hallmark of APL is the accumulation of malignant myeloid cells blocked at the stage of promyelocytes that are hypergranulated or contain multiple elongated Auer rods [20]. Also, specific immunophenotypic features has bee described in APL, including absence of HLA-DR expression, low frequency of CD34 and consistent expression of CD13, CD33 and MPO [21]. The t(15;17) leads to a fusion between the RARα and PML genes.The RARα is a 10-exon gene located on 17q21. It possesses two promoters that control the expression of two isoforms

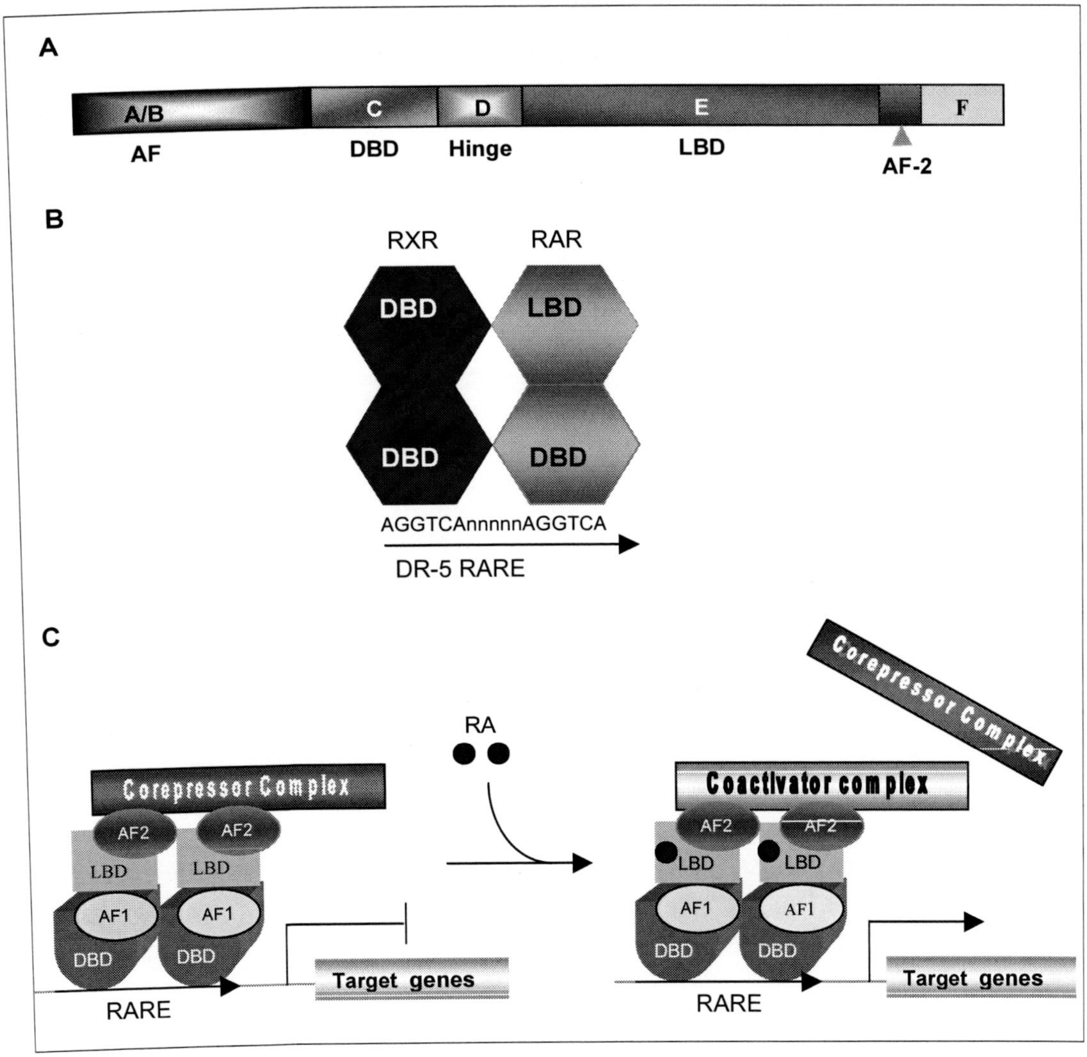

Figure 1. The structure and functional activity of the RARα molecule. (A) Major functional domains of retinoid receptors. The A/B N-terminus represents an autonomous activation domain, named activation function-1 (AF-1). The C segment is the DNA-binding domain (DBD). The D region acts as a flexible hinge. The E segment represents the Ligand-binding domain (LBD) and activation function-2 (AF-2) motif. The C-terminal tail (F) is of unknown function. (B) The heterodimer RARα-RXR binds to specific consensus sequence, the retinoic acid response element (RARE), on the promoter of target genes. This figure illustrates a DR5 RARE that contains a 6-bp direct repeats separated by 5 nucleotides. (C) A model for RARα-RXR acting as transcription repressor or transactivator. In the absence of RA (left), RARα-RXR binds to a corepressor complex leading to silencing of the target genes. When RA binds to the heterodimeric receptor, the latter releases the corepressor complex and recruits a co-activator complex leading to gene transactivation.

(RARα1 and RARα2) [22]. The RARα2 transcript differs from RARα1 by lacking the first and second exons. Although expression of RARα1 is ubiquitous, the RARα2 is expressed in an ATRA-dependent and tissue-specific pattern.

The RARα protein is a member of the retinoid receptor family that contains the retinoic acid receptors (RARα, β, and γ) and retinoid "X" receptors (RXRα, β, and γ). These receptors belong to the nuclear hormone receptor superfamily, which includes also the steroid receptors, the thyroid hormone receptors (TR), the vitamin D receptors (VDR) and the peroxisome-proliferator activated receptors (PPARs). The members of the nuclear receptor superfamily share specific functional domains, including the DNA-binding and ligand-binding domains, and function as ligand-dependant transcription factors [24]. Retinoid receptors are usually divided into 6 distinct structural domains (designated with letters from A to F) [25-27] (figure 1A). The N-terminal A/B regions contain a ligand-independent transactivation domain named also AF1. The ligand-independent transactivation domain is followed by the C segment, which contains two zinc-finger motifs responsible for DNA binding and is thus called the DNA-binding domain (DBD). At the base of the first finger, there is a box responsible for discriminating the DNA sequence, whereas the base of the second finger participates in receptor dimerization. Next to the DNA-binding domain, there is a flexible hinge region (segment D) that contains a nuclear localization signal. The E region is located near the C-terminus and possesses a ligand-dependant transactivation function and is thus named the ligand-binding domain (LBD). The 3D structure of this domain is folded to create a pocket for high-affinity ligand binding. The E domain is also responsible for heterodimerization and interaction with coregulator (coactivator and cosupressor) proteins. At the end of the LBD region, there is a small region called AF-2, which is responsible for ligand-induced conformational changes. Finally, at the C-terminal tail there is the (F) segment, which is of unknown function.

Normally, an RAR protein exists as heterodimer with a member of the RXRs (figure 1B). The RXRs can homodimerize or heterodimerize with other nuclear receptors, including RARs, VDR, TR and PPARs. The heterodimerization process of RAR and RXR molecules is critical to enhance the DNA-binding affinity and consequently to increase the transcriptional efficiency of RARs [28]. The ligand specificities of these receptors differ. RARs bind to *all-trans*-retinoic acid (ATRA) with high affinity and to *9-cis*-retinoic acid (9-cis-RA) with lower affinity, whereas RXRs bind mainly to 9-cis-RA [29]. The effects of vitamin A and its derivatives on gene transcription are presumably mediated by ATRA, since 9-cis-RA is undetectable in tissues *in vivo*. Furthermore, the activity of the heterodimers RAR-RXR is under the control of RAR molecules, because RXR-selective ligands do not activate transcription; a phenomenon called "RARs dominance" [30]. To regulate gene transcription, an RAR-RXR complex must bind to a specific retinoic acid responsive elements (RARE) in the regulatory regions of specific target genes. Both components of heterodimer are required for efficient DNA-binding activity. The RAREs are usually composed of direct repeats (DR) of the consensus half-site sequence AGGTCA separated by either 5 (DR5) or 2 (DR2) nucleotides [31] (figure 1B). This consensus sequence has been found in the regulatory elements of many genes that play important roles in hematopoiesis and are known to be regulated by retinoic acid, including C/EBPβ, C/EBPε, *p*21$^{WAF1/CIP1}$, STAT-1, STAT2, P48, INF, INF-responsive-1 (IRF-1), many members of the HOX family genes and the RARα2 isoform [32-40]. In absence of retinoic acid (RA), the RARα-RXR heterodimer recruits a repression complex that contains a nuclear-receptor corepressor (SMRT or N-CoR), a transcriptional corepressor (Sin3A or Sin3B) and a histone deacetylase (HDAC1 or HDAC2). This complex induces deacetylation of lysine residues in core histones leading to chromatin condensation and transcription repression. Upon ATRA binding, even at small doses (physiological doses), the receptor undergoes conformational change and the structure becomes much more compact. This conformational change results in alteration in protein-protein interaction due to change in AF-2 domain position leading to dissociation of the repression complex and the subsequent recruitment of a coactivation complex containing a nuclear receptor coactivator from the p160 family member and the histone acetyltransferase CBP/p300 (figure 1C). In addition, two other histone acetylases, ACTF and P/CAF, are recruited to this complex in a ligand-dependent manner. Together these three enzymatic units act cooperatively to acetylate lysine residues in histone proteins leading to chromatin unfolding and gene transactivation. It has been shown that the three RAR molecules have different ability to bind to corepressors. RARα binds strongly to SMRT, whereas RARβ and RARγ interact weakly with this corepressor. In fact, this difference may account for the different transcriptional properties of the three RAR molecules in the absence of hormone ligand. Under this condition RARα is a strong repressor of the target genes, whereas both RARβ and RARγ fail to repress and mediate substantial levels of hormone-independent transcriptional activity [41].

There is extensive evidence to postulate that retinoids and retinoid receptors play important roles in hematopoiesis. It has been shown, using colony assays, that retinoic acid enhances the granulocytic-progenitors growth and differentiation and decreases the production of colonies representing other hematopoietic lineages, particularly erythroid and monocytic [42-44]. This effect on granulocytic lineage seems to be selective to the cytokine used. For example, ATRA enhances the generation of IL-3 and GM-CSF dependent colonies, while it has no effect on colonies stimulated by G-CSF [45]. These findings suggest that ATRA regulate the proliferation and differentiation of relatively early stages of myelopoiesis. This is pertinent with the differentiation block seen upon inhibiting the retinoid receptors and with differentiation arrest seen in APL [46]. In liquid culture, it has been demonstrated that ATRA enhances the proliferation of human $CD34^+$ cells and the repopulating activity (CFU-S, short-term and long-term) of mouse stem cells (lin^-, c-kit+, $Sca\text{-}1^+$) [47-49]. Taken together, these findings suggest that retinoid receptors not only play an important role in the myeloid proliferation and differentiation but also in stem cells self-renewal capacity. In addition, the effect on hematopoiesis of selectively disrupted RA receptor has been explored *in vivo* using knock-out mouse models. In contrast to RARα and RARγ, the receptor RARβ is poorly expressed in myeloid cell. Therefore, knock-out studies assessing the role of the RAR mplecules in hematopoiesis have focused on the first two receptors. Interestingly, $RAR\alpha^{-/-}$ mice displayed early postnatal mortality, but with no apparent disruption of hematopoiesis [50]. Similarly, RARγ homozygous knock-out mice did not show overt hematopoietic abnormalities [51]. Furtheremore, mice that are double homozygous mutant ($RAR\alpha^{-/-}$/ $RAR\gamma^{-/-}$) die *in utero,* and their liver hematopoietic cells consisted of mostly mature granulocytes that display similar morphology and phenotype to wild type mice. Moreover, these animals did not display any enhanced RARβ expression that might compensate for the lack of RARα and RARγ [51]. The normal granulopoiesis observed in the fetal liver of $RAR\alpha^{-/-}$/ $RAR\gamma^{-/-}$ double mutant animals indicates that RARs are not essential and are, in fact, dispensable for granulopoiesis. While there is extensive evidence regarding the importance of the role that RARs play in physiological and pathological myeloid growth and differentiation, relatively little is known concerning the role of RXRs in these processes. The most abundant member of the RXR family of proteins in the myeloid cells is RXRα. As $RXR\alpha^{-/-}$ is lethal *in utero*, and its hematopoiesis is not assessable, the suppression in transgenic mice of the function of heterodimeric receptors composed of RXR and their counterparts was carried out in myeloid cells by using targeted expression of a dominant-negative form of RXRβ. This analysis resulted in severe maturation arrest at the promyelocytic stage in 1 of the 12 mice studied and to mild inhibition of myeloid differentiation in 3, indicating that RXRs may play role in myelopoiesis[52,53]

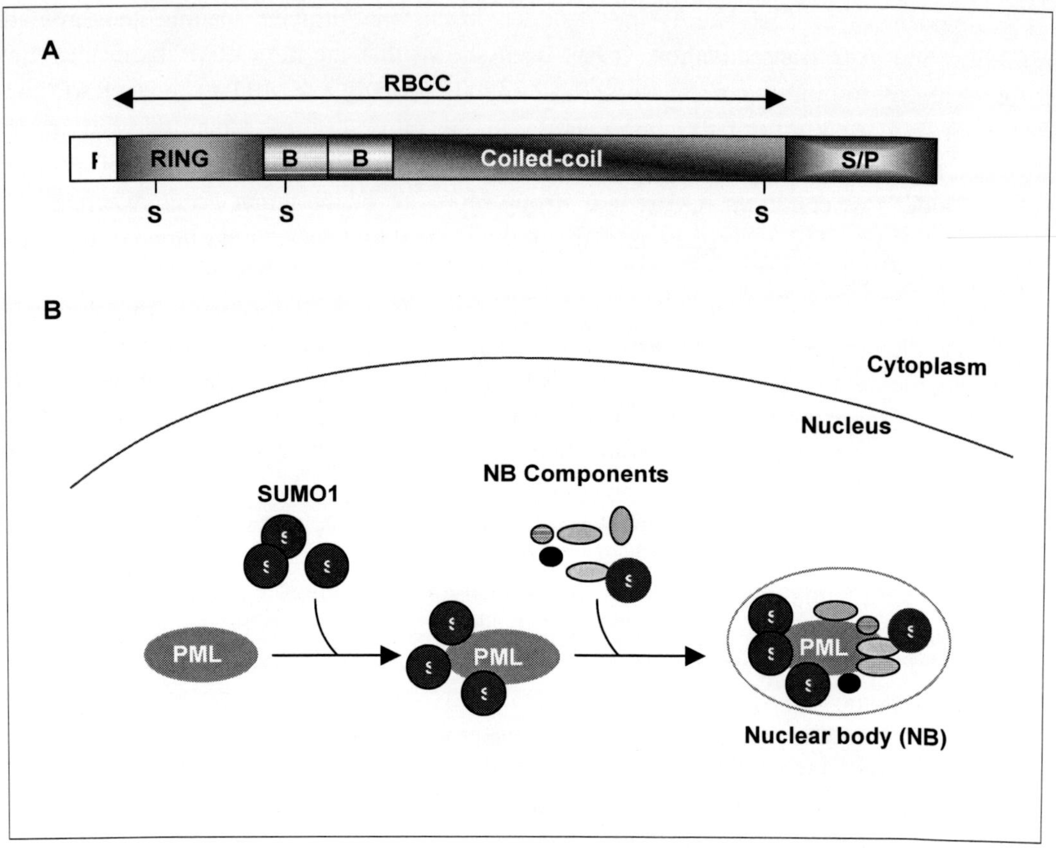

Figure 2. The structure of the PML protein and its contribution to the formation of the nuclear bodies (NBs). (A) Schematic presentation of PML protein structure. N-terminal proline-rich region (P), followed by a RING domain, two B-boxes, a coiled-coil domain and finally a serine/proline-rich region (S/P). The three S sites represent the three lysine residues by which PML binds to SUMO-1. (B) The sequential events that occur in the nucleus and lead to nuclear body (NB) formation. First step is the SUMOylation of PML. Then, SUMOylated PML interacts with the other NB components.

The PML gene is located on 15q22; It was originally identified by de The and his team due to its implication in the t(15;17) (q22;q12). De The and his collaborators in their intial report named this gene MYL (for melocytic leukemia) [55]; however, in a leter paper they changed the name to PML (for promyelocytic leukemia) [9]. This gene is composed of 9 exons with exons 7-9 having potential to generate by alternative splicing multiple isoforms that may have different biological functions [56-61].

The PML protein is a member of RING (Really Interesting New Gene) superfamily. In addition to the RING domain, the N-terminal of PML has a proline rich region and two cysteine/histidine-rich domains known as B-boxes (figure 2A). The B-boxes and the RING domain together form a putative transcriptional activation region. The central part of PML is occupied by α-helical leucine-rich coiled-coil domain. The RING/B box/coiled-coil (RBCC) domain is thought to be responsible for PML dimerization and possibly multimerization [60,62]. The C-terminal domain is a serine/proline-rich region that is variable in length due to transcript alternative splicing. The nuclear localization of sequence has been located in the carboxy terminal region. Although this sequence is sufficient for nuclear translocation, the RBCC domain is required for proper sunuclear localization of PML. In the nucleus, PML concentrates in speckled subnuclear organelles referred as PML nuclear bodies (NBs), Kremer bodies, ND10 (nuclear domains 10) or PODs (PML oncogenic domains). In fact, PML is invariably associated with the NBs in all cell types so far studied. In addition to PML, these bodies contain many other proteins, including SUMO-1, HAUSP, SP100, SP140, CBP, BLM, Daxx, pRb, p53, NDP55, elf-4, Sp-1, PRH and AML1 [61, 63-75]. PML is essential for the formation of the nuclear bodies. Homologous disruption of PML gene leads to aberrant localization pattern of the other components of NBs, including SUMO-1, Sp100, Sp140, CBP and Daxx, whereas lack of the other components of NBs does not affect their normal pattern of protein assembly [69,76]. Moreover, integrity of the RBCC domain is required for PML localization in the NBs. Mutations of this domain prevents PML from forming the NBs. In order to be able to recruit other molecules to form the NBs, PML needs to bind covalently with SUMO1 at three lysine residues [77-79] (figure 2 B). Unmodified PML is associated with the soluble nucleoplasmic fraction, whereas SOMUylated PML is tightly associated with NB [78]. Moreover, mutant PML that can no longer be modified by SUMO-1 fails to induce proper localization of NB components, including Sp100 and Daxx [69]. Other components of the NBs could be subjects for SUMOylation also. However, this process does not seem essential for their integration in the NBs [80,81].

The colocalization of PML with such variety of different proteins suggests that PML could be responsible for diverse cellular processes, including gene expression regulation, apoptosis and genomic stabilization. PML is a pro-apoptotic protein. Over-expression of PML in various cell lines increased apoptosis [82-84]. $PML^{-/-}$ mice and cells are protected from multiple caspase-dependent apoptotic processes [85,86]. In fact, the proapoptotic activity of PML requires intact RBCC domain [87]. Numerous *In vitro* and *in vivo* studies have indicated that PML has a tumor suppressor role. PML overexpression inhibits the transformation ability of rat embryo fibroblasts expressing Ha-ras and mutant p53 or Has-ras and c-Myc. However $PML^{-/-}$ mice do not develop spontaneously tumors, they are highly susceptible to physical or chemical carcinogens [86]. PML-NB regulates the function of transcription factors that are important for tumor suppression, including NB-dependent acetylation of p53 by p300/CBP and Daxx-regulated apoptosis [88-91]. Although PML doesn't bind to DNA, it has been found that to act as transcription regulator. PML is coactivator of the transcriptional complex RARα-RXRα. Inactivation of PML leads to impairment of the retinoic acid-induced myeloid differentiation, growth inhibition and alteration in the retinoic acid-dependent transcriptional activation [86]. Moreover, It has been recently found that PML binds to the transcription factor AML1 and regulates its transcriptional activity through enhancing its interaction with the co-activator P300 [61]. Also, while PML stimulates the transcriptional activity of P53, it inhibits the transcriptional activtivation carried out by pRB and SP-1, Moreover, PML inhibits the transcriptional repression activity of Daxx and Tax [88,92-94]. PML has been implicated, also, in genomic stability, DNA repair and the control of the checkpoint activation of the G1 phase of cell cycle. The co-localization PML and BLM in the NBs suggested a potential role of PML in genomic instability and DNA repair. BLM is muted and defective in Bloom Syndrome (BS), a disease associated with predisposition to malignancy; and chromosomal instability. In fact, the BLM protein functions as a 3'-5' DNA helicase and plays an important role in the genomic stability through the control of homologous recombination. Patients with BS display genomic instability with high rate of sister-chromatid exchange and chromosomal breakage (SCE). Interestingly, $PML^{-/-}$ murine fibroblast fails to form NBs and to concentrate BLM within them, and shows high degree of SCE [67,95]. These observations suggested a potential functional link between PML and genomic stability directly or mediated by BLM. Furthermore, BLM colocalize with PML in late S/G2 phase along with the recombination proteins RAD51 and replication factor (RP)-A. This colocalization became even more evident after γ-irradiation of normal human fibroblast, suggesting that this complex may be involved in recombinational repair after DNA damage in G2 phase [96]. The role of PML in cell cycle has been investigated in Hela cells synchronized at various phases of cell cycle. It has been found that PML is expressed at lower levels in S, G2 and M phases, whereas higher levels were found in G1 phase [97].

PML regulates hematopoiesis. It has been shown that PML is differentially expressed in various stages of hematopoietic cell differentiation. While it is barely expressed in the early quiescent hematopoietic stem cells, it is consistently expressed in erythroid and granulocytic progenitors. Thereafter, PML is downregulated in late granulocytic maturation, whereas it has sustainable expression throughout the eryrhroid pathway. Consistent with its expression pattern, inhibition of PML, using antisense oligomers, resulted in reduction of erythroid colonies and early granulocytic but not late granulocytic colonies [98]. Moreover, PML cooperates with pRb 105 to promote erythroid differentiation. Combined inhibition of PML and pRb 105 led to complete block of erythroid development. Overexpression of PML stimulates the G-CSF-induced myeloid differentiation of L-G cells *in vitro* [61]. When $PML^{-/-}$ mice were compared with $PML^{+/+}$, the homozygous PML^{-} animals showed marked reduction in granulocytes and monocytes terminal differentiation in peripheral blood and bone marrow. Surprisingly, the culture on semi-solid medium of hematopoietic cells obtained from $PML^{-/-}$ mice resulted in normal erythroid and myeloid colony formation. On the other hand, these hematopoietic cells showed impaired erythroid and myeloid- colony in response to RA, suggesting that PML is crucial for RA-induced myeloid differentiation [99]. This block of differentiation may be explained, at least partially, by the fact that PML is required for RA-mediated induction of $p21^{waf1}$. Other possible mechanisms through which PML can play role in hematopoiesis are its interacrions with PRH and AML1 [61,75]. Both AML1 and PRH are components of the nuclear bodies (NBs) and interact with PML through their NMTS and RING domains, respectively. PRH is a tumor suppressor gene that slows the proliferation of hematopoietic cells and prevents their leukemic transformation potenial [100]. AML1 is a transcription factor that plays a very important role in hematopoiesis; this role will be discussed in details in a subsequent section of this chapter. Finally, it is important to note that the full spectrum of PML activity and the role of its interaction with other transcription factors in hematopoiesis are still to be determined.

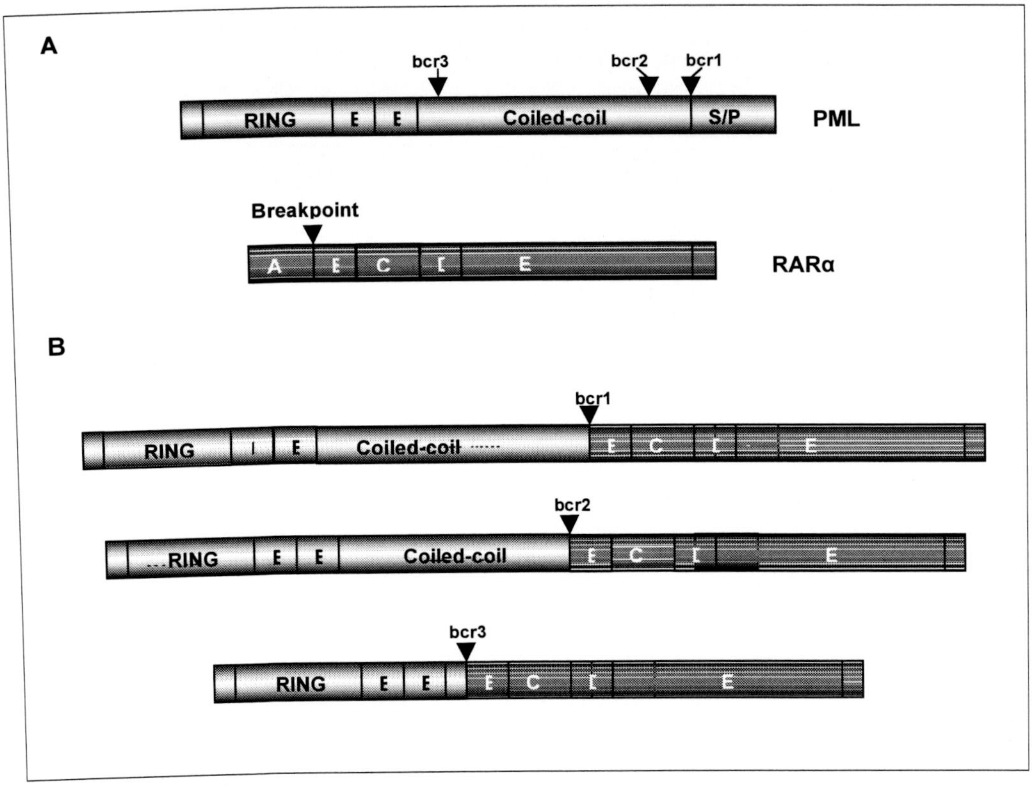

Figure 3. The structure of RARα, PML and PML-RARα fusion protein. (A) Schematic representation of PML and RARα genes. (B) Illustration of the 3 different PML- RARα isomers.

The t(15;17) results in the formation of two chimeric genes: RARα-PML on derivative 17 and PML-RARα on der(15). In fact, it is the PML-RARα transcript, but not RARα-PML, plays the crucial causal role in the pathogenesis of APL. This premise has been supported by gene expression data showing the PML-RARα transcript could be detected in all cases of APL with t(15;17), whereas the transcript RARα-PML could be only found in 80% of cases. Moreover, the experimental data ontained from transgenic mice have demonstrated that RARα-PML chimeric gene, but not PML-RARα, is required for the development of APL-like leukemia. The breakpoint in the RARα gene occurs consistently within the second intron, while in PML there are three breakpoint cluster regions (bcrs) located in intron 6 (bcr1), exon 6 (bcr2) and intron 3 (bcr3) [101, 102](figure 3A). In the majority of patients, the breakpoint in PML occurs in the bcr1 or bcr2 (70%) [103,104].These breakpoints result in a fusion product that contains the N-terminal of PML including the ring domain, the 2 B Boxes and the majority of the α-helical coiled coil domain, while the bcr3 breakpoint results in the deletion of the most coiled-coil domain [105]. In addition, alternative splicing of PML exons 3,4,5,6 may lead to more heterogeneity in the amount of PML included in the PML-RARα protein [106]. Whether these variation in PML moiety length lead to variation in function of the different isomers is still to be fully explored. However, It has been shown that different patients with the different PML breakpoints have distinct clinical, cytological and immunophenotypic characteristics. Patients with the bcr3 breakpoint show higher WBC, increased prevalence of severe hemorrhagic complications, more frequent M3v morphology and more likely to express CD34, CD2 and CD57 on their malignant cells [107-109]. Moreover, it has been shown that the PML-RARα with the bcr1 and PML-RARα with the bcr3 have different binding affinity to ATRA [105]; whether this difference in binding affinity translates into different *in vivo* response to treatment is still controversial [108-110].

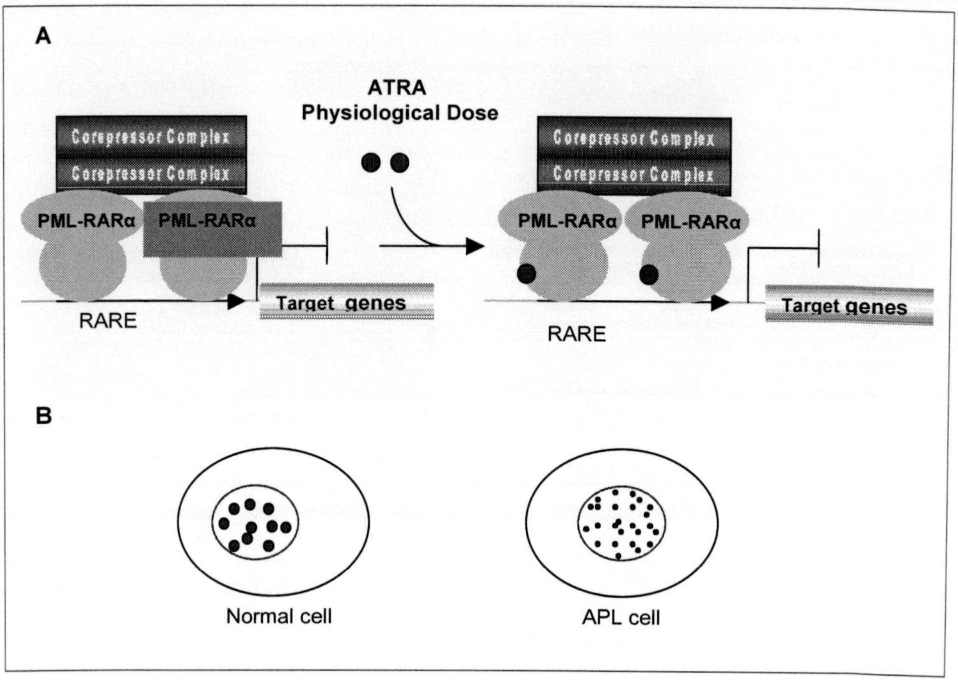

Figure 4. Major mechanisms by which the fusion orotein PML-RARα contributes to leukemogenesis. (A) The PML-RARα binds with high affinity to co-repressor complexes that are not released in the presence of physiological dose of ATRA, leading to supression of the RA induced transcrition. (B) Normal of nuclear bodies (Left). The disruption of (NBs) induced by PML-RARα in APL cells (Right).

As mentioned above, the epidemiological and experimental data support the implication of PML-RARα in the pathogenesis of APL. What are the molecular and cellular mechanisms by which this chimeric protein contributes to leukemogenesis? Although the molecular mechanisms by which PML-RARα exerts its oncogenic effects have been extensively investigated, they are not yet fully understood. In fact, both RARα and PML moieties are required for the oncogenic potential of PML-RARα [111-113]. Moreover, it has been proposed that the fusion protein disrupts the functions of RARs, RXRs and PML [110, 114-116]. In the PML-RARα chimeric protein, the N-terminal portion of PML that comprises the proline-rich, the RING finger, the B-boxes and the coiled coil domains is fused with the c-terminal of RARα from the beginning of B domain through to the end of the F region (figure 3B). Thus, PML-RARα contains nearly all the key functional domains of both molecules including the protein-protein interaction motifs PML and the DNA-binding, dimerization, ligand-binding and transcriptional activation domains of RARα (figure 3B). As predicted from this functional domain structure, PML-RARα can homodimerize, homooligomerize and heterodimerize with PML or RXRs; also, PML-RARα can bind to the DNA at the RAREs and recruits the same corespressors to which wild-type RARα usually binds. However, PML-

RARα displays higher affinity to the corepressors SMRT and N-CoR than wild-type RARs and recruits HDACs (in the absence of or under physiological doses of RA). This enhanced corepressor binding efficiency is due to the ability of PML-RARα to form homodimers or oligohomodime throurg the RBCC domain of its PML portion. This formation of homodimers and oligohomodimers results in altered stoichiometry of corepressor binding by the PML-RARα molecule that accounts for its super suppressor property [117-122]. Indeed, it has been shown that, unlike RAR-RXR heterodimers that interacts with one unit of the corepressor complex, the PML-RARα homodimers can recruit multiple units of the corepressor complex [118-122]. This high affinity of PML-RARα to the corepressor endows PML-RARα with the potential to act as a dominant inhibitor of RAR-RXR heterodimers, leading to a major aberrant repression of physiological transcriptional activity of RAR-RXR and providing an important ground for its oncogenic activity [115-123] (figure 4A). Interestingly, mutation of the N-CoR binding site of PML-RARα molecule abolishes the oncogenic potential. Furthermore, PML-RARα differs from RAR-RXR by its DNA-binding properties. PML-RARα can bind efficiently to a variety of target sequences as homodimers. These homodimers bind selectively to a series of response elements with a specific pattern that is different from that of RAR/RXR heterodimers, leading to suppression of transcription of genes that are not usually controlled by RAR/RXR heterodimers [115,123].

PML-RARα is able to sequester RXRs partially in the cytoplasm and to alter their mobility in the nucleus, leading to alteration of the high-affinity DNA-binding potential of other nuclear receptors whose activity is dependent on heterdimeric interaction with RXR, including VDR, TR, PPARs and residual RARs [115, 125, 126]. It has been shown that the sequestration of RXRs by PML-RARα prevents the binding of VDR to its target sequences in vitro, leading to inhibition of the vitamin-D3-dependent activation. Ectopic expression of PML-RARα protein in U937, a myeloid precursor cell line at is responsive to vitamin D treatment, abrogates their ability to differentiate under the action of vitamin D [122]. Whether this defect is due to RXRs sequestration is still to be determined.

While there is little doubt about the mechanisms by which PML-RARα disrupts the retinoid receptors functions, little is known about PML activity alterations in APL. One of the most striking characteristics of APL is the aberrant localization of PML [114,116,127,128]. As mentioned earlier, PML is found in normal cells organized with other proteins in subnuclear structures called NBs. One of the striking features of APL cells is PML-RARα disrupts the normal NBs structure by PML-RARα and displacement of PML into ill-defined nuclear microspeckles. In APL cells, the different component of NBs are colocalized with the aberrant protein PML-RARα through direct interaction with PML and the formation of PML-RARα/PML heterodimers (figure 4B). It seems that the microspeckles formation is crucial for at least some of the alterations induced by PML- RARα. While the formation of these microspeckles appears to be not important for the block of differentiation induced by PML-RARα [111], it is critical for its effect on apoptosis by targeting RARα to specific subnuclear and thus triggering abnormal transcriptional activity [112]. The impact of the alteration of NBs structure on PML function or alteration of the transcriptional activity of RARα and their contributions to the molecular pathogenesis of APL are not yet fully understood. Furthermore, PML-RARα transgenic mice experienced non-random chromosomal abnormalities, a manifestation that is seen in 30% of patients APL [129-132]. Whether this genomic instability is due to disruption of the NBs and functional alteration of PML is still to be examined.

At the cellular level, PML-RARα has selective effect on the survival of different hematopoietic cell lines. While it has a potent growth inhibitory effect on the majority of the hematopoietic cell lines due to increased apoptosis, it promotes the survival and reduce apoptosis of NB4 and U937 cell lines [112,126,133]. Moreover, it blocks the differentiation induced by different stimuli, including vitamin D3 and TGFβ, in U937 cell line [126]. APL cells are blocked at the level of promyelocytes that are $CD34^{low/-}$. However, expression of PML-RARα has been found in earlier hematopoietic progenitor cells (HPC) characterized by the phenotype $C34^{+}/CD38^{+}$ but not in the earlier hematopoietic stem cells/ hematopoietic progenitor cells (HSC/HPC) with the phenotype $C34^{+}/CD38^{-}$. These finding suggests that although the differentiation block occurs the stage of promyelocytes, the APL arises from earlier progenitor cells. Consistent with this hypothesis, ectopic expression of the chimeric gene PML-RARα in HPC/HSC with the phenotype $CD34^{+}/lin^{-}/CD71^{-}$ induced accumulation of promyeloctes through 3 mechanisms: rapid induction of HPC differentiation, reprogramming of HPC commitment with preferential granulocytic differentiation, and differentiation block at the stage of promyelocytes [134]. Retroviral transduction of murine hematopoitic progenitors with PML-RARα cDNA led to immortalization of these progenitors, to increase in the colony-forming efficiency and differentiation block at the stage of promyelocytes [135,136]. Inoculation of the (PML-RARα)-transduced murine hematopoietic progenitors into syngeneic, irradiated mice resulted in the development of retinoic-acid-sensitive promyelocytic leukemia at high frequency and short latency period (4 months).

In transgenic mice expressing PML-RARα under the control of the myeloid-specific promoter MRP-8 or the highly promyelocyte-specific promoter cathepsin G, the mice developed APL-like leukemia after a latency period (6-18 months) and at low penetrance (10%-30%) [137-139]; during this latency period, the mice experienced a so-called "preleukemic" condition characterized by impaired hematopoiesis with an expansion in the myeloid compartment, increase colony formation and incomplete block in the neutrophilic differentiation. Simultaneous expression of the bcr3 form of the chimeric gene RARα-PML with bcr1 form of fusion gene PML-RARα was shown to increase the frequency to 50% without affecting the length of latency period [140]. Taken together, these findings indicate that although the fusion between PML and RARα genes is required for the pathogenesis of APL, it is not sufficient for leukemogenesis. The addition of cooperating mutations that promote survival (BCL2) or proliferation (FLT3-ITD, FLT^{w51}, IL3, IL3R) led, in all cases, to complete penetrance (100%) and variable reduction in the latency period, with overexpression of IL3 or constitutive activation of IL3R showing the shortest latency period (<21 days) [130,132,141,142].

To end this section regarding the role of PML-RARα in the pathogenesis of APL, it is paramount to discuss the role of ATRA as an agent that specifically targets PML-RARα fusion protein in the treatment of this disease. Indeed, the use of ATRA has resulted in evolution of this disease from one that was historically of the most deadly subtype of AML with mortality of more than 70% to one that appears curable in 70%-80% of patients [143-145]. ATRA treatment is characterized by the induction of cellular differentiation rather than by cytotoxicity. ATRA drives APL blasts to undergo differentiation to neutrophils, leading to progressive replacement of the malignant immature cells to "benign" mature neutrophils. In fact, this effect of ATRA requires pharmacological concentrations of ATRA ($>10^{-6}$ M), but not a physiological one (10^{-9} M). Two molecular mechanisms account for the cellular changes induced ATRA: the abrogation of PML-RARα transcriptional repression and degradation of the chimeric protein. As discussed above, PML-RARα binds to the corepressors SMRT and N-CoR with higher affinity than the complex RXR-RARα. ATRA at physiological dose lead to the dissociation the co-repressors from the receptor RXR-RARα. and the recruitment of co-activator complexes. In contrast, the dissociation the corepressor complex from the chimeric protein requires higher doses (pharmacological doses). At pharmacological doses of ATRA, PML-RARα releases the HDAC complex and recruits co-activators to activate transcription, thus mimicking the physiological function of RXR-RARα [119,120,146] (figure 5A). In addition, many in vitro studies have shown that ATRA induces complete or partial degradation PML-RARα, thus not only restoring wild type RARα function but also that of PML [147-151]. Consistent with hypothesis, ATRA treatment of APL cells triggers re organization PML to generate normal appearing nuclear bodies [127] (figure 5B).

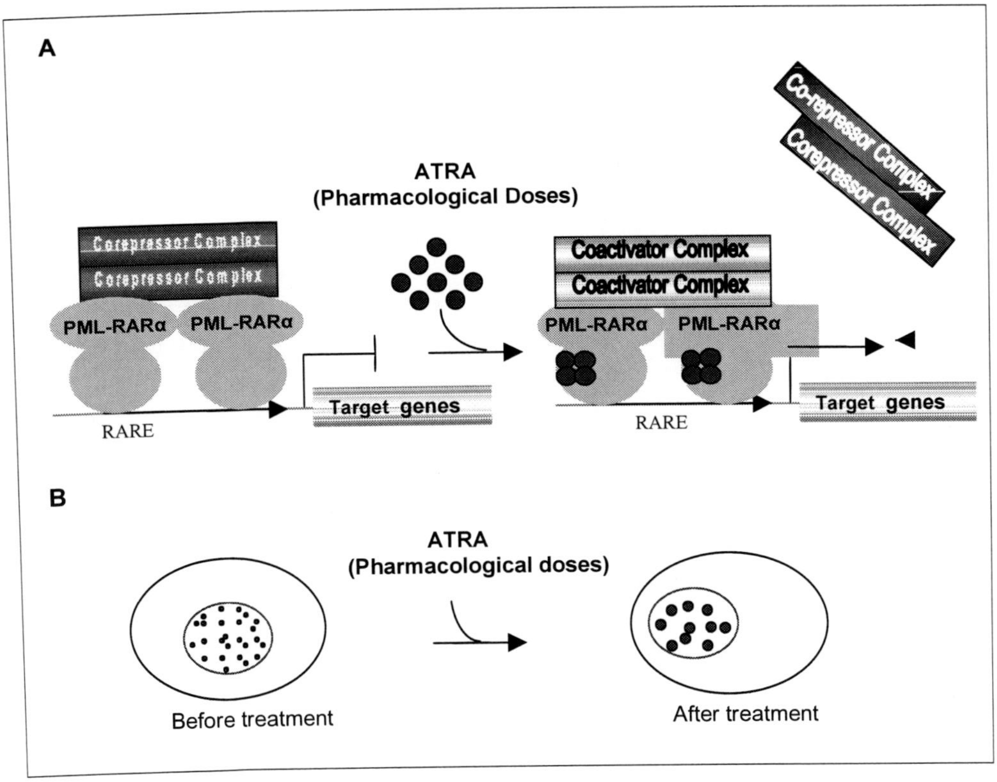

Figure 5. Mechanisms of action of ATRA treatment. (A) Treatment with ATRA (at pharmacological doses) results in the release of the corepressor complexes and the recruitment of coactivator complexes, leading to the restoration of the retinoic acid (RA) signaling. (B) ATRA treatment triggers reorganization PML to generate normal appearing nuclear bodies.

The t(8;21)(q22;q22)

AML cases with t(8;21) accounts for 5-12% of AML in adult and 40% of cases in children. Based on its clinical, morphological and immunophenotypic characteristics, AML with t(8;21) has been considered as distinct pathological entity. Clinically, high complete remission and long overall survival characterize it, particularly when treated with regimens containing high-dose Ara-c [152]. Morphologically, it is most likely to have M2 subtype according to FAB classification, type III myeloblasts, frequent Auer rods, increased eosinophils and dysplastic granulopoiesis [153]. Moreover, AML with t(8;21) have been associated with specific immunophenotypic features including frequent expression of CD19 and CD56, high intensity of CD34, HLADR and MPO expression, and low expression level of CD13 and CD33 [154,155]. This translocation involves the genes AML1 and ETO.

AML1 is located on 21q22 and has 9 exons. AML1 was first cloned due to its involvement in the t(8;21). In addition to its implication in the t(8;21), it is involved in the following genomic abnormalities: t(3;21)(q26;q22) seen in the CML blast crisis and therapy-related AML, t(16;21) found in rare cases of AML, t(12;21) the most frequent translocation in pediatric ALL and AML1 germline mutation or intragenic deletion associated with familial thrombocytopenia and propensity for AML development. The expression of AML1 is a subject of alternative splicing process and two independent promoters regulation, leading to the expression of many different isomers [156-159]. Of these, the most frequent is AML1B on which we will focus in this chapter and we will refer to as simply AML1. Some isomers, like AML1A, lack the transcription activation domain and act as dominant negative of full-length AML1. AML1 protein is a member of the RUNT homology domain (RHD) superfamily that includes also AML2 and AML3 protein [160,161]. Each of these genes shows restrictive pattern of expression. A high level of expression of AML1 is limited to cells of the hematopoietic lineage and to ganglion cells of the developing nervous system [162,163].

AML1 protein is composed of 479 amino acids and contains many functional domains. Besides the RUNT domain, AML1 comprises a transactivation domain (TA), a proline/serine/threonine-rich stretch, a nuclear matrix signal (NMTS), and 3 inhibitory domains (figure 6A). The RUNT domain provides AML1 the ability to bind to DNA, to heterodimerize with CBFβ and to interact with other transcription factors, including C/EBPα, PU1, ETS1, MEF, Pax5 and GATA1. The transcriptional activity of AML1 is mediated by the transactivation domain through which AML1 can bind to the transactivators p300, CBP and YAP1 [164]. Moreover, the transactivation domain can interact with ALY; ALY, in turn, can stabilizes the interaction between AML1 and YAP1. AML1 is phosphorylated within the PST stretch. The NMTS region directs AML1 to its specific matrix-associated subnuclear localization. AML1 has three repression domains. The first repression domain is located immediately C-terminal to the RUNT domain and interacts with the transcription repressor Ear2. The second is located between the NMTS and the TA domains; this domain binds to the repressors the mammalian Sin3 homologue A (mSin3A) nuclear rceptor that, in tutrn, recruits histone deacetylase (HDAC). The last repressor damain is located at the C-terminus and composed of 5 (VWRPY) residues that interact with the repressor proteins the human homologues of Groucho (TLE 1-4).

AML1 exerts its biological effects as a part of the heterodimeric complex AML1/CBFβ that also called the core-binding factor (CBF). The heterodimer binds to DNA through the DNA-binding domain of AML1. The CBFβ does not have any DNA-binding capacity; however its heterodimarization with AML1 is crucial high-affinity DNA-binding. Moreover, AML1 is a phosphoprotein whose level of phosphorylation may lead to alterations in its ability to interact with CBFβ and DNA [165]. Phosphoryaltion of AML1 occurs at the PST-

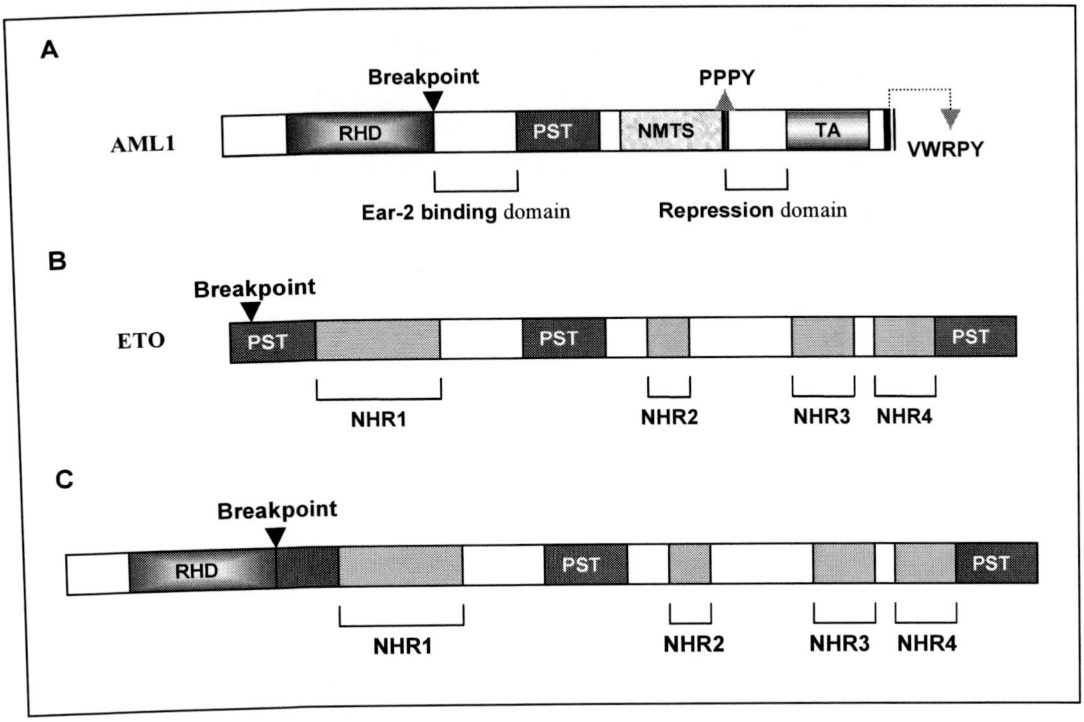

Figure 6. The structure of AML1, ETO and the AML1-ETO fusion protein. (A) RUNX1 functional domains. RHD is the Runt homolgy domain by which RUNX1 binds to CBFβ, DNA and other transcription factors such as C/EBPα. It is followed by the Ear2-binding repression domain. The proline-serine-threonine region is the phosphorylation site. The nuclear matrix signal (NMTS) guides RUNX1 to its specific nuclear sites. The PPPY sequence binds to the co-activator YAP. This is followed by the second repression domain. The Transactivation (TA) domain mediates transcativation and binds to the coactivators p300, CBP, YAP1 and ALY. The VWRPY is the binding site of the TLE proteins. (B) ETO functional domain with particularly the four nervy homolgy regions (NHR 1-4). The NHRs binds to co-repressors including HDACs, N-Cor and SMRT. The NHR2 contains the hydrophobic heptad repeats and responsible for dimerization. (C) RUNX1-ETO protein contains the RHD of RUNX1 and almost all of ETO.

rich region. It is achieved through the action of MAP kinase cascade and binding of the ERK protein. The AML1/CBFβ transcription complex recognizes the enhancer core sequence (TGTGGT) on the promoters of the target genes. It regulates many genes that are crucial for normal hematopoietic development, including myeloperoxidase (MPO), elastase, granzyme B, T-cell receptor (TCR), IL-3, GM-CSF, M-CSF receptor and G-CSF receptor, Cycline D3 [166-175]. In addition to its ability to bind to DNA and to heterodimerize with CBFβ, the transcriptional activity of AML1 depends on its ability to interact with a variety of other nuclear proteins including lineage specific transcription factors, transcription co-activators and in some cases transcription co-repressor. Certain transcriptional activities of AML1 require the presence of adjacent binding sites for lineage restricted transcription factors. These transcriptional regulatory sequences recruit transcription factors including c-Myb, C/EBPα, PU1, TCF/LEF1, MEF, Pax5, GATA1 and ETS family members to allow them to have physical interaction with the RUNT domain of AML1, leading to enhanced DNA-binding and generating an activation surface that promotes the interaction with coactivators

[176-187]. This observation suggests that the heterodimer AML1/CBFβ functions as a transcriptional organizer that recruits specific factors into nucleoprotein complex, or enhancesome, that stimulates lineage-specific transcription [188] (figure 7A). This paradigm is substantiated by the findings showing that AML1 synergizes with Ets1 to activate TCRα and TCRβ transcription, with CEBPα to promote the transcription of NP-3, with both CEBPα and PU1 to initiate M-CSF receptor transcription, and with CEBPα and ETS2 to activate NF1 [177,151,189-195]. Moreover, The AML1 can recruit through its transactivation domains transcriptional coactivators with intrinsic, including HAT activity CBP, P300 and P/CAF, leading to enhanced transcription. Also, AML1 can interact with the protein ALY. ALY does not have an intrinsic activation domain but it is involved in stabilizing the interaction of AML1 and the transcription factors TCF/LEF on the promoter of TCRα. Moreover, ALY strengthens the interaction between AML1 and the yes-associated protein (YAP-1). YAP-1 interacts directly with PPPY motif of AML1 located next to the NMTS region. Indeed, YAP-1 contains a strong intrinsic tranactivation domain and may act to convert AML-1 into a strong activator (figure 7A). In most cellular contexts, AML1/CBFβ has transcriptional activation potential. However in some other contexts, AML1 can interact with repressors such as Ear2, Sin3A and B, and the human Groucho homologues, the transducin-like enhancer of split 1-4 (TLE 1-4). Ear-2 an orphan member of the nuclear receptor superfamily that binds with an 80 amino acids domain immediately C-terminal to RUNT domain and inhibits the transcriptional cooperation between AML1 and CEBPα [196]. Furthermore, AML1 acts as a transcriptional repressor of the cyclin-dependent kinase inhibitor $p21^{Waf1/Cip1}$ promoter [197]. This repression is mediated by interaction of AML1 with the mammalian Sin3 homologue A (mSin3A) nuclear rceptor that can recruit histone deacetylase (HDAC). Another group of repressors that can interact with AML1, through the c-terminal motif VWRPY, are the transducin-like enhancer of split 1-4 (TLE 1-4) [198,199]. Coexpression of TLE and AML1 can inhibit the ability of AML1 to activate the promoters of the M-CSF gene and the neutrophil elastase gene. On the other hand, AML1 can interact with HES-1 that inhibits the interaction between AML1 and TLE [200,201]. In addition to these complex protein interactions, the proper transcriptional activity of the complex AML1/CBFβ relies on the integrity of its NMTS sequence. The NMTS domain is requird for AML1 association with the nuclear matrix and for its routing toward its proper subnuclear localization. Indeed, in these subnuclear localizations reside physiological transcriptional targets of AML1. Alterations of the NMTS sequence by point mutation or by deletion lead to abnormal subnuclear localization of the heterodimer AML1/CBFβ and result in aberrant transcriptional activity of the CBF complex with disruption of myeloid differentiation [202-205]. Furthermore, it it is a specific tyrosine residue in the NMTS sequence that is critical for the proper routing of the AML1 molecules to its appropriate targets.

AML1 shows a restrictive pattern of expression. A high level of expression of AML1 is limited to cells of the hematopoietic lineage and to the developing nervous system. AML1 is crucial for embryonic hematopoiesis. Mice with homolgous knock-out of AML1 die *in utero* during the midpoint development due to lack of fetal liver hematopoiesis and massive hemorrhage, mainly, in central nervous system [206,207]. Although primitive yolk sac-derived erythropoiesis appeared normal, definitive hematopoietic progenitors were absent. The defect resulting from the loss of AML1 was restricted to hematopoietic cells, and no disruption in hepatic microenvironment was seen. Consistent with these findings, analysis of

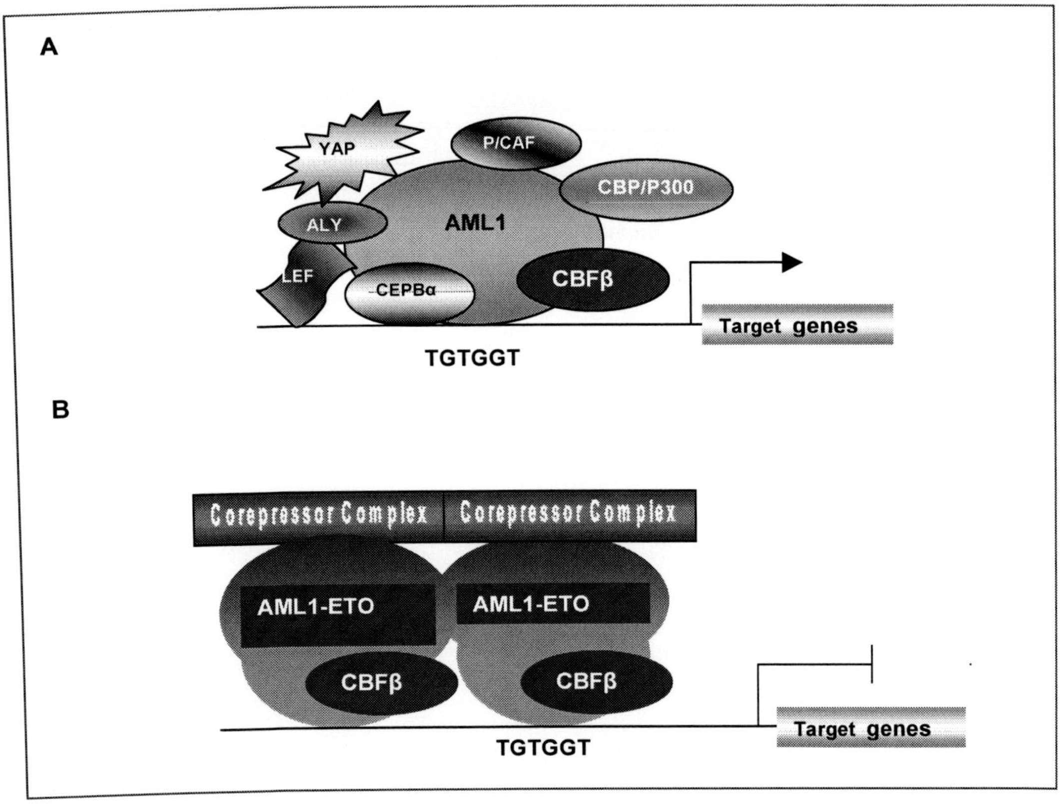

Figure 7. AML1 interaction with other proteins to induce transcription and how this transcriptional activity abrogated by the fusion protein AML1-ETO. (A) AML1/CBFβ functions as a transcriptional organizer at the core enhancer sequence, TGTGGT, that recruits coactivators (CBP/P300, P/CAF, YAP and ALY) and cooperating transcription factor(s) (such as C/EBPα, PU1, ETS1, LEF, C-Myb and others) into nucleoprotein complex, or enhancesome, that stimulates the transcription of target genes. This figure represents the enhancesome formed at the TCRα enhancer. (B) AML1-ETO can homodimerize or heterodimerize with ETO family members leading to the assembly of multiple transcription co-repressor complexes on AML1 targets.

AML1-deficient embryonic stem (ES) cells demonstrated that this gene is essential for the generation of normal numbers of blast colony-forming cells (BL-CFCs). AML1$^{-/-}$ Embryonic bodies (EBs) generated consistently 10- to 20-fold fewer blast colonies than wild-type control and displayed a complete block in definitive hematopoiesis. As observed in yolk sac AML1-deficient embryos, AML1$^{-/-}$ EBs showed little or no defect in primitive erythropoiesis. Moreover, induced expression of AML1 can restore *in vitro* definitive hematopoiesis arising from AML1-defficient embryonic stem cells. Taken together, these data indicate the while AML1 is crucial for definitive embryonic hematopoietic reconstitution; it is not required for early primitive embryonic hematopoiesis. In contrast with its crucial function during development, AML1 has less critical role for sustaining hematopoiesis in adult mice. Using conditional gene-targeting strategy, it has been shown that AML1 is dispensable for the hematopoietic stem cells, granulocytic progenitors and erythroid-progenitors maintenance; however significant reduction in long-term repopulating activity was seen. Moreover, it appears that AML1 is very important for efficient lymphoid lineage development at all stages and also for megakaryocytic maturation. AML1-defficient mice developed marked reduction in B-lymphocytes, T-lymphocytes and platelets (80% reduction). On the other hand, the mice exhibited mild myeloproliferative phenotype characterized by an increased myeloid progenitors and peripheral blood neutrophils with extramedullary hematopoiesis composed of maturing myeloid and erythroid elements [208]. The biological functions of AML1 appear to be exquisitely sensitive to its intracellular levels. AML1 haploinsufficiency affects tempral and spatial emergence of HSCs during murine embryogenesis, leading to a precocious appearance in HSCs with later reduction in their expansion and maintenance accompanies by decrease in the committed progenitors [209-211]. In ES cells, AML1 monoallelic loss resulted in acceleration of hemangioblast development with accelerated hematopoietic development in the EBs [212]. Adult mice with heterozygous loss of AML1 displayed decrease in the long-term repopulating hematopoietic stem cells with increase in the multilineage progenitors, in addition to moderate reduction in $CD4^+$ T calls and mild decrease in platelets (15%).

ETO (Eight Twenty One) gene was initially cloned due to its involvement in the t(8;21). It is also called (MTG8) and (CBF2T1). Subsequently, two other members of this family have been found MTGR1 and MTG16. MTGR1 was originally cloned due to its ability to heterodimerize with ETO [207,208]. A second ETO family member, MTG16, was discovered by screening of murine and human cDNA library [215,216]. Interestingly, MTG16 was identified as a fusion partner of AML1 in the t(16;21) found in rare cases of AML and MDS [217]. Furthermore, the t(16;21) produces the chimeric protein AML1-MTG16 which is similar in structure to AML1-ETO. ETO gene is located at chromosome 8q 22 and consists of 13 exons. Different transcripts resulting from alternative promoter and alternative splicing have been described. Alternative promoters yield two isomers that differ in 27 amino acids [218]. An alternatively spliced transcript of this gene that contains a 155-bp exon (9a) was identified. The inclusion of this exon results in the introduction of a premature stop codon leading to a truncated protein that lacks the 177 C-terminal residues. The significance of these different isoforms is still to be determined. ETO is evolutionarily conserved. It is the mammalian homologue of the *Drosophila* nervy gene, a target of the *Drosophila* ultra bithorax gene. In *Drosophila,* Nervy appears to be involved in the development of the CNS of *Drosophila*. Initial analysis of the ETO protein sequence provided very few hints about its function. The comparison of its sequence with the *Drosophila* nervy protein revealed the presence of four conserved regions called nervy homology regions (NHR1-4) [214,219] (figure 6B). NHR1 has similarity to several TATA-binding protein-associated factors including TAF105, TAF130 and *Drosophila* TAF110 [218,220]. The homology to TAFs provided an early hint that ETO could function as transcription factor [214]. However, it was found that this domain might play a role in subnuclear localization of ETO [221]. The NHR2 has alpha helical structure with a hydrophobic heptad repeat (HHR) of amino acids [222]. The NHR3 has a small helical structure and lacks similarity with other proteins to provide hints about its potential function. The fourth region, termed also MYND (MTG8, Nervy, Deformed), has two non-DNA-binding zinc fingers similar to those found in BS69, BOP and DEAF1. This similarity suggests that MYND domain can interact with transcription corepressors. In addition to the four NHR domains, ETO has proline, serine and threonine rich sequences in N- and C-terminus that could affect the protein stability. ETO functions mainly as transcription corepressor. Furthermore, ETO interacts with many corepressor or correpreesion complexes simultaneously. ETO can recruit HDAC1, 2 and 3 through direct interaction with the NHR2 and NHR3 domain [222,227-229]. In addition, ETO interacts with the complexes N-coR/SMRT/HDAC3 through binding with the MYND domain and with Sin3/(HDAC1/2) [228,230,231]. Moreover, it has been shown recently that many of the non-conserved areas may contribute to the interaction of ETO with co-repressors [232]. Furthermore, ETO interacts with PZLF and Gfi proteins. PZLF itself is implicated in the t(11;17). PLZF functions a repressive transcription factor that blocks myeloid growth and differentiation and promote apoptosis [233]. The region of ETO that interacts with PZLF is located between NHR1 and NHR2. ETO behaves as a corepressor in the transcription inhibition activity of PLZF on IL3 promoter [234,235]. The Gfi-1 and Gfi-1b proto-oncogenes are zinc-finger DNA binding transcription repressors that interact with ETO via binding to its NHR1 and NHR2 to facilitate the recruitement of HDAC [236]. The Gfi proteins are highly expressed in hematopoietic cells and are essential for normal hematopoiesis [237-240]. Additionally, ETO has the ability to homodimerize or heterodimerize with other member of the ETO family through the HHR domain [214]. The formation of high molecular weight oligomers enhances the ability of these molecules to interact with N-CoR and SMRT. Thus, the ETO family members are likely to form multi-subunit complexes that function in transcriptional regulation. ETO is expressed in a variety of tissues but is most abundant in heart, brain, lung and testis [216, 241]. In normal hematopoietic cells, ETO is expressed in $CD34^+$ cells but is not present in more differentiated cells [242]. Its normal physiological function in these tissues is still unknown. However, a targeted homozygous mutation of the mouse Mtg8 resulted in greatly reduced postnatal viability and severe growth impairment, which was likely due to a gross disruption of the gut architecture, indicating a crucial role of ETO in the gastrointestinal system [243]

The t(8;21) results in in-frame fusion between the first five exons of AML1 gene and nearly of all the ETO gene [244,245]. The product of this rarrangement is a chimeric protein called AML1-ETO. AML1-ETO protein contains the first 177aa of AML1 and all the amino acids of ETO [246] (figure 6C). Thus, AML1-ETO retains the ability of AML1 to heterodimerize with CBFβ, to bind to DNA and to interact with other transcription factors. Moreover, AML1-ETO conserves the ability of ETO to recruit co-repression partners and to interact with the transcription factors like PLZF and Gfi. Furthermore, AML1-ETO can form homodimers and homooligomers and has the ability to form heterodimers or heteromultimers with ETO family members throught the NHR2 domain. This property of AML1-ETO to dimerize and to oligomerize endowed the chimeric protein with the potential to assemble multiple transcription repressor complexes on AML1 regulatory targets [247,248](figure 7B). Furthermore, AML1-ETO can heterodimerize more efficiently with CBFβ than wild-type AML1, suugesting that AML-ETO may have higher DNA-binding capacity than wild type AML1. Moreover, this high affinity of AML-ETO to CBFβ leads to sequestration of CBFβ away from wild type AML1 [249]. All these mechanisms provided AML1-ETO the possibility to act as potent suppressor of AML1 on its target genes. In fact, AML1-ETO reduces 25 times the expression level of TCRβ reporter, when AML1-ETO was cotransfected with AML1 vector. Subsequent experiments on GM-CSF promoter showed that AML1-ETO not only blocks the ability of AML1 to activate the promoter but also decreased its expression below the baseline level [175]. These data strongly suggest that AML1-ETO is dominant repressor form of AML1 that did not simply work by competing for AML1 binding site but actively repress transcription of AML1 target. Mutagenic analysis of the AML1-ETO protein revealed that all of the RHD region of AML1, the NH4 of ETO and also the dimerization domain (NH2) of ETO are required for this suppressive effect on AML1 target. Transient transfection reporter gene studies have shown that AML1-ETO suppress different AML1 target genes including regulation of TCRβ enhancer, NF1, the GM-CSF, p14^{ARF}, immunoglobulin α promoter, MDR1, and NP3 [175,195,223,250,251]. Surprisingly, AML1-ETO up-regulates the M-CSF receptor, a target of AML1, synergistically with wild-type AML1 in a dosage-dependent manner. Such effect of AML1-ETO requires the NHR1 and NHR2 domains of ETO [168]. In addition, AML1-ETO upregulates BCL2 by binding to the enhancer core sequence localized in BCL2 promoter, whereas AML1 has no effect on the promoter [252,253]. Both RHD and NHR4 regions seems to be important for such BCL2 transactivation. Also, AML1-ETO activates the G-CSF receptor independently of the presence of AML binding site, but it is dependent on the C/EBP binding site in the upstream region of G-CSF receptor gene. Moreover, AML1-ETO can interact with other transcription factors including C/EBPα, PU1 and MEF affecting negatively their transcriptional activity by direct binding to them or by blocking their AML1-adjacent DNA-binding sites. For example, AML1-ETO competes with c-Jun in binding to PU1, suppressing the coactivation of PU1 by c-jun [254]. Furthermore, AML1-ETO inhibits the C/EBPα by blocking its adjacent binding site or suppressing its positive auto regulation [255]. Also, the interaction of AML1-ETO and MEF leads to inhibition of the transcriptional activity of MEF [256]. Finally, AML1-ETO interacts with PLZF and Gfi proteins. In fact, by binding to PZLF, AML1-ETO reduces its DNA-binding capacity, inhibits its repressor activity and disrupts its compartmentalization in the nuclear matrix [234]. On the other hand, the effect of AML1-ETO on the Gfi proteins is still to be defined. Another mechanism that may contribute to AML1-ETO abnormal transcriptional activity is its aberrant nuclear localization. It has been shown, that AML1-ETO resides in subnuclear sites that differ from those of AML1. This misrouting of AML1-ETO is due to loss of NMTS of AML1 and the acquisition of new subnuclear localization site, the NHR1 domain of ETO. This abnormal localization of AML1-ETO leads to the presence of AML1-ETO near the regulatory sites that are not usually under the regulation of AML1 [257-260].

At the cellular level, it has been shown that AML1-ETO affects proliferation, differentiation and apoptosis programs. Forced expression AML1-ETO blocked the myeloid and erythroid differentiation in many murine and human cell lines including 32D, L-G, K562, MEL, U937 cells [196,214,261-265]. In retrovirally transduced human CD34^{+} cells, AML1-ETO promotes hematopoietic stem/progenitors cells (HSPC) self-renewal and proliferation [266,267]. These cells retain the ability to differentiate to more mature progenitors of myeloid and lymphoid phenotype, however with decreased granulocytic and erythroid colony-forming efficiency and reduced granulocytic terminal differentiation [267-269]. Inoculation of CD34^{+} cells in NOD/SCID mice lead to low engraftment and does not leukemia. Mice with AML1-ETO$^{+/-}$ genotype generated by knock-in strategy display severe block in liver hematopoiesis and die in midgestation by CNS hemorrhage. A phenotype that is similar to that seen by AML1$^{-/-}$ mice. However, in contrast to AML$^{-/-}$ mice, yolk sac cells from AML1-ETO$^{+/-}$ mice can give rise to monocytic colonies in CFU assays and their fetal liver contained rare hematopoietic progenitors. Moreover, when these fetal hepatic progenitors were cultured in vitro, they showed increase self-renewal capacity with ability to form immortalized cell line and dysplatic multilineage differentiation. These data indicate that AML1-ETO functions at least partially as dominant negative inhibitor of normal AML1/CBFβ activity. When AML1-

ETO was expressed in mice using an inducible system that turned on after birth, the mice lived a normal life span without development of leukemia, Moreover, transgenic mice theat express AML1-ETO under the control of MRP8 promoter developed AML, with a frequency of 45% and a latency period of 4-7 months, only upon treatment with the alkalating agent N-ethyl-N-nitrosurea (ENU) [270]. Taken together, these findings indicate that however AML1-ETO is critical for leukemic transformation, one or more additional mutations are required for overt leukemogenesis. This paradigm was confirmed by experimental studies showing that AML1-ETO can induce leukemia in cooperation with aberrant enhanced expression of FLT3-LM, upregulation of WT1, concomitant expression of the fusion protein TEL-PDFGβR or downregulation of ISCBP [271-274].

To end this section regarding the role of the chimeric protein AML1-ETO in the molecular pathogenesis of AML with t(8;21), it is important to discuss a recent efforts aimed to develop new molecules that target specifically this fusion protein. Based on the premise that oligohomomerization of AML1-ETO molecules, through the interaction of their NHR2 domains, is crucial for its super suppressor activity and its oncogenic potential (figures 6C & 7B), Wichmann et al. developed a polypeptide composed of 128 amino acids that is aimed to inhibits the NHR2 domain and consequently prevents AML1-ETO from forming high molecular weight complexes [17]. Treatment of Kasumi-1 cell line, a cell line that possess the t(8;21) with NC128 led to restoration of the transcriptional activity of AML1 on its target gene. Moreover, NC128 completely overcame the maturation block mediated by AML1-ETO seen in Kasumi-1 cell line and U937 transfected with AML1-ETO leadin these cells to differentiate expressing mature monocytic markers. Furthermore, NC128 induced a cell cycle arrest at G1 phase in Kasumi-1 and SKNO-1 cell lines. Finally, treatment of human $CD34^{+}$ immortalized with AML1-ETO resulted in loss of CD34, growth arrest and increased apoptosis [17]. Indeed, this selective interference with the NHR2-mediated oligommerization of AML1-ETO could offer a real promise that could target specifically this oncogenic molecule.

The inv(16)(p13q22)/t(16;16)(p13q22)

Inv (16)(p13q22)/t(16;16) has a prevalence of 10-15% in AML patients. It has recognized as distinct clinicopathological entity. The majority of patients with cytogenetic abnormality have AML-4Eo subtype, which is characterized by blasts of myelomonocytic differentiation and increased eosinophils with distinctive basophilic granules [275]. The immunophenotypic analysis revealed a distinctive pattern with high intensity CD34 and CD13 and low intensity of CD33, in addition to frequent expression of CD2 [276,277]. Patients with inv(16)/t(16;16) tend to have a good response to chemotherapy and had a similar prognosis to patients with t(8;21) AML [278] . The genes involved in inv(16)/t(16;16) are CBFβ with the MYH11.

The human CBFβ gene maps to chromosome 16q22. It contains 6 exons with alternatively spliced transcripts that yield 4 isoforms that encodes 4 different isomeric proteins with varied length (187, 182, 155 and 148 amino acids). CBFβ (187) and CBFβ (182) are generated by two different splice donors (31 nucleotides apart) at the 3' end of exon 5, generating proteins that differ from amino acid 166 their C-termini. CBFβ (148) and CBFβ (155) are genrtated by splicing skipping exons 3 and 5, respectively.

CBFβ is a transcription cofactor that does not bind directly to DNA, but it interacts with RUNX transcription factors (AML1, AML2, AML3) to form transcription complexes called core-binding factors (CBFs). In fact, it is in this complexed form that the RUNX transcription factors are able to regulate their target genes. Therefore, the RUNX proteins can be designated as the α subunits (CBFα) of the transcription complexes CBF. CBFβ is composed one known functional domain located at the N-terminus (amino acids) 1-135 and a cytoplasmic tail of unknown function (figure 8A). By its N-terminal domain, CBFβ binds to the RUNT domain of the RUNX proteins. The main functions of CBFβ are to enhance the DNA binding affinity of CBFα subunits and to protect them from ubiquitin –proteosome mediated degradation. The RUNX proteins alone bind very weakly to DNA. On the other hand, by heterodimirizing with the longer CBFβ isomers (that are composed of 187 or 182 aa, respectively), their affinity to the consensus core site (TGTGGT) is increased at least 5 folds [279,280]. The CBFβ core domain interacts with RUNT domain of the RUNX proteins at a site distant from the RUNX-DNA interface. The interaction between CBFα and CBFβ stabilizes the flexible C-portion of the CBFα runt domain that interacts with the minor groove of DNA [280-283]. This stabilization forms the allosteric regulation of CBFα-DNA complex formation by CBFβ. The stoichiometry of the CBFα-CBFβ-DNA is 1:1:1 [284,285]. On the other hand, the shorter CBFβ isomers (isomer 148, isomer155) do not bind stably with the RUNT domain [279]. Interestingly, it has been shown, by deletion analysis, that the truncated form of CBFβ containing the amino acids 1-141 is able to heterodimerize with the RUNX proteins and to enhance their DNA-binding affinity as efficiently as the long isomers (187 or 182 amino acids) of CBFβ [279,280,286]. However, very short truncations of the N-terminus (5-6 amino acids) or longer truncations of the C-terminus up to the amino acid 127 abrogate the ability of CBFβ to bind to RUNT proteins [286-288]. In addition to increasing its DNA binding capacity, the dimerization of CBFβ with CBFα protects CBFα subunits from degradation. It has been found that AML1 is continuously subject to proteolytic degradation mediated by ubiquitin-proteasome pathway. Heterodimerization of CBFβ with AML1 abrogates AML1 ubiquitylation. In mice with homozygous knock-out of Cbfb, AML1 is barely detectable [289]. In contrast to the individual AML1 protein, CBFβ is ubiquitously in human cells and cell lines derived from varied tissues [285,287]. In adult mice Cbfb was found the in hematopoietic stem cells, hematopoietic progenitors, mature myeloid and lymphoid, but not during the terminal stages of erythropoiesis [290]. CBFβ is predominantly present in the cytoplasm, whereas the RUNX proteins are always in the nucleus. RUNX protein has a nuclear localization signal and translocates CBFβ to the nucleus in a dose dependent manner. Without dimerizing with CBFα subunits, CBFβ remains in the cytoplasm [291-293]. AML1 and AML3 proteins require CBFβ to regulate hematopoiesis. As in mice with Runx-1 $^{-/-}$, Cbfb$^{-/-}$ murine embryos die at 11.5-13.5 dpc due to hemorrhage in the CNS and severe impairment in the fetal liver hematopoiesis [294-296]. Mouse ES cells deficient for Cbfb can differentiate into primitive erythroid colonies *in vitro*, but are severely impaired in their ability to form definitive erythroid and myeloid colonies; and construct containing 1-141 amino acids of Cbfb can rescue definitive hematopoiesis of Cbfb$^{-/-}$ ES cells, suggesting that this region is sufficient for Cbfb function *in vivo* [297].

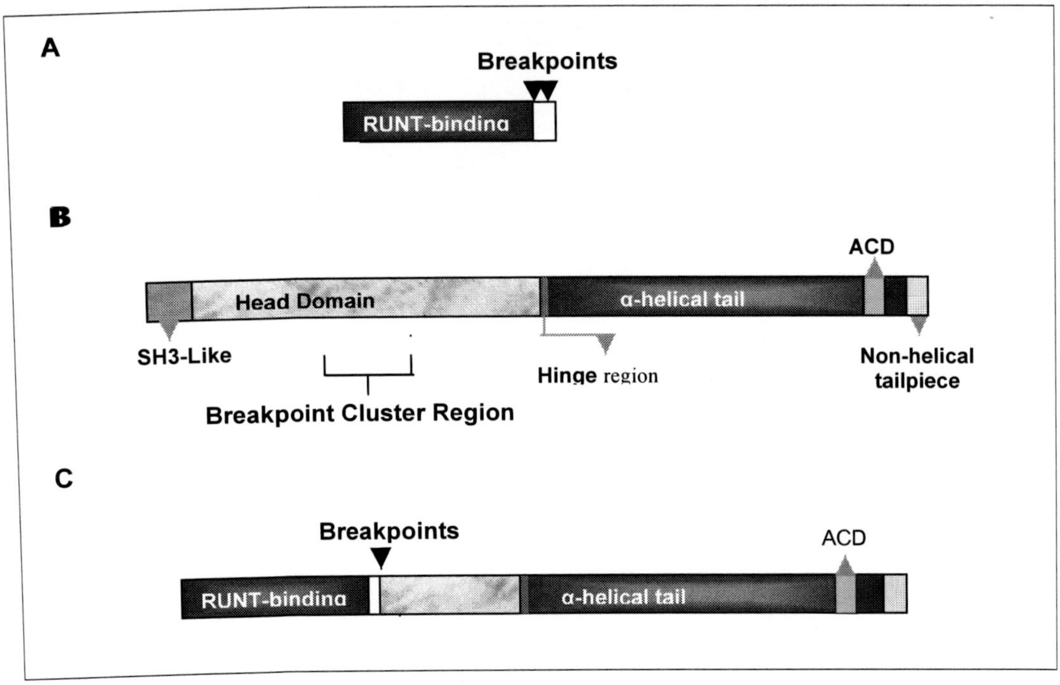

Figure 8. The Structure of CBFβ, MYH11 and the fusion protein CBFβ-MYH11. (A) schematic representation of CBFβ protein. It has only one functional domain, the AML1-binding domain. (B) Schematic representation of MYH11 protein. It has main two functional domains: the head domain and α-helical domain. The ACD motif provides the MYH11 the ability to multidimerize. (C) The most froquent isomer of CBFβ-MYH11 contains the RUNT-binding domain in addition to a part of the globular head and the α-helical tail.

The smooth muscle myosin heavy chain (SMMHC), called also myosin heavy chain 11 (MYH11), is mapped to the band 16p13 and has 42 exons. Alternative splicing of 39bp at the 3' end produces two isoforms that are termed SM1 and SM2 [298]. MYH11 protein is composed of 4 domains: a short N-terminal region that has an SH3-like fold of unknown function, a very large head domain, a short hinge composed of calmodulin-binding motif, large tail domain and finally at the C-terminus the spectrin repeats that form the uncoiled tailpiece (figure 8B). Analysis of the 3D structure reveals the head to be globular in structure, whereas the tail is in a rod-like form [299]. The head has an actin-binding and ATPase activities. The rod-like tail sequence is highly repetitive, showing cycles of a 28-residue repeat pattern composed of 4 heptapeptides, characteristic for alpha-helical coiled coils [300-302]. The coiled-coil is composed of the tail from two molecules of myosin. These can then assemble into the macromolecular thick filament [302]. The coiled-coil region provides the structural backbone of the thick filaments, and a sequence named assembly competence domain (ACD) composed of 29 residues located near the C-terminus (residues 1874-1902) is required these thick filaments formation. The SM1 (1972 amino acids) and SM2 (1945 amino acid) isomers are identical over 1424 nucleotides, but differ in C-terminal region. Although the additional 39 nucleotides in the SM2 transcript result in the insertion of 9 amino, they lead to a shorter carboxy terminus due the formation of an early stop codon. The differences in structure between the SM1 and the SM2 isomers are located in the end of the tail region that form non-helical tailpiece. This variability in two isomers affects differentially their filament assembly properties; rendering SM2 rods are less able to assemble into stable filaments than SM1 or tailless rods [303].

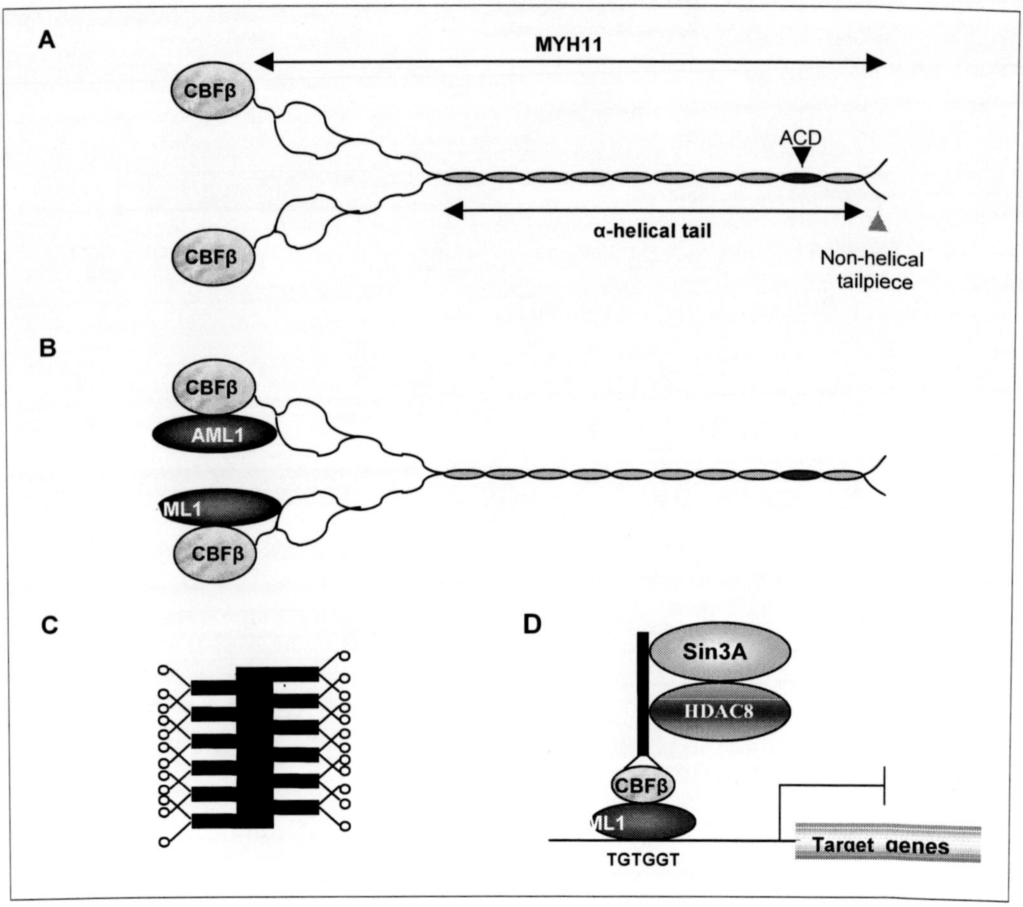

Figure 9. Schematic representations of 3D structure of the fusion protein CBFβ-MYH11, its interaction with AML1, multidimerization of the wild type MYH11 and suppression of AML1 transcriptional activity by the fusion protein CBFβ-MYH11. (A) Schematic representation of the fusion protein CBFβ-MYH11. CBFβ-MYH11 always exits as a homodimer. ACD motif is responsible for the multihomodimerization. (B) Model of interaction of CBFβ-MYH11. (C) Filamentous multihomodimer. The multihomodimers form deposits associated with the actin cytoskeleton or aggregates in the cytoplasm, or speckles or rod-like structures in the nucleus. (D) Repression of AML1 transcriptional activity by CBFβ-MYH11.

In the fusion gene CBFβ-MYH11 the breakpoints occur in the fourth or fifth introns of CBFβ, whereas at least 8 breakpoints have been recognized in MYH11 [304,305] (figure 8A&B). Moreover, alternative splicing of exon 5 of CBFβ and C-terminus of MYH11 may make the heterogeneity of the CBFβ-MYH11 isoforms even more complex [306,307]. It is not yet known whether these molecular variants have different clinical or biological characteristics. However, all these variants contain at least the first 133 amino acids of CBFβ and 1313 or 1346 of MYH11 generated by alternative splicing. The most frequent breakpoint combination (80% of cases) lead for a chimeric protein that contains the amino acids 1-165 of

N-terminus of CBFβ and 1313 or 1346 amino acids of the C-terminus of MYH11 (figure 8C). Structural analysis, revealing that of all isomers contain most of the RUNT-binding domain of CBFβ in addition to part of globular head and α-helical tail including ACD domain of MYH11, would indicate that they are all able to interact efficiently to RUNX proteins, to bind to actin and to homodimerize and multihomodimerize (figure 9 A,B&C). Moreover, biochemical studies revealed that CBFβ-MYH11 has increased affinity to AML1 protein when compared to wild type CBFβ, a phenomenon that is called "hyper-heterodimerization". NMR studies revealed that the RUNT domain of AML1 interacts with both CBFβ and MYH11 [308] (figure 9B). By functional studies using serial deletion analysis, it has been shown that the N-terminal of MYH11 segment that is involved in the fusion protein is required for this hyper heterodimerization [309]. Isometric titration calorimetry analysis yielded a stoichiometry of two Runt domain proteins for one CBFβ-MYH11 molecule [302]. It appears that CBFβ-MYH11 is able to stabilize AML1 more efficiently than wild-type CBFβ [289]. Whether this is due to its higher affinity to AML1 and the implication of the resulting accumulation of RUNX1 are still unknown. Furthermore, CBFβ-MYH11 can recruit corepressors including Sin3A and HDAC8 via the C-terminal 163- amino acid region of the myosin tail [310,311] (figure 9D). In contrast to wild type CBFβ, the chimeric protein CBFβ-MYH11 is localized predominantly in the cytoplasm in form of deposits that are associated with the actin cytoskeleton or in form of aggregates [288,312,313]. A direct consequence of the predominant cytoplasmic localization of CBFβ-MYH11 is the sequestration of CBFα subunits (including AML1) in its abnormal cytoplasmic structures [291,314]. This sequestration is dependent on the N-terminal region of CBFβ that interacts with the CBFα subunits and the ACD motifs of coiled-coil domains of MYH11 that are implicated in myosin homodimerization and multihomodimerization (figure 9B&C). Although most of the CBFβ-MYH11 protein remains in the cytoplasm, a small fraction can enter to the nucleus through binding with the CBFα subunits [315]. In the nucleus it form speckles or at higher concentration rod-like structures, which contain also the CBFα subunits [316]. It has been proposed that CBFβ-MYH11 acts as dominant negative inhibitor of AML1. Many mechanisms have been suggested to explain the disruption of AML1 function by CBFβ-MYH11. First, the cytoplasmic sequestration of AML1 by CBFβ-MYH11 would make it unavailable to bind to its target-gene regulatory domains [288,291]. Second, the CBFβ-MYH11 the multidimerized intranuclear AML1-β/SMMH complexes by binding to the co-repressors Sin3A and HDAC8 and interacting with the regulatory elements of AML1 target genes would lead to inhibition of transcription of these genes [304,305] (figure 9D). Third, the increased ability of CBFβ-MYH11 to hyperheterodimerize with AML1 would reduce the amount of AML1 available to heterodimerize with CBFβ [308,309]. Finally, sequestration of increased AML1 amount would lead to sequestration of putative activators that interact with AML1 [289]. It has been shown *in vitro* that nuclear extract from cells transfected with CBFβ-MYH1 construct reduces endogenous AML1/CBFβ complex to bind to DNA [317]. CBFβ-MYH1 inhibits the biological activity of AML1/CBFβ complex. This has been demonstrated by showing that CBFβ-MYH1 reduces the transition from G1 to S phase of the cell cycle by inhibiting AML1-induced cyclin D3 or cdk4 expression [318-321]. Moreover, it has been shown that patients with inv(16) has lower levels of C/EBPα, a downstream target of the transcription factor AML1, when they compared with a control group [322]. Studying mouse embryos heterozygous for the fusion gene (Cbfb-MYH11) provided *in vivo* evidence supporting the hypothesis that CBFβ-MYH1 acts as dominant negative inhibitor of AML1. These embryos died around day 12.5 by CNS hemorrhage and blockage of the definitive hematopoiesis, a phenotype that is very similar to that seen in AML1 $^{-/-}$ or Cbfb$^{-/-}$ mice [323].

At the cellular level, CBFβ-MYH11 slows the cell proliferation by blocking the progression from G1 to S phase, however it reduces apoptosis by down-regulating p53 via p21-independent mechanism [317,324,325]. Induction of CBFβ-MYH1 in the murine granulocytic cell line 32D cl3 and B-lymphoid Ba/F3 reduced their proliferation capacity by reducing G1 to S phase transition, whereas the G-CSF-induced differentiaion of 32D cells and their myeloperoxiase RNA expression were not affected [325]. Retroviral transduction of CBFβ-MYH11 of human CD34^{+} or murine progenitor cells led to reduced proliferation in liquid culture. Moreover, CBFβ-MYH11 decreased the number and the size of myeloid, but not erythroid, colonies in semi-solid culture. However, CBFβ-MYH11 does not affect granulocytic or myeloid differentiation [321]. To study the role of Cbfb-MYH11 during adult hematopoiesis and its role to leukemogenesis, Chimeras with Cbfb$^{+/cbfb\text{-}MYH11}$ ES cells were generated. These chimeras did not develop any AML in the first year of their life; However, they had defect in myeloid and lymphoid, but not erythroid, differentiation *in vivo* and *in vitro*; a phenotype that is also similar to that seen in cbfb$^{-/--}$ chimeras [297,326]. However, The ES heterozygous for the Cbfb-MYH11 allele produced significantly more definitive hematopoietic progenitors than either cbfb$^{-/-}$ or AML1$^{-/-}$ ES cells *in vitro* [291]. Moreover, Cbfb-MYH11 led to immortalization of hematopoietic stem cells [297]. Moreover, when these chimeras were exposed to sublethal dose of an alkylating agent (EUN), 84% of the animals developed acute leukemia within 2-6 month [326]. Furthermore, the leukemia developed in these mice has some phenotypic similarity with human AML with inv(16) including the myelomonocytic differentiation and the excess of eosinophils [326]. In transgenic mice with a CBFβ-MYH11 allele expressed under the hMRP8 promoter, the mice did not show any leukemia with a normal life span, however partial impairment in neutrophilic differentiaion with dysplastic features was seen [327]. In a conditional CBFβ-MYH11 knock-in mice model, CBFβ-MYH11 led to the expansion of the hematopoietic stem cell compartment, but their multilineage repopulation capacity was reduced. Analysis of myeloid colonies revealed an increase in CFU-GEMM with reduction in CFU-GM. Moreover, while CBFβ-MYH11 blocked definitive B-lymphoid and megakaryocytic differentiation, it increased CFU-Meg progenitors, indicating a block in CFU-Meg to megacaryocye transition [328]. Taken together, these results indicate that CBFβ-MYH11 leads to expansion of the HSCs and some progenitors in addition to disruption of hematopoietic differentiation, but it is not sufficient for leukemia. When transgenic mice with a CBFβ-MYH11 allele expressed under the hMRP8 promoter crossed with transgenic mice that express constitutively activated NRAS (NRAS^{G12D}) under the control of The resulting double transgenic mice developed increased immature neutrophilic cells with severe neutrophilic dysplasia, however without overt AML [327]. To identify potential cooperating genes, newborn CBFβ-MYH11 knock-in chimeric mice were treated with retrovirus to induce changes in the genome by insertional mutagenesis. This retroviral exposure resulted in the development of AML in 63% of cases, but none of the control wild-type mice exposed to the virus. The latency period was 3-6 months and the phenotype was similar to those seen in ENU-treated chimeras. Analysis of the isertional sites suggested non-random integration with six genes were inserted more than once, and they are Plag1, Plagl2, D6Mm, H2T24, Myb and AML2 [329]. When Plag-1 or Plagl-2 were retrovirally transduced in bone marrow cells of transgenic mice expressing the chimeric protein CBFβ-MYH11, they developed leukemia 3-12 weeks after transplantation, suggesting that Plag1 and plgl2 independently cooperate with CBFβ-MYH11 to induce AML *in vivo*. PLAG1 is an oncogene that has been associated with many human tumors, including pleomorphic adenoma of the salivary glands, lipoblastoma and hepatoblastom. Interestingly, overexpression of PLAG1 and PLAGL2 was seen in 20% of AML patients with preferential expression of PLGL2 in AML with inv(16). Finally, it has been shown also that loss of the cell cycle regulator including p16INK4a and p19ARF cooperates with CBFβ-MYH11 to induce leukemia in mice [318].

The Abnormalities of 11q23

Chromosomal translocations involving 11q23 affect a gene called the mixed-lineage leukemia or myeloid/lymphoid leukemia gene, MLL (Also named HTRX, HRX, TRX1 or ALL1). Translocations invoving the MLL gene have been reported in 5-10% of cases of AML. Moreover, chromosomal aberrations involving 11q23 have, been found in approximately 20% acute lymophoblastic leukemia (ALL) and also in acute biphenotypic leukemia.These abnormalities are more common in children. Two clinical groups of patients have high frequency of 11q23 aberrations and AML: the first is infants with acute leukemia and the second is patients with therapy-related AML (t-AML). MLL translocations have been found in 80% of infants with AML and ALL [330]. In infant leukemia, MLL translocations are leukemia are likely to occur in utero, however full leukmic transformation is usually seen between the age of 5-20 months. Epidemeological studies suggested that maternal exposure tp pesticide, marijuana or excess of flavonoides might be associated with acute leukemia in infants. In therapy-related acute myeloid leukemia (t-AML), MLL aberrations occur in 25% of cases, particularly following treatment by topoisomerase II inhibitors. In general, the translocations in therapy-related leukemias are the same as those occurring in "de novo" leukemia, including t(9;11), t(10; 11) and t(11;19), although the location of breakpoint in MLL gene might differ. AML cases with translocations involving the MLL gene are likely to have morophology of acute monoblastic leukemia (AML-M5 of FAB classification) or acute myelomonocytic leukemia (AML-M4 of FAB classification). Moreover, they have characteristic immunophenotypic features, including high frequency of CD56, CD4 and rare expression of CD34. AML with 11q23 translocation are usally considered, by most authers, of intermediate prognosis.

MLL maps on chromosome 11, at 11q23.3. It was initially cloned by the virtue of its involvement in the t(4;11) found mainly in ALL and translocations (9;11) and (11;19) found in AML. The MLL genomic structure covers over 100 kb and consists of 36 exons. It produces a mRNA of 12Kb. In addition to chromosomal rearrangements, MLL gene undergoes the so-called partial tendem duplication (MLL-PTD) [331]. This abnormality occurs in approximately 5% of AML cases, but particularly in cases with normal karyotype or trisomy 11. These duplications consist of an in-frame fusion of a 5' segment of the MLL gene with itself and lead to a potentially translatable sequence.

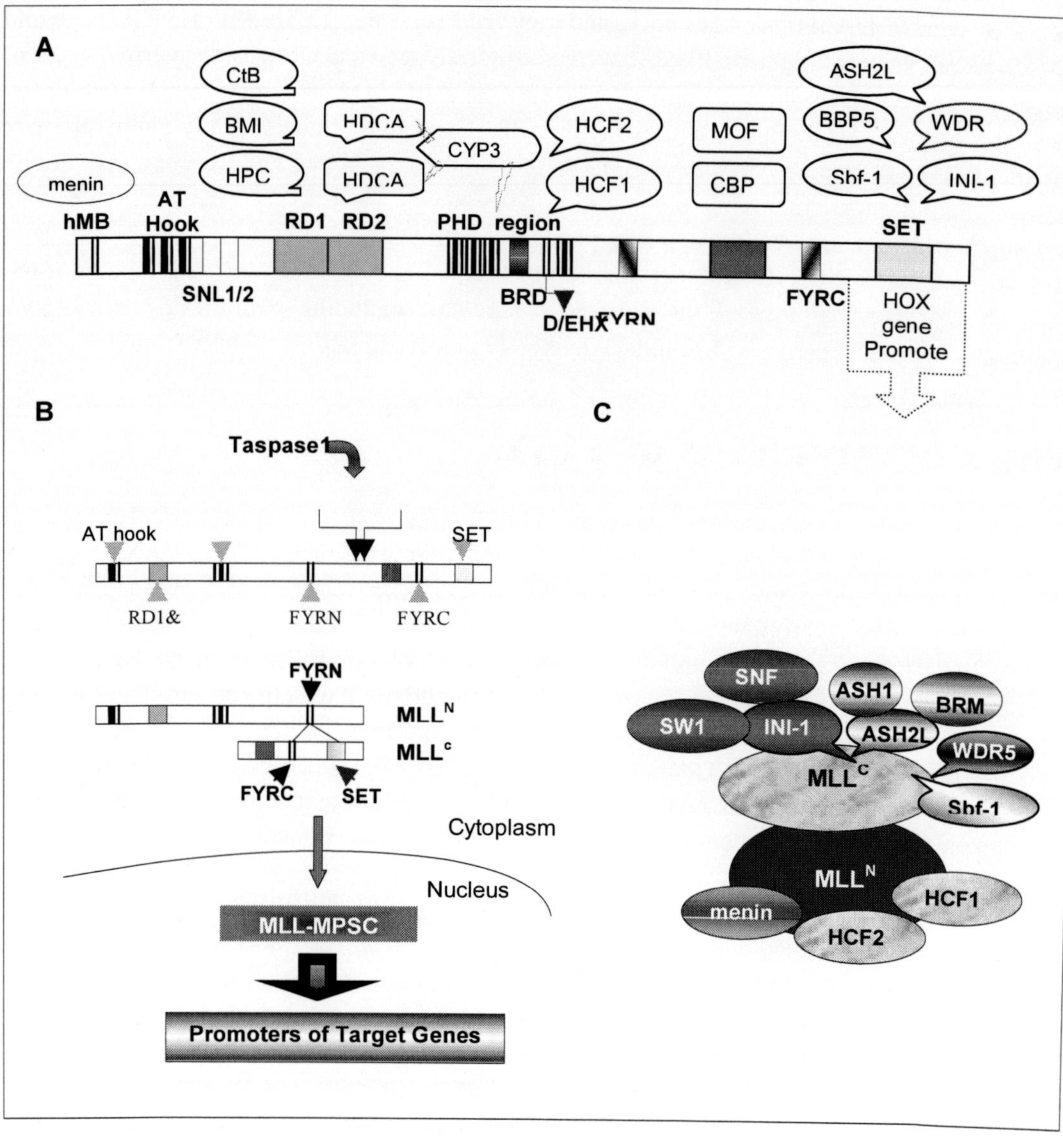

Figure 10. Stucture of the MLL protein, its interaction with other molecules, its cleavage and its formation of MLL multiprotein supercomplex (MPSC). (A) MLL functional domains with its major interacting proteins. Menin interacts with the hMBM domain. The supressors CtBP, BMI-1, HPC2 interacts with the RD1 domain. HDAC1&2 interact with RD2. Cyp33 interacts with third finger of PHD and stabilizes the interaction of HDACs with the RD2. HCF1&2 interact with the D/EHXY sequence. MOF and CBP interact with the TAD. ASHL2, BBP5, Sbf-1, WDR5 and INI-1 interact with SET. (B) Cleavage of MLL in cytoplasm into two major segments MLL^N and MLL^c that bind to each other through the interaction of FYRN motif with FYRC motif and the SET domain. Then, the assembled MLL fragments enter the nucleus to serve as a basic structure on which the MLL multiprotein supercomplex (MLL-MPSC) is built. (C) Schematic presentation of MLL-MPSC. MLL^N interacts with multiple proteins including menin, HCF1 and HCF2. MLL^C interacts with INI-1, AS2L, WRD5 and Sbf-1. INI-1, in turn, binds to SNF and SW1. In addition, ASH2L binds to ASH1 and BRM.

The MLL protein is a large molecule composed of 3969 amino acids and has a molecular weight of 431kDa. The predicted amino acid sequence indicates that it is homologous to the trithorax gene of D. melnogaster [324, 325] (figure 10 A). The native MLL protein undergoes post-translational modification that consists of cleavage at two conserved amino acids motifs, namely CS1 (D/GADD) located at the amino residues 2666/2667 and CS2 (D/GVDD) located at 2718/2719 (figure 10B). This process is carried out by a novel threonine protease called taspase 1 (an endopeptidase with an aspraginase 2 homolgy domain) [334, 335]. The products of this proteolysis are two fragments: an N-terminal 320kDa (amino acids 1-2666; MLL^{N}) with both transcription activation and repression functional domains, and C-terminal 180kDa (amino acids 2719-3969; MLL^{C}) with strong transactivation property [334]. Processed MLL^{N} and MLL^{C} fragments bind noncovalently to each other via N-terminal (1253-2254 amino acids) and C-terminal (3602-3742 amino acids) intramolecular interaction domains [334](figure 10B). This proteolytic cleavage and association seems to confer protein stability and proper subnuclear localization and gene expression [335,336]. MLL is a nuclear protein with several putative functional domains, including the hMBM motif, the AT hook region, the SNL1 and SNL2 regions, RD1 and RD2 domains, PHD zing-finger domain, bromodomain, the FYRN domain, the TAD domain, the FYRNC domain and the SET domain (figure 10A). The high-affinity menin-binding motif (hMBM) is composed of the consensus sequence (RXRFP) located near the N- terminal (amino acids 6-10) and required for menin binding [337,338]. Menin is the product of the tumor suppressor gene MEN1, which is muted in the heritable and sporadic multiple endocrine neoplasia syndrome type 1. By interacting with MLL, menin plays an important role in the transactivation of Hox and cyclin-dependant kinase inhibitor genes [338,339]. The AT hook region consists of three AT hook motifs. These motifs, with homology to the high-mobility group I (HMG-I) protein, are composed of repeats of the sequence Arg-Ly-Srg-Pro and mediate binding to the AT-rich regions of the minor groove of the DNA through recognizing DNA structure rather specific consensus sequence [340,341]. The speckled nuclear localization signals 1 and 2 (SNL1 and SNL2) guide the MLL protein to the nucleus. MLL is localized in a punctate pattern of distribution close to the nuclear envelope and the periphery of the nucleolus in the resting cells and surrounds the condensed chromosomes in the mitotically activated cells [342,343]. The repression domain 1 (RD1) contains the DNMT1 DNA methyltansferase (MT) homology region including the cystein-rich CXXC zinc finger domain [342,344]. It has DNA-binding activity to unmethylated CpG sequence and exerts its transcriptional repression activity by recruiting the polycomb group (PcG) repressor proteins human polycomb2 (HPC2) and BMI-1, and the corepressor C-terminal-binding protein (CtBP) [345,346]. The repression domain 2 (RD2) mediates repression through recruitment of the histone deacytelases HDCA1 and HDCA2, which can interact with RD1. The central part of MLL harbors the plant homeodomain (PHD) zing fingers with a nested bromodomain. The PHD region has 4 zinc-finger motifs. Little is known about the functions of these motifs. Many if not all PHD proteins are nuclear and implicated in the control of chromatin or transcription [347]. The PHD region of MLL did not have transcriptional activating or repression function. However, it binds to Cyclophilin 33 (Cyp33) that can contribute to the transcriptional activity of MLL. Cyp33 is a member of the cyclophilin family with an RNA recognition motif (RRM). It binds to the third PHD zinc-finger domain through its RRM domain and enhances the repression activity of the HDACs by increasing its attachment to the RD2 domain. Overexpression Cyp 33 leads to down-regulation of HOXC8 and HOXC9 RNA [348,349]. The exact function of the BRD domain in the particular case of MLL has not yet been identified. However, it is still possible to speculate regarding its potential role. Indeed, the bromodomain has been found in many transcription factors and chromatin regulator. Several functions of the bromodomain have been recognized, including chromatin acetylation by HATs, acetylation-dependent nucleosome assembly and remodeling, chromatin reorganization and interaction with acetylated non-histone proteins. Association between the PHD motifs and the bromodomain (BRD) has been recognized in many proteins. The PHD fingers seem to cooperate with the adjacent BRD to constitute a functional nucleosome-binding module designed to target or to tether chromatin-remodeling complexes to acetylated stretches of the nucleosomes. Between the BRD and the fourth Zinc finger motif is located the consensus sequence D/EHXY (residues 1799-1803) that interacts with the transcriptional coregulators HCF1 and HCF2 proteins [337]. Little is known about HCF2. On the other hand, it has been shown that HCF1 is a regulator of cell cycle and cell growth by regulating transcription, both positively and negatively [350]. The molecular and cellular consequences of the interaction between MLL and HCF proteins are not known yet .MLL has a transcriptional activation domain (TAD) that binds to the co-activators CBP and MOF. CBP, in turn, binds to phosphorylated CREB to induce the acetylation of H3 and H4 and promote the transcription of HOXA9 and HOXC8 [351]. MOF mediates H3K4 methylation and H4K16 acetylation to induce transactivation of MLL targets [352]. Near the C terminus there is a SET [Su(var)3-9, enhancer-of-zeste trithorax] domain, which was shown recently to have an intrinsic histone methyltransferase activity promoting transactivation of Hox genes through methylation of histone3 at Lys 4 [353]. Moreover, SET interacts with the human proteins ASH2L, WDR5, RBBP5, INI-1 and Sbf-1 (SET binding factor 1) [337, 354-356]. INI-1 is a component of the SW1/SNF multi protein complex, which acts to remodel chromatin and consequently to activate transcription. Sfb-1 functions as SET-dependent regulator of growth-inducing kinase signaling pathways and appears to be important for B-cell-progenitors regulation. ASH2L, in turn interacts with ASH1 and BRM. The FYRN (FY-rich, N-terminal) domain (amino acids 1979-2130) located in MLL^N (N320) interacts with the FYRC (FY-rich, C-terminal) and SET domains (amino acids 3656-3969) located in MLL^c (C180), thus building the basic structure for MLL multiprotein supercomplex (MLL-MPSC) that consists of the two MLL fragments and at least 29 additional proteins [329,348] (figure 10C). The majority of these proteins are components of the human transcription complexes that include SW1/SNF, NuRD, Sin3A, TFIID, SAGA and hSNF2H [349,350]. As concluded from its composition and from the experimental data, MLL-MPSC has roles in chromatin remodeling and histone modification [357].

Many studies have demonstrated the importance of the MLL protein for Hox gene expression, including several Hox A cluster genes (Hoxa5, Hoxa7, Hoxa9, Hoxa10), several Hox B cluster genes (Hoxb4, Hoxb5, Hoxb6, Hoxb8), several Hox C cluster genes (Hoxc6, Hoxc8, Hoxc9) [359-362]. In fact, MLL has no role in the initiation of Hox gene transcription, but in its maintenance [363]. Regulation of Hox genes by wild-type MLL involves both the SET domain –mediated histone H3lys4 methylation [345,348] and recruitment of HATs by the TAD domain [351,352]. In addition, there is evidence that MLL activates cyclin-dependent kinase inhibitors, including $p27^{kip1}$ and $P18^{Ink4c}$, through its interaction with menin [331]. Other genes, also, have been shown to be downstream targets of Mll including the many DLX genes (DLX2, DLX3, DLX5, DLX6 and DLX7) and the Ptx2 gene [364,365]. Moreover, gene expression profiling of $Mll^{+/+}$ and $Mll^{-/-}$ murine cell lines revealed 197 potential target genes of Mll [366].

Animal studies have been shown that MLL gene has very important role in morphogenesis, particularly that of axial-skeletal morphogenesis and hemoatopoietic system development. In contrast to heterozygous mice, animal with homozygous loss of Mll die at 10.5 dpc [359,367,368] and leads to defects in yolk sac and liver hematiopoesis [359,360]. In vitro analysis of yolk sac and fetal progenitors taken from mice the were $MLL^{+/+}$, $Mll^{+/-}$ or $Mll^{-/-}$ revealed that cells that are $Mll^{-/-}$ give rise to early colonies that were fewer in number, smaller in size and took longer time to develop. Analysis of the lineage-restricted colonies showed marked decrease in CFU-GEMM, CFU-M and BFU-E, while CFU-GM, CFU-G and CFU-E were unaffected. Moreover, liquid culture of $Mll^{-/-}$ cells did not reveal any maturation arrest [367,368]. Analysis of $Mll^{-/-}$chimera showed that Mll-deficient cells failed to contribute to fetal liver hematioietic stem cells/progenitor population. AGM cells from these chimeras lacked HSCs activity despite their ability to generate hematioietic progeny *in vitro* [362]. However, $Mll^{-/+}$ mice were viable, they exhibit, in addition growth retardation and skeletal abnormalities, anemia and thrombocytopenia without morphological changes. Moreover, these mice had decreased number of B-cell population [359]. Taken together, MLL seems to be crucial for the generation and growth early hematopoietic stem/progenitor cells while it does not affect their late differentiation potential.

In AML, more than 39 partners have been cloned, and the most frequently implicated sites include 6q27, 9p22, 10p12, 17q21, 19p13.1 and 19p13.3, involving the AF6, AF9, AF10, AF17, ELL and ENL, respectively (table 1). MLL translocations to the previously described sites account for more than 85% of all AML cases with 11q23 abnormailties. Less frequently involved sites include AF1Q at 1q22 at, AF1P at 1p32, AF3p21 at 3p21, AF4 at 4q21, AF5q31 at 5q31, ARHGAP26 at 5q31, FOXO3A at 6q21, CDK6 at 7q21-q22, MLLT3 at 9q22, FBP17 at 9q34, CBL at 11q23.3, GEPHYRIN at 14q24, AF15q14 at 14q14, CBP at 16p13.3, GAS7 at 17p13, MLLT6 at 17q21, SEPT9 at 17q25, SEPT5 at 22q11.2, MLLT7 at Xq13.1 and SEPT6 at Xq24. All MLL translocations lead to the formation of oncogenic fusion genes that are located on der(11) and consist of the 5' portion of MLL fused to the 3' of the partner gene. The breakpoints on the MLL gene localized in a breakpoint cluster region of 8.5 kb located between exons 5 and 11 (figure 11A&B).

Table 1. The most frequent MLL fusion partners.

Gene	Locus	Putative protein function	Subcellular Location
AF9	9p22	Transcription factors	Nuclear
AF10	10p12	Transcription factors	Nuclear
ENL	19p13.3	Transcription factor	Nuclear
EEL	19p13.1	RNA polymerase II transcription elongation factor	Nuclear
AF6	6q27	Maintenance of cell-cell junctions and cell polarity	Cytoplasmic
AF17	17q21	Transcription factor	Nuclear

Evidence supporting the importance of the reciprocal product for leukemogenesis is two-fold. First, knock-in mice with Mll fusions under the control of the Mll promoter in correct chromatin environment do develop leukemia only if Mll is fused to a partner but not if it is truncated e.g. by the addition of a simple myc-tag [369-371]. Second, many series of retroviral transduction experiments have shown that only the combination Mll with its partners are able to transform hematopoietic precursors whereas the non-dimerized N-terminal of Mll is not oncogenic [372-376]. Indeed, the large number of different partners implicated in MLL translocations has made it challenging to define the exact mechanisms by which the chimeric proteins contribute to leukemogenesis.

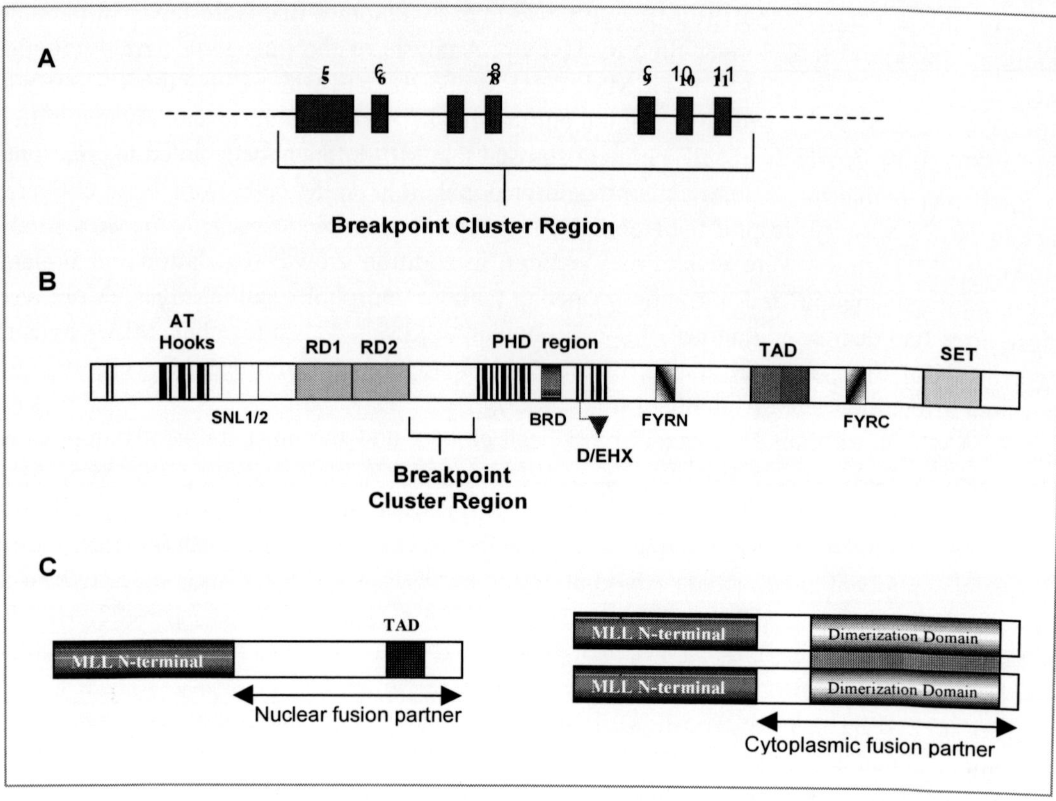

Figure 11. The Breakpoint Cluster region on MLL and the different types of fusion protein formed by MLL translocations. (A) Representation of MLL breakpoint cluster region located between exons 5-11. (B) The location of the breakpoint cluster region on the MLL protein. (C) Two forms of MLL chimeric proteins: with nuclear proteins leading to remplacement of TAD of MLL by a new TAD that pertains to the fusion partner (left) and with cytoplasmic partners leading to homodimerization of the fusion protein (right).

However, the consensus at present is that they act by a gain-of-function mechanism by converting the truncated MLL to a potent transcriptional activator. The classification of MLL fusion into two groups according to the function and cellular localization of the fusion partner may help establish relatively simplified paradigm to understand the mechanism by which the different fusion proteins enhance the transcriptional activity of MLL (figure 11C). The partners of the first class are the most frequent and include nuclear proteins with transcriptional activation function (table 1). These proteins are transcription factors (e.g. AF9, AF10, ENL, AF17, AFX and FKHR-L), transcriptional co-activators (such as p300 and CBP), or proteins such as ELL and ABI1 that can recruit transcription activators. The transcription factors that are fused with MLL have endogenous transctivation domains (TADs) on their C-terminus that replaces the TAD of MLL and converts its N-terminal region to an active oncoprotein [375,377]. This hypothesis was substantiated by the finding that the transcativating domain ENL is necessary and sufficient to confer transforming activity to the chimeric protein MLL-ENL [367]. The mechanism by which the transcription activators CMP and P300 contribute to the leukemic transformation potential of their respective fusions with MLL is not known yet. However, it seems that both the bromodomain and the HAT domain of CBP are required and sufficien for full *in vitro* transformation when fused with the N-terminal of MLL to induce leukemic phenotype *in vivo* [378]. Although ELL does not have a transactivation domain, it can interact with proteins that have transcriptional activity. It has been shown the fusion gene MLL-ELL colocalizes with p53 in the nucleus and interacts, and these colocalization and interaction are dependent on the extreme c-terminus of ELL, which is needed for leukemic transformation ability MLL-ELL, and the transcativating domain of p53 [379]. The exact role of this interaction in the leukemic transformation is yet to be defined. The second class of MLL translocations comprises fusion in which the partners are cytoplasmic proteins. It has been noticed that although these proteins lack any functional similarity, they have common structural feature of possessing coiled-coil domain. This coiled-coil domain is important for the dimerization of the MLL fision products. It has been shown that the dimerization domains, like the coiled-coil domains of MLL-GAS7 and MLL-AF1p, are required for their transformation potential [376]. More importantly, the dimerization of MLL N-terminus predisposes for leukemia. This fact has been corroborated by studies showing that Mll-LacZ knock-in mice succumbed to leukemia, albeit with reduced penetrance and long latency period [370]. LacZ does not possess any transcriptional activity but forms tetramers in solution, suggesting that multidimerization may be the key mechanism of this transformation. Furthermore, an inducible dimerization of N-terminal MLL was shown to cause hematopoietic differentiation blockade, which was reversible by withdrawing the pharmacological dimerization agent [380]. Dimerized MLL is an extremely potent transactivator that has higher binding-affinity to Hox promoters than monomeric MLL leading to upregulation of Hoxa7, Hoxa9 and Meis1 accompanied by enhanced self-renewal of hematopoietic progenitors with myeloid block [376, 380-382]. The mechanism by which the dimerization of truncated MLL alters gene transactivation and disrupts normal hematpoietic growth and differentiation is unknown. However, it has been shown that intact methyltransferase homology, RERE/snRNP and Ala/Ser-rich (AS-rich) domains are required for the transformation ability of MLL dimers [376]. The latter are similar to those required for the transforming activity of MLL-ENL, a fusion with nuclear MLL partner, indicating that both classes of fusion proteins may share similar oncogenic pathways. As discussed above, MLL is a component of MLL-MPSC that exerts histone/chromatin-modifying/remodeling activity leading transactivation or transcription repression activity. In MLL translocations, the normal epigenetic function of MLL is disrupted leading to abnormal gene transcription regulation. Some translocation partners, such as ENL and AF9, recruit SWI-SNF chromatin-remodeling complex. This complex contains also INI-1 and BRG-1. The MLL-ENL transcriptional activity seems to be dependent on the integrity SWI-SNF complex, since transactivation of Hoxa7 promoter by MLL-ENL is inhibited by dominant negative Brg-1 [383]. The MLL fusion partners AF10 and AF17 bind to GAS41that recruits the SWI-SNF complex [384]. The transformation ability of both later fusion proteins requires a conserved leucine-zipper region that binds to GAS41, suggesting that these fusion molecules exert oncogenic chromatin remodeling activity through SWI-SNF recruitment. Other translocations involve partners with known HAT activity, such as CBP and p300, suggesting that abnormal histone acetylation may have a role in leukemogenic potential of these translocations. Recruitment of CBP has been implicated in MLL-AFX and MLL-FKHRL1 fusion genes.

Since HOX genes are positively regulated by MLL protein, it is legitimate to hypothesize that fusion proteins produced by MLL translocations lead to dysregulation of HOX genes. Overexpression of HOXA7, HOXA9 and the homeobox cofactor MEIS1 was consistently seen in patients with MLL abnormalities in several cDNA microarray studies [385-387]. These findings were confirmed by experimental models showing overexpression of Hoxa5, Hoxa7, Hoxa9, Hoxa10 and Meis1 as major consequence of the forced expression of different MLL fusion proteins in murine cells [388,389]. In normal hematopoiesis, HOX gene expression is high in hematopoietic stem cells and decrease gradually during differentiation. Overexpression of HOX genes in hematopoietic cells can dramatically disrupt the differentiation of different cell lineages and contribute to leukemogenesis.

In a knock-in mouse design using 3' human AF9 fragment fused into 5' mouse Mll introduced into ES cells under the control of Mll promoter, the chimeric and heterozygous mice developed an early myeloproliferative-like condition, resulting in the accumulation of Mac1/Gr-1 double positive cells, followed by an overt AML with frequency of 90% and a latency period of from 4.5-11.5 months [369,370]. Identical results were obtained using different mouse model, including Cre-LoxP-mediated *de novo* translocations and retroviral transduction/bone marrow transplantation [390]. For instance, the transplantation of MLL-CBP retrovirally-transduced bone marrow cells in mice induced leukemia of myelomonocytic phenotype in 100% of animals preceded by myelodysplastic syndrome of long duration (4-6 months) [378]. Likewise, similar pathological syndromes were obtained with transplantation of other retrovirally-trasnduced MLL fusion genes, including MLL-AF10, MLL-ELL, MLL-AF1P, MLL-AFX and MLL-FKHRLI. In contrast, bone marrow cells tranduced with the fusion gene MLL-ENL and MLL-GAS7 and transplanted in mice led, consistent with their phenotype in human leukemia, to acute myeloid leukemia, acute lymphoid leukemia or acute biphenotypic leukemia.

Conclusion

Although our knowledge regarding the mechanisms that govern the leukemic transformation has largely increased over the last 15 years, it is clear that much still remains to be learned in this field. Indeed, the cloning and characterization of the major gene rearrangemnts in AML has provided important insight into the regulatory pathways that control normal and leukemic hematopoiesis. Moreover, these studies have revealed two important aspects of the pathogenesis of AML. First, abnormal chromatin remodeling, particularly histone deacetylation, leading to abnormal gene transcription seems to be a recurrent theme in the pathogenic pathway arising from the chimeric genes. Second, leukemogenesis is a multistep process. The expression of the fusion proteins induces molecular and cellular changes that contribute to the leukemic transformation, but is not sufficient for leukemogenesis. In fact, additional molecular aberrations are needed for the development of an overt leukemic phenotype. The identification of these additional mutations appears to be a difficult task.

Based on the encouraging results obtained from the use of ATRA in APL, it is plausible that a detailed understanding of the molecular genetic basis will lead to the discovery of new agents that could improve treatment efficacy and reduce toxicity. It has been possible to compile a list of rational targets against which novel therapy can be developed, including histone deacetylase (HDAC) inhibitors and tyrosine kinase antagonists that target flt3 [391-396]. These agents have shown promise in laboratory studies and small clinical studies. However, more studies are needed to confirm these initial encouraging results.

References

[1] Song WJ, Sullivan MG, Legare RD, Hutchings S, Tan X, Kufrin D, Ratajczak J,Resende IC, Haworth C, Hock R, Loh M, Felix C, Roy DC, Busque L, Kurnit D,Willman C, Gewirtz AM, Speck NA, Bushweller JH, Li FP, Gardiner K, Poncz M,Maris JM, Gilliland DG. Haploinsufficiency of CBFA2 causes familial thrombocytopenia with propensity to develop acute myelogenous leukaemia. *Nat Genet.* 1999; 23(2):166-75.

[2] Buijs A, Poddighe P, van Wijk R, van Solinge W, Borst E, Verdonck L, Hagenbeek A, Pearson P, Lokhorst H. A novel CBFA2 single-nucleotide mutation in familial platelet disorder with propensity to develop myeloid malignancies. *Blood.* 2001; 98(9):2856-8.

[3] Osato M, Yanagida M, Shigesada K, Ito Y. Point mutations of the RUNx1/AML1 gene in sporadic and familial myeloid leukemias. *Int J Hematol.* 2001; 74(3):245-51.

[4] Frohling S, Scholl C, Gilliland DG, Levine RL. Genetics of myeloid malignancies: pathogenetic and clinical implications. *J Clin Oncol.* 2005; 23(26):6285-95.

[5] Verhaak RG, Goudswaard CS, van Putten W, Bijl MA, Sanders MA, Hugens W, Uitterlinden AG, Erpelinck CA, Delwel R, Lowenberg B, Valk PJ. Mutations in nucleophosmin (NPM1) in acute myeloid leukemia (AML): association with other gene abnormalities and previously established gene expression signatures and their favorable prognostic significance. *Blood.* 2005; 106(12):3747-54.

[6] Marcucci G, Mrozek K, Bloomfield CD. Molecular heterogeneity and prognostic biomarkers in adults with acute myeloid leukemia and normal cytogenetics. *Curr Opin Hematol.* 2005; 12(1):68-75.

[7] Look AT. Oncogenic transcription factors in the human acute leukemias. *Science.* 1997; 278(5340):1059-64.

[8] Harris NL, Jaffe ES, Diebold J, Flandrin G, Muller-Hermelink HK, Vardiman J, Lister TA, Bloomfield CD. The World Health Organization classification of hematological malignancies report of the Clinical Advisory Committee Meeting, Airlie House, Virginia, November 1997. *Mod Pathol.* 2000; 13(2):193-207.

[9] de The H, Lavau C, Marchio A, Chomienne C, Degos L, Dejean A. The PML-RAR alpha fusion mRNA generated by the t(15;17) translocation in acute promyelocytic leukemia encodes a functionally altered RAR. *Cell.* 1991 23; 66(4):675-84.

[10] Miyoshi H, Shimizu K, Kozu T, Maseki N, Kaneko Y, Ohki M. t(8;21) breakpoints on chromosome 21 in acute myeloid leukemia are clustered within a limited region of a single gene, AML1. *Proc Natl Acad Sci U S A.* 1991; 88(23):10431-4.

[11] Ziemin-van der Poel S, McCabe NR, Gill HJ, Espinosa R III, Patel Y, Harden A, Rubinelli P, Smith SD, LeBeau MM, Rowley JD, et al. Identification of a gene, MLL, that spans the breakpoint in 11q23 translocations associated with human leukemias. *Proc Natl Acad Sci U S A.* 1991; 88(23):10735-9.

[12] Liu P, Tarle SA, Hajra A, Claxton DF, Marlton P, Freedman M, Siciliano MJ, Collins FS. Fusion between transcription factor CBF beta/PEBP2 beta and a myosin heavy chain in acute myeloid leukemia. *Science.* 1993; 261(5124):1041-4.

[13] Alcalay M, Orleth A, Sebastiani C, Meani N, Chiaradonna F, Casciari C, Sciurpi MT, Gelmetti V, Riganelli D, Minucci S, Fagioli M, Pelicci PG. Common themes in the pathogenesis of acute myeloid leukemia. *Oncogene.* 2001; 20(40):5680-94.

[14] Gilliland DG, Jordan CT, Felix CA. The molecular basis of leukemia. *Hematology (Am Soc Hematol Educ Program).* 2004: 80-97.

[15] Tallman MS. Acute promyelocytic leukemia as a paradigm for targeted therapy. *Semin Hematol.* 2004; 41(2 Suppl 4):27-32.

[16] Jing Y. The PML-RARalpha fusion protein and targeted therapy for acute promyelocytic leukemia. *Leuk Lymphoma.* 2004; 45(4):639-48.

[17] Wichmann C, Becker Y, M Grez The NRH2 domain of AML1/ETO as a Novel Therapeutic Target Structure in t(8;21)- Positive Leukemia. *Blood* 2007; 110:210a

[18] McKenzie SB. Advances in understanding the biology and genetics of acute myelocytic leukemia. *Clin Lab Sci.* 2005; 18(1):28-37.

[19] Jabbour EJ, Estey E, Kantarjian HM. Adult acute myeloid leukemia. *Mayo Clin Proc.* 2006; 81(2):247-60

[20] Lengfelder E, Saussele S, Weisser A, Buchner T, Hehlmann R. Treatment concepts of acute promyelocytic leukemia. *Crit Rev Oncol Hematol.* 2005; 56(2):261-74.

[21] Liso V, Bennett J. Morphological and cytochemical characteristics of leukaemic promyelocytes. *Best Pract Res Clin Haematol.* 2003; 16(3):349-55.

[22] Paietta E. Expression of cell-surface antigens in acute promyelocytic leukaemia. *Best Pract Res Clin Haematol.* 2003; 16(3):369-85.

[23] Leroy P, Nakshatri H, Chambon P. Mouse retinoic acid receptor alpha 2 isoform is transcribed from a promoter that contains a retinoic acid response element. *Proc Natl Acad Sci U S A.* 1991; 88(22):10138-42.

[24] Li YP, Andersen J, Zelent A, Rao S, Paietta E, Tallman MS, Wiernik PH, Gallagher RE. RAR alpha1/RAR alpha2-PML mRNA expression in acute promyelocytic leukemia cells: a molecular and laboratory-clinical correlative study. *Blood.* 1997; 90(1):306-12.

[25] Novac N, Heinzel T. Nuclear receptors: overview and classification. *Curr Drug Targets Inflamm Allergy.* 2004; 3(4):335-46.

[26] Tenbaum S, Baniahmad A. Nuclear receptors: structure, function and involvement in disease. *Int J Biochem Cell Biol.* 1997; 29(12):1325-41.

[27] Moras D, Gronemeyer H. The nuclear receptor ligand-binding domain: structure and function. *Curr Opin Cell Biol.* 1998; 10(3):384-91.

[28] Renaud JP, Moras D. Structural studies on nuclear receptors. *Cell Mol Life Sci.* 2000; 57(12):1748-69.

[29] Minucci S, Leid M, Toyama R, Saint-Jeannet JP, Peterson VJ, Horn V, Ishmael JE, Bhattacharyya N, Dey A, Dawid IB, Ozato K. Retinoid X receptor (RXR) within the RXR-retinoic acid receptor heterodimer binds its ligand and enhances retinoid-dependent gene expression. *Mol Cell Biol.* 1997; 17(2):644-55.

[30] Soprano DR, Qin P, Soprano KJ. Retinoic acid receptors and cancers. *Annu Rev Nutr.* 2004; 24:201-21.

[31] Kurokawa R, DiRenzo J, Boehm M, Sugarman J, Gloss B, Rosenfeld MG, Heyman RA, Glass CK. Regulation of retinoid signalling by receptor polarity and allosteric control of ligand binding. *Nature.* 1994; 371(6497):528-31.

[32] Perez A, Kastner P, Sethi S, Lutz Y, Reibel C, Chambon P. PMLRAR homodimers: distinct DNA binding properties and heteromeric interactions with RXR. *EMBO J.* 1993; 12(8):3171-82.

[33] Gianni M, Terao M, Fortino I, LiCalzi M, Viggiano V, Barbui T, Rambaldi A, Garattini E. Stat1 is induced and activated by all-trans retinoic acid in acute promyelocytic leukemia cells. *Blood.* 1997; 89(3):1001-12.

[34] Kolla V, Weihua X, Kalvakolanu DV. Modulation of interferon action by retinoids. Induction of murine STAT1 geneexpression by retinoic acid. *J Biol Chem.* 1997; 272(15):9742-8.

[35] Matikainen S, Lehtonen A, Sareneva T, Julkunen I. Regulation of IRF and STAT gene expression by retinoic acid. *Leuk Lymphoma.* 1998; 30(1-2):63-71.

[36] Pelicano L, Li F, Schindler C, Chelbi-Alix MK. Retinoic acid enhances the expression of interferon-induced proteins: evidence for multiple mechanisms of action. *Oncogene.* 1997; 15(19):2349-59.

[37] Langston AW, Thompson JR, Gudas LJ. Retinoic acid-responsive enhancers located 3' of the Hox A and Hox B homeobox gene clusters. Functional analysis. *J Biol Chem.* 1997; 272(4):2167-75.

[38] Langston AW, Gudas LJ. Identification of a retinoic acid responsive enhancer 3' of the murine homeobox gene Hox-1.6. Mech Dev. 1992; 38(3):217-27.

[39] Ogura T, Evans RM. Evidence for two distinct retinoic acid response pathways for HOXB1 gene regulation. *Proc Natl Acad Sci U S A.* 1995; 92(2):392-6.

[40] Popperl H, Featherstone MS. Identification of a retinoic acid response element upstream of the murine Hox-4.2 gene. *Mol Cell Biol.* 1993; 13(1):257-65.

[41] Duprez E. A new role for C/EBPbeta in acute promyelocytic leukemia. *Cell Cycle.* 2004; 3(4):389-90.

[42] Hauksdottir H, Farboud B, Privalsky ML. Retinoic acid receptors beta and gamma do not repress, but instead activate target gene transcription in both the absence and presence of hormone ligand. *Mol Endocrinol.* 2003; 17(3):373-85.

[43] Gratas C, Menot ML, Dresch C, Chomienne C. Retinoid acid supports granulocytic but not erythroid differentiation of myeloid progenitors in normal bone marrow cells. *Leukemia.* 1993; 7(8):1156-62.

[44] Labbaye C, Valtieri M, Testa U, Giampaolo A, Meccia E, Sterpetti P, ParoliniI, Pelosi E, Bulgarini D, Cayre YE, et al. Retinoic acid downmodulates erythroid differentiation and GATA1 expression in purified adult-progenitor culture. *Blood.* 1994; 83(3):651-6.

[45] Van Bockstaele DR, Lenjou M, Snoeck HW, Lardon F, Peetermans ME. Effect of retinoic acid (RA) on myeloid progenitors in normal bone marrow (BM). *Leukemia.* 1994; 8(1):214-5.

[46] Tocci A, Parolini I, Gabbianelli M, Testa U, Luchetti L, Samoggia P, Masella B, Russo G, Valtieri M, Peschle C. Dual action of retinoic acid on human embryonic/fetal hematopoiesis: blockade of primitive progenitor proliferation and shift from multipotent/erythroid/monocytic to granulocytic differentiation program. *Blood.* 1996; 88(8):2878-88.

[47] Tsai S, Bartelmez S, Sitnicka E, Collins S. Lymphohematopoietic progenitors immortalized by a retroviral vector harboring a dominant-negative retinoic acid receptor can recapitulate lymphoid, myeloid, anderythroid development. *Genes Dev.* 1994; 8(23):2831-41.

[48] Lardon F, Snoeck HW, Haenen L, Lenjou M, Nijs G, Weekx SF, Van Ranst PC, Berneman ZN, Van Bockstaele DR. The combined effects of all-trans retinoic acid and TGF-beta on the initial proliferation of normal human bone marrow progenitor cells. *Leukemia.* 1996; 10(12):1937-43.

[49] Purton LE, Bernstein ID, Collins SJ. All-trans retinoic acid enhances the long-term repopulating activity of cultured hematopoietic stem cells. *Blood.* 2000; 95(2):470-7.

[50] Purton LE, Morris JC, Bernstein ID, Collins SJ, Kiem HP. All-trans retinoic acid facilitates oncoretrovirus-mediated transduction of hematopoietic repopulating stem cells. *J Hematother Stem Cell Res.* 2001; 10(6):815-25.

[51] Lufkin T, Lohnes D, Mark M, Dierich A, Gorry P, Gaub MP, LeMeur M, Chambon P. High postnatal lethality and testis degeneration in retinoic acid receptor alpha mutant mice. *Proc Natl Acad Sci U S A.* 1993; 90(15):7225-9.

[52] Lohnes D, Mark M, Mendelsohn C, Dolle P, Dierich A, Gorry P, Gansmuller A, Chambon P. Function of the retinoic acid receptors (RARs) during development (I). Craniofacial and skeletal abnormalities in RAR double mutants. *Development.* 1994; 120(10):2723-48.

[53] Sucov HM, Dyson E, Gumeringer CL, Price J, Chien KR, Evans RM. RXR alpha mutant mice establish a genetic basis for vitamin A signaling in heart morphogenesis. *Genes Dev.* 1994; 8(9):1007-18.

[54] Kastner P, Grondona JM, Mark M, Gansmuller A, LeMeur M, Decimo D, Vonesch JL, Dolle P, Chambon P. Genetic analysis of RXR alpha developmental function: convergence of RXR and RAR signaling pathways in heart and eye morphogenesis. *Cell.* 1994; 78(6):987-1003.

[55] de The, H.; Chomienne, C.; Lanotte, M.; Degos, L.; Dejean, A. The t(15;17) translocation of acute promyelocytic leukaemia fuses the retinoic acid receptor alpha gene to a novel transcribed locus. *Nature* 1990; 347: 558-561.

[56] Salomoni P, Bellodi C. New insights into the cytoplasmic function of PML. *Histol Histopathol.* 2007; 22(8):937-46.

[57] D, Grignani F, Pelicci PG. Alternative splicing of PML transcripts predicts coexpression of several carboxy-terminally different protein isoforms. *Oncogene.* 1992; 7(6):1083-91.

[58] Fogal V, Gostissa M, Sandy P, Zacchi P, Sternsdorf T, Jensen K, Pandolfi PP, Will H, Schneider C, Del Sal G. Regulation of p53 activity in nuclear bodies by a specific PML isoform. *EMBO J.* 2000; 19(22):6185-95

[59] Hofmann H, Sindre H, Stamminger T. Functional interaction between the pp71 protein of human cytomegalovirus and the PML-interacting protein human Daxx. *J Virol.* 2002; 76(11):5769-83.

[60] Jensen K, Shiels C, Freemont PS. ML protein isoforms and the RBCC/TRIM motif. *Oncogene.* 2001; 20(49):7223-33

[61] Nguyen LA, Pandolfi PP, Aikawa Y, Tagata Y, Ohki M, Kitabayashi I. Physical and functional link of the leukemia-associated factors AML1 and PML. *Blood.* 2005; 105(1):292-300.

[62] Maul GG, Negorev D, Bell P, Ishov AM. Review: properties and assembly mechanisms of ND10, PML bodies, or PODs. *J Struct Biol.* 2000; 129(2-3):278-87.

[63] Negorev D, Ishov AM, Maul GG. Evidence for separate ND10-binding and homo-oligomerization domains of Sp100. *J Cell Sci.* 2001; 114(Pt 1):59-68.

[64] Hodges M, Tissot C, Howe K, Grimwade D, Freemont PS. Structure, organization, and dynamics of promyelocytic leukemia protein nuclear bodies. *Am J Hum Genet.* 1998; 63(2):297-304.

[65] LaMorte VJ, Dyck JA, Ochs RL, Evans RM. Localization of nascent RNA and CREB binding protein with the PML-containing nuclear body. *Proc Natl Acad Sci U S A.* 1998; 95(9):4991-6.

[66] Alcalay M, Tomassoni L, Colombo E, Stoldt S, Grignani F, Fagioli M, Szekely L, Helin K, Pelicci PG. The promyelocytic leukemia gene product (PML) forms stable complexes with the retinoblastoma protein. *Mol Cell Biol.* 1998; 18(2):1084-93.

[67] Zhong S, Hu P, Ye TZ, Stan R, Ellis NA, Pandolfi PP. A role for PML and the nuclear body in genomic stability. *Oncogene.* 1999; 18(56):7941-7.

[68] Zhong S, Muller S, Ronchetti S, Freemont PS, Dejean A, Pandolfi PP. Role of SUMO-1-modified PML in nuclear body formation. *Blood.* 2000; 95(9):2748-52.

[69] Zhong S, Salomoni P, Ronchetti S, Guo A, Ruggero D, Pandolfi PP. Promyelocytic leukemia protein (PML) and Daxx participate in a novel nuclear pathway for apoptosis. *J Exp Med.* 2000; 191(4):631-40

[70] Ishov AM, Sotnikov AG, Negorev D, Vladimirova OV, Neff N, Kamitani T, Yeh ET, Strauss JF 3rd, Maul GG. PML is critical for ND10 formation and recruits the PML-interacting protein daxx to this nuclear structure when modified by SUMO-1. *J Cell Biol.* 1999; 147(2):221-34.

[71] Muller S, Matunis MJ, Dejean A. Conjugation with the ubiquitin-related modifier SUMO-1 regulates the partitioning of PML within the nucleus. *EMBO J.* 1998; 17(1):61-70.

[72] Everett RD, Meredith M, Orr A, Cross A, Kathoria M, Parkinson J. A novel ubiquitin-specific protease is dynamically associated with the PML nuclear domain and binds to a herpesvirus regulatory protein. EMBO J. 1997; 16(3):566-77. Corrected and republished in: *EMBO J.* 1997; 16(7):1519-30.

[73] Seeler JS, Marchio A, Sitterlin D, Transy C, Dejean A. Interaction of SP100 with HP1 proteins: a link between the promyelocytic leukemia-associated nuclear bodies and the chromatin compartment. *Proc Natl Acad Sci U S A.* 1998; 95(13):7316-21.

[74] Bloch DB, de la Monte SM, Guigaouri P, Filippov A, Bloch KD. Identification and characterization of a leukocyte-specific component of the nuclear body. *J Biol Chem.* 1996; 271(46):29198-204.

[75] Topcu Z, Mack DL, Hromas RA, Borden KL. The promyelocytic leukemia protein PML interacts with the proline-rich homeodomain protein PRH: a RING may link hematopoiesis and growth control. *Oncogene.* 1999 Nov; 18(50):7091-100.

[76] Ishov AM, Sotnikov AG, Negorev D, Vladimirova OV, Neff N, Kamitani T, Yeh ET, Strauss JF 3rd, Maul GG. PML is critical for ND10 formation and recruits the PML-interacting protein daxx to this nuclear structure when modified by SUMO-1. *J Cell Biol.* 1999; 147(2):221-34.

[77] Boddy MN, Howe K, Etkin LD, Solomon E, Freemont PS. PIC 1, a novel ubiquitin-like protein which interacts with the PML component of a multiprotein complex that is disrupted in acute promyelocytic leukaemia. *Oncogene.* 1996; 13(5):971-82.

[78] Muller S, Miller WH Jr, Dejean A. Trivalent antimonials induce degradation of the PML-RAR oncoprotein and reorganization of the promyelocytic leukemia nuclear bodies in acute promyelocytic leukemia NB4 cells. *Blood.* 1998; 92(11):4308-16.

[79] Chang, K.-S.; Fan, Y.-H.; Andreeff, M.; Liu, J.; Mu, Z.-M. The PML gene encodes a phosphoprotein associated with the nuclear matrix. *Blood* 1995, 85: 3646-3653.

[80] Kamitani T, Kito K, Nguyen HP, Wada H, Fukuda-Kamitani T, Yeh ET. Identification of three major sentrinization sites in PML. *J Biol Chem.* 1998; 273 (41):26675-82.

[81] Sternsdorf T, Jensen K, Will H. Evidence for covalent modification of the nuclear dot-associated proteins PML and Sp100 by PIC1/SUMO-1. *J Cell Biol.* 1997; 139(7): 1621-34.

[82] Sternsdorf T, Puccetti E, Jensen K, Hoelzer D, Will H, Ottmann OG, Ruthardt M. PIC-1/SUMO-1-modified PML-retinoic acid receptor alpha mediates arsenic trioxide-induced apoptosis in acute promyelocytic *leukemia. Mol Cell Biol.* 1999; 19(7):5170-8.

[83] Borden KL, CampbellDwyer EJ, Salvato MS. The promyelocytic leukemia protein PML has a pro-apoptotic activity mediated through its RING domain. *FEBS Lett.* 1997; 418(1-2):30-4.

[84] Li L, He D, He H, Wang X, Zhang L, Luo Y, Nan X. Overexpression of PML induced apoptosis in bladder cancer cell by Caspase-dependent pathway. *Cancer Lett.* 2006; 236(2):259-68.

[85] Son SH, Yu E, Choi EK, Lee H, Choi J. Promyelocytic leukemia protein-induced growth suppression and cell death in liver cancer cells. *Cancer Gene Ther.* 2005; 12(1):1-11.

[86] Quignon F, De Bels F, Koken M, Feunteun J, Ameisen JC, de The H. PML induces a novel caspase-independent death process. *Nat Genet.* 1998; 20(3):259-65.

[87] Wang ZG, Ruggero D, Ronchetti S, Zhong S, Gaboli M, Rivi R, Pandolfi PP. PML is essential for multiple apoptotic pathways. *Nat Genet.* 1998; 20(3):266-72.

[88] Fagioli M, Alcalay M, Tomassoni L, Ferrucci PF, Mencarelli A, Riganelli D, Grignani F, Pozzan T, Nicoletti I, Grignani F, Pelicci PG. Cooperation between the RING + B1-B2 and coiled-coil domains of PML is necessary for its effects on cell survival. *Oncogene.* 1998; 16(22):2905-13.

[89] Li H, Chen JD. PML and the oncogenic nuclear domains in regulating transcriptional repression. *Curr Opin Cell Biol.* 2000; 12(5):641-4.

[90] Li H, Leo C, Zhu J, Wu X, O'Neil J, Park EJ, Chen JD. Sequestration and inhibition of Daxx-mediated transcriptional repression by PML. *Mol Cell Biol.* 2000; 20(5):1784-96.

[91] Bernardi R, Scaglioni PP, Bergmann S, Horn HF, Vousden KH, Pandolfi PP.PML regulates p53 stability by sequestering Mdm2 to the nucleolus. *Nat Cell Biol.* 2004; 6(7):665-72.

[92] Pearson M, Carbone R, Sebastiani C, Cioce M, Fagioli M, Saito S, Higashimoto Y, Appella E, Minucci S, Pandolfi PP, Pelicci PG. PML regulates p53 acetylation and premature senescence induced by oncogenic Ras. *Nature.* 2000; 406(6792):207-10.

[93] Vallian S, Chin KV, Chang KS. The promyelocytic leukemia protein interacts with Sp1 and inhibits its transactivation of the epidermal growth factor receptor promoter. *Mol Cell Biol.* 1998; 18(12):7147-56.

[94] Gostissa M, Hengstermann A, Fogal V, Sandy P, Schwarz SE, Scheffner M, Del Sal G. Activation of p53 by conjugation to the ubiquitin-like protein SUMO-1. *EMBO J.* 1999; 18(22):6462-71.

[95] Rodriguez MS, Desterro JM, Lain S, Midgley CA, Lane DP, Hay RT. SUMO-1 modification activates the transcriptional response of p53. *EMBO J.* 1999; 18(22): 6455-61.

[96] Yankiwski V, Noonan JP, Neff NF. The C-terminal domain of the Bloom syndrome DNA helicase is essential for genomic stability. *BMC Cell Biol.* 2001; 2:11.

[97] Sanz MM, Proytcheva M, Ellis NA, Holloman WK, German J. BLM, the Bloom's syndrome protein, varies during the cell cycle in its amount, distribution, and co-localization with other nuclear proteins. *Cytogenet Cell Genet.* 2000; 91(1-4):217-23.

[98] Labbaye C, Valtieri M, Grignani F, Puglisi R, Luchetti L, Masella B, Alcalay M, Testa U, Peschle C. Expression and role of PML gene in normal adult hematopoiesis: functional interaction between PML and Rb proteins in erythropoiesis. *Oncogene.* 1999; 18(23):3529-40.

[99] Wang ZG, Delva L, Gaboli M, Rivi R, Giorgio M, Cordon-Cardo C, Grosveld F,Pandolfi PP. Role of PML in cell growth and the retinoic acid pathway. *Science.* 1998; 279(5356):1547-51.

[100] Jayaraman PS, Frampton J, Goodwin G. The homeodomain protein PRH influences the differentiation of haematopoietic cells. *Leuk Res.* 2000; 24(12):1023-31.

[101] Diverio D, Lo Coco F, D'Adamo F, Biondi A, Fagioli M, Grignani F, Rambaldi A, Rossi V, Avvisati G, Petti MC, et al. Identification of DNA rearrangements at the retinoic acid receptor-alpha (RAR-alpha) locus in all patients with acute promyelocytic leukemia (APL) and mapping of APL breakpoints within the RAR-alpha second intron. Italian Cooperative Study Group "GIMEMA". *Blood.* 1992; 79(12):3331-6.

[102] Biondi A, Rambaldi A, Pandolfi PP, Rossi V, Giudici G, Alcalay M, Lo Coco F, Diverio D, Pogliani EM, Lanzi EM, et al. Molecular monitoring of the myl/retinoic acid receptor-alpha fusion gene in acute promyelocytic leukemia by polymerase chain reaction. *Blood.* 1992; 80(2):492-7.

[103] Geng JP, Tong JH, Dong S, Wang ZY, Chen SJ, Chen Z, Zelent A, Berger R, Larsen CJ. Localization of the chromosome 15 breakpoints and expression of multiple PML-RAR alpha transcripts in acute promyelocytic leukemia: a study of 28 Chinese patients. *Leukemia.* 1993; 7(1):20-6.

[104] Borrow J, Shipley J, Howe K, Kiely F, Goddard A, Sheer D, Srivastava A, Antony AC, Fioretos T, Mitelman F, et al. Molecular analysis of simple variant translocations in acute promyelocytic *leukemia. Genes Chromosomes Cancer.* 1994; 9(4):234-43.

[105] Gallagher RE, Li YP, Rao S, Paietta E, Andersen J, Etkind P, Bennett JM, Tallman MS, Wiernik PH. Characterization of acute promyelocytic leukemia cases with PML-RAR alpha break/fusion sites in PML exon 6: identification of a subgroup with decreased in vitro responsiveness to all-trans retinoic acid. *Blood.* 1995; 86(4):1540-7.

[106] Pandolfi PP, Alcalay M, Fagioli M, Zangrilli D, Mencarelli A, Diverio D,Biondi A, Lo Coco F, Rambaldi A, Grignani F, et al. Genomic variability and alternative splicing generate multiple PML/RAR alpha transcripts that encode aberrant PML proteins and PML/RAR alpha isoforms in acute promyelocytic leukaemia. *EMBO J.* 1992; 11(4):1397-407.

[107] Biondi A, Luciano A, Bassan R, Mininni D, Specchia G, Lanzi E, Castagna S,Cantu-Rajnoldi A, Liso V, Masera G, et al. CD2 expression in acute promyelocytic leukemia is associated with microgranular morphology (FAB M3v) but not with any PML gene breakpoint. *Leukemia.* 1995; 9(9):1461-6.

[108] Gallagher RE, Willman CL, Slack JL, Andersen JW, Li YP, Viswanatha D, Bloomfield CD, Appelbaum FR, Schiffer CA, Tallman MS, Wiernik PH. Association of PML-RAR alpha fusion mRNA type with pretreatment hematologic characteristics but not treatment outcome in acute promyelocytic leukemia: an intergroup molecular study. *Blood.* 1997; 90(4):1656-63.

[109] Khoury H, Nevill TJ, Dalal BI, Barnett MJ, Shepherd JD, Hogge DE, Toze CL, Conneally EA, Sutherland HJ, Nantel SH, Horsman DE. "Clinical and Biological Relevance of PML Breakpoints in Acute Promyelocytic Leukemia". *Blood.* 2001; 98: 108a.

[110] Fukutani H, Naoe T, Ohno R, Yoshida H, Miyawaki S, Shimazaki C, Miyake T,Nakayama Y, Kobayashi H, Goto S, et al. Isoforms of PML-retinoic acid receptor alpha fused transcripts affect neither clinical features of acute promyelocytic leukemia nor prognosis after treatment with all-trans retinoic acid. The Leukemia Study Group of the Ministry of Health and Welfare (Kohseisho). *Leukemia.* 1995; 9(9):1478-82.

[111] Grignani F, Testa U, Rogaia D, Ferrucci PF, Samoggia P, Pinto A, Aldinucci D, Gelmetti V, Fagioli M, Alcalay M, Seeler J, Grignani F, Nicoletti I, PeschleC, Pelicci PG. Effects on differentiation by the promyelocytic leukemia PML/RARalpha protein depend on the fusion of the PML protein dimerization and RARalpha DNA binding domains. *EMBO J.* 1996; 15(18):4949-58.

[112] Ferrucci PF, Grignani F, Pearson M, Fagioli M, Nicoletti I, Pelicci PG. Cell death induction by the acute promyelocytic leukemia-specific PML/RARalpha fusion protein. *Proc Natl Acad Sci U S A.* 1997; 94(20):10901-6.

[113] Kogan SC, Hong SH, Shultz DB, Privalsky ML, Bishop JM. Leukemia initiated by PMLRARalpha: the PML domain plays a critical role while retinoic acid-mediated transactivation is dispensable. *Blood.* 2000; 95(5):1541-50.

[114] Kastner P, Perez A, Lutz Y, Rochette-Egly C, Gaub MP, Durand B, Lanotte M,Berger R, Chambon P. Structure, localization and transcriptional properties of two classes of retinoic acid receptor alpha fusion proteins in acute promyelocytic leukemia(APL): structural similarities with a new family of oncoproteins. *EMBO J.* 1992; 11(2):629-42.

[115] Perez A, Kastner P, Sethi S, Lutz Y, Reibel C, Chambon P. PMLRAR homodimers: distinct DNA binding properties and heteromeric interactions with RXR. *EMBO J.* 1993; 12(8):3171-82.

[116] Weis K, Rambaud S, Lavau C, Jansen J, Carvalho T, Carmo-Fonseca M, Lamond A, Dejean A. Retinoic acid regulates aberrant nuclear localization of PML-RAR alpha in acute promyelocytic leukemia cells. *Cell.* 1994; 76(2):345-56.

[117] Lin RJ, Nagy L, Inoue S, Shao W, Miller WH Jr, Evans RM. Role of the histone deacetylase complex in acute promyelocytic leukaemia. *Nature.* 1998; 391(6669): 811-4.

[118] He LZ, Guidez F, Tribioli C, Peruzzi D, Ruthardt M, Zelent A, Pandolfi PP. Distinct interactions of PML-RARalpha and PLZF-RARalpha with co-repressors determine differential responses to RA in APL. *Nat Genet.* 1998; 18(2):126-35.

[119] Grignani F, De Matteis S, Nervi C, Tomassoni L, Gelmetti V, Cioce M, Fanelli M, Ruthardt M, Ferrara FF, Zamir I, Seiser C, Grignani F, Lazar MA, Minucci S, Pelicci PG. Fusion proteins of the retinoic acid receptor-alpha recruit histone deacetylase in promyelocytic leukaemia. *Nature.* 1998; 391(6669):815-8.

[120] Guidez F, Ivins S, Zhu J, Soderstrom M, Waxman S, Zelent A. Reduced retinoic acid-sensitivities of nuclear receptor corepressor binding to PML- and PLZF-RARalpha underlie molecular pathogenesis and treatment of acute promyelocytic leukemia. *Blood.* 1998; 91(8):2634-42.

[121] Lin RJ, Evans RM. Acquisition of oncogenic potential by RAR chimeras in acute promyelocytic leukemia through formation of homodimers. *Mol Cell.* 2000; 5(5): 821-30.

[122] Minucci S, Maccarana M, Cioce M, De Luca P, Gelmetti V, Segalla S, Di Croce L, Giavara S, Matteucci C, Gobbi A, Bianchini A, Colombo E, Schiavoni I, Badaracco G, Hu X, Lazar MA, Landsberger N, Nervi C, Pelicci PG. Oligomerization of RAR and AML1 transcription factors as a novel mechanism of oncogenic activation. *Mol Cell.* 2000; 5(5):811-20

[123] Hauksdottir H, Privalsky ML. DNA recognition by the aberrant retinoic acid receptors implicated in human acute promyelocytic leukemia. *Cell Growth Differ.* 2001; 12(2):85-98.

[124] Dong S, Stenoien DL, Qiu J, Mancini MA, Tweardy DJ. Reduced intranuclear mobility of APL fusion proteins accompanies their mislocalization and results in sequestration and decreased mobility of retinoid X receptor alpha. *Mol Cell Biol.* 2004; 24(10):4465-75.

[125] Grimwade D, Solomon E. Characterisation of the PML/RAR alpha rearrangement associated with t(15;17)acute promyelocytic leukaemia. *Curr Top Microbiol Immunol.* 1997; 220:81-112.

[126] Grignani F, Ferrucci PF, Testa U, Talamo G, Fagioli M, Alcalay M, Mencarelli A, Grignani F, Peschle C, Nicoletti I, et al. The acute promyelocytic leukemia-specific PML-RAR alpha fusion protein inhibits differentiation and promotes survival of myeloid precursor cells. *Cell.* 1993; 74(3):423-31.

[127] Dyck JA, Maul GG, Miller WH Jr, Chen JD, Kakizuka A, Evans RM. A novel macromolecular structure is a target of the promyelocyte-retinoic acid receptor oncoprotein. *Cell.* 1994; 76(2):333-43.

[128] Koken MH, Puvion-Dutilleul F, Guillemin MC, Viron A, Linares-Cruz G, Stuurman N, de Jong L, Szostecki C, Calvo F, Chomienne C, et al. The t(15;17) translocation alters a nuclear body in a retinoic acid- reversible fashion. *EMBO J.* 1994; 13(5):1073-83.

[129] Zimonjic DB, Pollock JL, Westervelt P, Popescu NC, Ley TJ. Acquired, nonrandom chromosomal abnormalities associated with the development of acute promyelocytic leukemia in transgenic mice. *Proc Natl Acad Sci U S A.* 2000; 97(24):13306-11.

[130] Kogan SC, Brown DE, Shultz DB, Truong BT, Lallemand-Breitenbach V, Guillemin MC, Lagasse E, Weissman IL, Bishop JM. BCL-2 cooperates with promyelocytic leukemia retinoic acid receptor alpha chimeric protein (PMLRARalpha) to block neutrophil differentiation and initiate acute leukemia. *J Exp Med.* 2001; 193(4):531-43.

[131] Le Beau MM, Bitts S, Davis EM, Kogan SC. Recurring chromosomal abnormalities in leukemia in PML-RARA transgenic mice parallel human acute promyelocytic leukemia. *Blood.* 2002 ; 99(8):2985-91.

[132] Le Beau MM, Davis EM, Patel B, Phan VT, Sohal J, Kogan SC. Recurring chromosomal abnormalities in leukemia in PML-RARA transgenic mice identify cooperating events and genetic pathways to acute promyelocytic leukemia. *Blood.* 2003; 102(3):1072-4.

[133] Nason-Burchenal K, Allopenna J, Begue A, Stehelin D, Dmitrovsky E, Martin P. Targeting of PML/RARalpha is lethal to retinoic acid-resistant promyelocytic leukemia cells. *Blood.* 1998; 92(5):1758-67.

[134] Grignani F, Valtieri M, Gabbianelli M, Gelmetti V, Botta R, Luchetti L,Masella B, Morsilli O, Pelosi E, Samoggia P, Pelicci PG, Peschle C. PML/RAR alpha fusion protein expression in normal human hematopoietic progenitors dictates myeloid commitment and the promyelocytic phenotype. *Blood.* 2000; 96(4):1531-7.

[135] Minucci S, Monestiroli S, Giavara S, Ronzoni S, Marchesi F, Insinga A, Diverio D, Gasparini P, Capillo M, Colombo E, Matteucci C, Contegno F, Lo-Coco F, Scanziani E, Gobbi A, Pelicci PG. PML-RAR induces promyelocytic leukemias with high efficiency following retroviral gene transfer into purified murine hematopoietic progenitors. *Blood.* 2002; 100(8):2989-95.

[136] Du C, Redner RL, Cooke MP, Lavau C. Overexpression of wild-type retinoic acid receptor alpha (RARalpha) recapitulates retinoic acid-sensitive transformation of primary myeloid progenitors by acute promyelocytic leukemia RARalpha-fusion genes. *Blood.* 1999; 94(2):793-802.

[137] He LZ, Tribioli C, Rivi R, Peruzzi D, Pelicci PG, Soares V, Cattoretti G, Pandolfi PP. Acute leukemia with promyelocytic features in PML/RARalpha transgenic mice. *Proc Natl Acad Sci U S A.* 1997; 94(10):5302-7.

[138] Grisolano JL, Wesselschmidt RL, Pelicci PG, Ley TJ. Altered myeloid development and acute leukemia in transgenic mice expressing PML-RAR alpha under control of cathepsin G regulatory sequences. *Blood.* 1997; 89(2):376-87.

[139] Brown D, Kogan S, Lagasse E, Weissman I, Alcalay M, Pelicci PG, Atwater S, Bishop JM. A PMLRARalpha transgene initiates murine acute promyelocytic *leukemia. Proc Natl Acad Sci U S A.* 1997; 94(6):2551-6.

[140] Pollock JL, Westervelt P, Kurichety AK, Pelicci PG, Grisolano JL, Ley TJ. A bcr-3 isoform of RARalpha-PML potentiates the development of PML-RARalpha-driven acute promyelocytic *leukemia. Proc Natl Acad Sci U S A.* 1999; 96(26):15103-8.

[141] Kelly LM, Kutok JL, Williams IR, Boulton CL, Amaral SM, Curley DP, Ley TJ,Gilliland DG. PML/RARalpha and FLT3-ITD induce an APL-like disease in a mouse model. *Proc Natl Acad Sci U S A.* 2002; 99(12):8283-8.

[142] Phan VT, Shultz DB, Truong BT, Blake TJ, Brown AL, Gonda TJ, Le Beau MM, Kogan SC. Cooperation of cytokine signaling with chimeric transcription factors in leukemogenesis: PML-retinoic acid receptor alpha blocks growth factor-mediated differentiation. *Mol Cell Biol.* 2003; 23(13):4573-8

[143] Warrell RP Jr, Frankel SR, Miller WH Jr, Scheinberg DA, Itri LM, Hittelman WN, Vyas R, Andreeff M, Tafuri A, Jakubowski A, et al. Differentiation therapy of acute promyelocytic leukemia with tretinoin (all-trans-retinoic acid).*N Engl J Med.* 1991; 324(20):1385-93.

[144] Degos L, Dombret H, Chomienne C, Daniel MT, Miclea JM, Chastang C, Castaigne S, Fenaux P. All-trans-retinoic acid as a differentiating agent in the treatment of acute promyelocytic leukemia. *Blood.* 1995; 85(10):2643-53.

[145] Fenaux P, Chastang C, Chevret S, Sanz M, Dombret H, Archimbaud E, Fey M, Rayon C, Huguet F, Sotto JJ, Gardin C, Makhoul PC, Travade P, Solary E, Fegueux N, Bordessoule D, Miguel JS, Link H, Desablens B, Stamatoullas A, Deconinck E, Maloisel F, Castaigne S, Preudhomme C, Degos L. A randomized comparison of all transretinoic acid (ATRA) followed by chemotherapy and ATRA plus chemotherapy and the role of maintenance therapy in newly diagnosed acute promyelocytic *leukemia.* The European APL Group. *Blood.* 1999; 94(4):1192-200.

[146] He LZ, Guidez F, Tribioli C, Peruzzi D, Ruthardt M, Zelent A, Pandolfi PP. Distinct interactions of PML-RARalpha and PLZF-RARalpha with co-repressors determine differential responses to RA in APL. *Nat Genet.* 1998; 18(2):126-35.

[147] Raelson JV, Nervi C, Rosenauer A, Benedetti L, Monczak Y, Pearson M, Pelicci PG, Miller WH Jr. The PML/RAR alpha oncoprotein is a direct molecular target of retinoic acid in acute promyelocytic leukemia cells. *Blood.* 1996; 88(8):2826-32.

[148] Jing Y, Wang L, Xia L, Chen GQ, Chen Z, Miller WH, Waxman S. Combined effect of all-trans retinoic acid and arsenic trioxide in acute promyelocytic leukemia cells in vitro and in vivo. *Blood.* 2001; 97(1):264-9.

[149] Yoshida H, Kitamura K, Tanaka K, Omura S, Miyazaki T, Hachiya T, Ohno R, Naoe T. Accelerated degradation of PML-retinoic acid receptor alpha (PML-RARA) oncoprotein by all-trans-retinoic acid in acute promyelocytic leukemia: possible role of the proteasome pathway. *Cancer Res.* 1996; 56(13):2945-8.

[150] Duprez E, Lillehaug JR, Naoe T, Lanotte M. cAMP signalling is decisive for recovery of nuclear bodies (PODs) during maturation of RA-resistant t(15;17) promyelocytic leukemia NB4 cells expressing PML-RAR alpha. *Oncogene.* 1996; 12(11):2451-9.

[151] Zhu J, Gianni M, Kopf E, Honore N, Chelbi-Alix M, Koken M, Quignon F, Rochette-Egly C, de The H. Retinoic acid induces proteasome-dependent degradation of retinoic acid receptor alpha (RARalpha) and oncogenic RARalpha fusion proteins. *Proc Natl Acad Sci U S A.* 1999; 96(26):14807-12.

[152] Byrd JC, Dodge RK, Carroll A, Baer MR, Edwards C, Stamberg J, Qumsiyeh M, Moore JO, Mayer RJ, Davey F, Schiffer CA, Bloomfield CD. Patients with t(8;21)(q22;q22) and acute myeloid leukemia have superior failure-free and overall survival when repetitive cycles of high-dose cytarabine are administered. *J Clin Oncol.* 1999; 17(12):3767-75.

[153] Nucifora G, Dickstein JI, Torbenson V, Roulston D, Rowley JD, Vardiman JW. Correlation between cell morphology and expression of the AML1/ETO chimeric transcript in patients with acute myeloid leukemia without the t(8;21). *Leukemia.* 1994; 8(9):1533-8.

[154] Khoury H, Dalal BI, Nevill TJ, Horsman DE, Barnett MJ, Shepherd JD, Toze CL,Conneally EA, Sutherland HJ, Hogge DE, Nantel SH. Acute myelogenous leukemia with t(8;21)--identification of a specific immunophenotype. *Leuk Lymphoma.* 2003; 44(10):1713-8.

[155] Khoury H, Dalal BI, Nantel SH, Horsman DE, Lavoie JC, Shepherd JD, Hogge DE, Toze CL, Song KW, Forrest DL, Sutherland HJ, Nevill TJ. Correlation between karyotype and quantitative immunophenotype in acute myelogenous leukemia with t(8;21). *Mod Pathol.* 2004;17(10):1211-6.

[156] Miyoshi H, Ohira M, Shimizu K, Mitani K, Hirai H, Imai T, Yokoyama K, Soeda E, Ohki M. Alternative splicing and genomic structure of the AML1 gene involved in acute myeloid leukemia. *Nucleic Acids Res*. 1995; 23(14):2762-9.

[157] Levanon D, Bernstein Y, Negreanu V, Ghozi MC, Bar-Am I, Aloya R, Goldenberg D, Lotem J, Groner Y. A large variety of alternatively spliced and differentially expressed mRNAs are encoded by the human acute myeloid leukemia gene AML1. *DNA Cell Biol*. 1996; 15(3):175-85.

[158] Pozner A, Goldenberg D, Negreanu V, Le SY, Elroy-Stein O, Levanon D, Groner Y. Transcription-coupled translation control of AML1/RUNX1 is mediated by cap- and internal ribosome entry site-dependent mechanisms. *Mol Cell Biol*. 2000; 20(7):2297-307.

[159] Levanon D, Glusman G, Bangsow T, Ben-Asher E, Male DA, Avidan N, Bangsow C,Hattori M, Taylor TD, Taudien S, Blechschmidt K, Shimizu N, Rosenthal A, Sakaki Y, Lancet D, Groner Y. Architecture and anatomy of the genomic locus encoding the human leukemia-associated transcription factor RUNX1/AML1. *Gene*. 2001; 262(1-2):23-33.

[160] Ogawa E, Maruyama M, Kagoshima H, Inuzuka M, Lu J, Satake M, Shigesada K,Ito Y. PEBP2/PEA2 represents a family of transcription factors homologous to the products of the Drosophila runt gene and the human AML1 gene. *Proc Natl Acad Sci U S A*. 1993; 90(14):6859-63.

[161] Levanon D, Negreanu V, Bernstein Y, Bar-Am I, Avivi L, Groner Y. AML1, AML2, and AML3, the human members of the runt domain gene-family: cDNA structure, expression, and chromosomal localization. *Genomics*. 1994; 23(2):425-32.

[162] Corsetti MT, Calabi F. Lineage- and stage-specific expression of runt box polypeptides in primitive and definitive hematopoiesis. *Blood*. 1997; 89(7):2359-68.

[163] Levanon D, Brenner O, Negreanu V, Bettoun D, Woolf E, Eilam R, Lotem J, Gat U, Otto F, Speck N, Groner Y. Spatial and temporal expression pattern of Runx3 (Aml2) and Runx1 (Aml1)indicates non-redundant functions during mouse embryogenesis. *Mech Dev*. 2001; 109(2):413-7.

[164] Kitabayashi I, Yokoyama A, Shimizu K, Ohki M. Interaction and functional cooperation of the leukemia-associated factors AML1 and p300 in myeloid cell differentiation. *EMBO J*. 1998; 17(11):2994-3004.

[165] Tanaka T, Kurokawa M, Ueki K, Tanaka K, Imai Y, Mitani K, Okazaki K, Sagata N, Yazaki Y, Shibata Y, Kadowaki T, Hirai H. The extracellular signal-regulated kinase pathway phosphorylates AML1, an acute myeloid leukemia gene product, and potentially regulates its transactivation ability. *Mol Cell Biol*. 1996; 16(7):3967-79.

[166] Prosser HM, Wotton D, Gegonne A, Ghysdael J, Wang S, Speck NA, Owen MJ. A phorbol ester response element within the human T-cell receptor beta-chain enhancer. *Proc Natl Acad Sci U S A*. 1992; 89(20):9934-8.

[167] Redondo JM, Pfohl JL, Hernandez-Munain C, Wang S, Speck NA, Krangel MS. Indistinguishable nuclear factor binding to functional core sites of the T-cell receptor delta and murine leukemia virus enhancers. *Mol Cell Biol*. 1992; 12(11):4817-23.

[168] Rhoades KL, Hetherington CJ, Rowley JD, Hiebert SW, Nucifora G, Tenen DG,Zhang DE. Synergistic up-regulation of the myeloid-specific promoter for the macrophage colony-stimulating factor receptor by AML1 and the t(8;21) fusion protein may contribute to leukemogenesis. *Proc Natl Acad Sci U S A*. 1996; 93(21):11895-900.

[169] Zhang DE, Fujioka K, Hetherington CJ, Shapiro LH, Chen HM, Look AT, Tenen DG. Identification of a region which directs the monocytic activity of the colony-stimulating factor 1 (macrophage colony-stimulating factor) receptor promoter and binds PEBP2/CBF (AML1). *Mol Cell Biol.* 1994; 14(12):8085-95.

[170] Britos-Bray M, Friedman AD. Core binding factor cannot synergistically activate the myeloperoxidase proximal enhancer in immature myeloid cells without c-Myb. *Mol Cell Biol.* 1997; 17(9):5127-35.

[171] Nuchprayoon I, Meyers S, Scott LM, Suzow J, Hiebert S, Friedman AD. PEBP2/CBF, the murine homolog of the human myeloid AML1 and PEBP2 beta/CBF beta proto-oncoproteins, regulates the murine myeloperoxidase and neutrophil elastase genes in immature myeloid cells. *Mol Cell Biol.* 1994; 14(8):5558-68.

[172] Wargnier A, Legros-Maida S, Bosselut R, Bourge JF, Lafaurie C, Ghysdael CJ, Sasportes M, Paul P. Identification of human granzyme B promoter regulatory elements interacting with activated T-cell-specific proteins: implication of Ikaros and CBF binding sites in promoter activation. *Proc Natl Acad Sci U S A.* 1995; 92(15):6930-4.

[173] Cameron S, Taylor DS, TePas EC, Speck NA, Mathey-Prevot B. Identification of a critical regulatory site in the human interleukin-3 promoter by in vivo footprinting. *Blood.* 1994; 83(10):2851-9.

[174] Takahashi A, Satake M, Yamaguchi-Iwai Y, Bae SC, Lu J, Maruyama M, Zhang YW, Oka H, Arai N, Arai K, et al. Positive and negative regulation of granulocyte-macrophage colony-stimulating factor promoter activity by AML1-related transcription factor, PEBP2. *Blood.* 1995; 86(2):607-16.

[175] Frank R, Zhang J, Uchida H, Meyers S, Hiebert SW, Nimer SD.The AML1/ETO fusion protein blocks transactivation of the GM-CSF promoter by AML1B. *Oncogene.* 1995;1 1(12):2667-74.

[176] Zhang DE, Hetherington CJ, Meyers S, Rhoades KL, Larson CJ, Chen HM, Hiebert SW, Tenen DG. CCAAT enhancer-binding protein (C/EBP) and AML1 (CBF alpha2) synergistically activate the macrophage colony-stimulating factor receptor promoter. *Mol Cell Biol.* 1996; 16(3):1231-40.

[177] Petrovick MS, Hiebert SW, Friedman AD, Hetherington CJ, Tenen DG, Zhang DE. Multiple functional domains of AML1: PU.1 and C/EBPalpha synergize with different regions of AML1. *Mol Cell Biol.* 1998; 18(7):3915-25.

[178] Hanai J, Chen LF, Kanno T, Ohtani-Fujita N, Kim WY, Guo WH, Imamura T,Ishidou Y, Fukuchi M, Shi MJ, Stavnezer J, Kawabata M, Miyazono K, Ito Y. Interaction and functional cooperation of PEBP2/CBF with Smads. Synergistic induction of the immunoglobulin germline Calpha promoter. *J Biol Chem.* 1999; 274(44):31577-82.

[179] Ogihara H, Kanno T, Morii E, Kim DK, Lee YM, Sato M, Kim WY, Nomura S, Ito Y, Kitamura Y. Synergy of PEBP2/CBF with mi transcription factor (MITF) for transactivation of mouse mast cell protease 6 gene. *Oncogene.* 1999; 18(32):4632-9.

[180] Morii E, Ogihara H, Kanno T, Kim DK, Nomura S, Ito Y, Kitamura Y. Identification of the region of mi transcription factor which is responsible for the synergy with PEBP2/CBF. *Biochem Biophys Res Commun.* 1999; 261(1):53-7.

[181] Kim WY, Sieweke M, Ogawa E, Wee HJ, Englmeier U, Graf T, Ito Y. Mutual activation of Ets-1 and AML1 DNA binding by direct interaction of their autoinhibitory domains. *EMBO J.* 1999; 18(6):1609-20.

[182] Libermann TA, Pan Z, Akbarali Y, Hetherington CJ, Boltax J, Yergeau DA, Zhang DE. AML1 (CBFalpha2) cooperates with B cell-specific activating protein (BSAP/PAX5) in activation of the B cell-specific BLK gene promoter. *J Biol Chem.* 1999; 274(35):24671-6.

[183] Mao S, Frank RC, Zhang J, Miyazaki Y, Nimer SD. Functional and physical interactions between AML1 proteins and an ETS protein, MEF: implications for the pathogenesis of t(8;21)-positive leukemias. *Mol Cell Biol.* 1999; 19(5):3635-44.

[184] Gu TL, Goetz TL, Graves BJ, Speck NA. Auto-inhibition and partner proteins, core-binding factor beta (CBFbeta) and Ets-1, modulate DNA binding by CBFalpha2 (AML1). *Mol Cell Biol.* 2000; 20(1):91-103.

[185] Goetz TL, Gu TL, Speck NA, Graves BJ. Auto-inhibition of Ets-1 is counteracted by DNA binding cooperativity with core-binding factor alpha2. *Mol Cell Biol.* 2000; 20(1):81-90.

[186] Elagib KE, Racke FK, Mogass M, Khetawat R, Delehanty LL, Goldfarb AN. RUNX1 and GATA-1 coexpression and cooperation in megakaryocytic differentiation. *Blood.* 2003; 101(11):4333-41.

[187] Li FQ, Person RE, Takemaru K, Williams K, Meade-White K, Ozsahin AH, Gungor T, Moon RT, Horwitz M. Lymphoid enhancer factor-1 links two hereditary leukemia syndromes through core-binding factor alpha regulation of ELA2. *J Biol Chem.* 2004; 279(4):2873-84.

[188] Carey M. The enhanceosome and transcriptional synergy. *Cell.* 1998; 92(1):5-8.

[189] Hernandez-Munain C, Krangel MS. Regulation of the T-cell receptor delta enhancer by functional cooperation between c-Myb and core-binding factors. *Mol Cell Biol.* 1994; 14(1):473-83.

[190] Giese K, Kingsley C, Kirshner JR, Grosschedl R. Assembly and function of a TCR alpha enhancer complex is dependent on LEF-1-induced DNA bending and multiple protein-protein interactions. *Genes Dev.* 1995; 9(8):995-1008.

[191] Zaiman AL, Lenz J. Transcriptional activation of a retrovirus enhancer by CBF (AML1) requires a second factor: evidence for cooperativity with c-Myb. *J Virol.* 1996; 70(8):5618-29.

[192] Zaidi SK, Sullivan AJ, van Wijnen AJ, Stein JL, Stein GS, Lian JB. Integration of Runx and Smad regulatory signals at transcriptionally active subnuclear sites. *Proc Natl Acad Sci U S A.* 2002; 99(12):8048-53.

[193] Erman B, Cortes M, Nikolajczyk BS, Speck NA, Sen R. ETS-core binding factor: a common composite motif in antigen receptor gene enhancers. *Mol Cell Biol.* 1998; 18(3):1322-30.

[194] Wotton D, Ghysdael J, Wang S, Speck NA, Owen MJ. Cooperative binding of Ets-1 and core binding factor to DNA. *Mol Cell Biol.* 1994; 14(1):840-50.

[195] Yang G, Khalaf W, van de Locht L, Jansen JH, Gao M, Thompson MA, van der Reijden BA, Gutmann DH, Delwel R, Clapp DW, Hiebert SW. Transcriptional repression of the Neurofibromatosis-1 tumor suppressor by the t(8;21) fusion protein. *Mol Cell Biol.* 2005; 25(14):5869-79.

[196] Ahn MY, Huang G, Bae SC, Wee HJ, Kim WY, Ito Y. Negative regulation of granulocytic differentiation in the myeloid precursor cell line 32Dcl3 by ear-2, a mammalian homolog of Drosophila seven-up, and a chimeric leukemogenic gene, AML1/ETO. *Proc Natl Acad Sci U S A.* 1998; 95(4):1812-7.

[197] Lutterbach B, Westendorf JJ, Linggi B, Isaac S, Seto E, Hiebert SW. A mechanism of repression by acute myeloid leukemia-1, the target of multiple chromosomal translocations in acute *leukemia. J Biol Chem.* 2000; 275(1):651-6.

[198] Imai Y, Kurokawa M, Tanaka K, Friedman AD, Ogawa S, Mitani K, Yazaki Y,Hirai H. TLE, the human homolog of groucho, interacts with AML1 and acts as a repressor of AML1-induced transactivation. *Biochem Biophys Res Commun.* 1998; 252(3):582-9.

[199] Levanon D, Goldstein RE, Bernstein Y, Tang H, Goldenberg D, Stifani S, Paroush Z, Groner Y. Transcriptional repression by AML1 and LEF-1 is mediated by the TLE/Groucho corepressors. *Proc Natl Acad Sci U S A.* 1998; 95(20):11590-5.

[200] McLarren KW, Theriault FM, Stifani S. Association with the nuclear matrix and interaction with Groucho and RUNX proteins regulate the transcription repression activity of the basic helix loop helix factor Hes1. *J Biol Chem.* 2001; 276(2):1578-84.

[201] McLarren KW, Lo R, Grbavec D, Thirunavukkarasu K, Karsenty G, Stifani S. The mammalian basic helix loop helix protein HES-1 binds to and modulates the transactivating function of the runt-related factor Cbfa1. *J Biol Chem.* 2000; 275(1):530-8.

[202] McNeil S, Zeng C, Harrington KS, Hiebert S, Lian JB, Stein JL, van Wijnen AJ, Stein GS. The t(8;21) chromosomal translocation in acute myelogenous leukemia modifies intranuclear targeting of the AML1/CBFalpha2 transcription factor. *Proc Natl Acad Sci U S A.* 1999; 96(26):14882-7.

[203] Vradii D, Zaidi SK, Lian JB, van Wijnen AJ, Stein JL, Stein GS. Point mutation in AML1 disrupts subnuclear targeting, prevents myeloid differentiation, and affects a transformation-like phenotype. *Proc Natl Acad Sci U S A.* 2005; 102(20):7174-9.

[204] Li X, Vradii D, Gutierrez S, Lian JB, van Wijnen AJ, Stein JL, Stein GS, Javed A. Subnuclear targeting of Runx1 is required for synergistic activation of the myeloid specific M-CSF receptor promoter by PU.1. *J Cell Biochem.* 2005; 96(4):795-809.

[205] Stein GS, Lian JB, van Wijnen AJ, Stein JL, Javed A, Montecino M, Choi JY, Vradii D, Zaidi SK, Pratap J, Young D.Organization of transcriptional regulatory machinery in nuclear microenvironments: implications for biological control and cancer. *Adv Enzyme Regul.* 2007; 47:242-50.

[206] Okuda T, van Deursen J, Hiebert SW, Grosveld G, Downing JR. AML1, the target of multiple chromosomal translocations in human leukemia, is essential for normal fetal liver hematopoiesis. *Cell.* 1996; 84(2):321-30.

[207] Sasaki K, Yagi H, Bronson RT, Tominaga K, Matsunashi T, Deguchi K, Tani Y, Kishimoto T, Komori T. Absence of fetal liver hematopoiesis in mice deficient in transcriptional coactivator core binding factor beta. *Proc Natl Acad Sci U S A.* 1996; 93(22):12359-63.

[208] Growney JD, Shigematsu H, Li Z, Lee BH, Adelsperger J, Rowan R, Curley DP, Kutok JL, Akashi K, Williams IR, Speck NA, Gilliland DG. Loss of Runx1 perturbs adult hematopoiesis and is associated with a myeloproliferative phenotype. *Blood.* 2005; 106(2):494-504

[209] Cai Z, de Bruijn M, Ma X, Dortland B, Luteijn T, Downing RJ, Dzierzak E. Haploinsufficiency of AML1 affects the temporal and spatial generation of hematopoietic stem cells in the mouse embryo. *Immunity.* 2000; 13(4):423-31.

[210] Mukouyama Y, Chiba N, Hara T, Okada H, Ito Y, Kanamaru R, Miyajima A, Satake M, Watanabe T. The AML1 transcription factor functions to develop and maintain

hematogenic precursor cells in the embryonic aorta-gonad-mesonephros region. *Dev Biol.* 2000; 220(1):27-36.

[211] North TE, de Bruijn MF, Stacy T, Talebian L, Lind E, Robin C, Binder M, Dzierzak E, Speck NA. Runx1 expression marks long-term repopulating hematopoietic stem cells in the midgestation mouse embryo. *Immunity.* 2002; 16(5):661-72.

[212] Lacaud G, Kouskoff V, Trumble A, Schwantz S, Keller G. Haploinsufficiency of Runx1 results in the acceleration of mesodermal development and hemangioblast specification upon in vitro differentiation of ES cells. *Blood.* 2004; 103(3):886-9

[213] Fracchiolla NS, Colombo G, Finelli P, Maiolo AT, Neri A. EHT, a new member of the MTG8/ETO gene family, maps on 20q11 region and is deleted in acute myeloid leukemias. *Blood.* 1998; 92(9):3481-4.

[214] Kitabayashi I, Ida K, Morohoshi F, Yokoyama A, Mitsuhashi N, Shimizu K,Nomura N, Hayashi Y, Ohki M. The AML1-MTG8 leukemic fusion protein forms a complex with a novel member of the MTG8(ETO/CDR) family, MTGR1. *Mol Cell Biol.* 1998; 18(2):846-58.

[215] Calabi F, Cilli V. CBFA2T1, a gene rearranged in human leukemia, is member of a multigene family. *Genomics.* 1998; 52(3):332-41.

[216] Davis JN, Williams BJ, Herron JT, Galiano FJ, Meyers S. ETO-2, a new member of the ETO-family of nuclear proteins. *Oncogene.* 1999; 18(6):1375-83.

[217] Gamou T, Kitamura E, Hosoda F, Shimizu K, Shinohara K, Hayashi Y, Nagase T,Yokoyama Y, Ohki M. The partner gene of AML1 in t(16;21) myeloid malignancies is a novel member of the MTG8(ETO) family. *Blood.* 1998; 91(11):4028-37.

[218] Erickson PF, Robinson M, Owens G, Drabkin HA. The ETO portion of acute myeloid leukemia t(8;21) fusion transcript encodes a highly evolutionarily conserved, putative transcription factor. *Cancer Res.* 1994; 54(7):1782-6.

[219] Feinstein PG, Kornfeld K, Hogness DS, Mann RS. Identification of homeotic target genes in Drosophila melanogaster including nervy, a proto-oncogene homologue. *Genetics.* 1995; 140(2):573-86.

[220] Kozu T, Komori A, Sueoka E, Fujiki H, Kaneko Y, Matsui T, Uehara T, Seino Y,Ishii M. Significance of MTG8 in leukemogenesis. *Leukemia.* 1997; 11 Suppl 3:297-8.

[221] Odaka Y, Mally A, Elliott LT, Meyers S. Nuclear import and subnuclear localization of the proto-oncoprotein ETO (MTG8). *Oncogene.* 2000; 19(32):3584-97.

[222] Lutterbach B, Westendorf JJ, Linggi B, Patten A, Moniwa M, Davie JR, Huynh KD, Bardwell VJ, Lavinsky RM, Rosenfeld MG, Glass C, Seto E, Hiebert SW. ETO, a target of t(8;21) in acute leukemia, interacts with the N-CoR and mSin3 corepressors. *Mol Cell Biol.* 1998; 18(12):7176-84.

[223] Lutterbach B, Sun D, Schuetz J, Hiebert SW. The MYND motif is required for repression of basal transcription from the multidrug resistance 1 promoter by the t(8;21) fusion protein. *Mol Cell Biol.* 1998; 18(6):3604-11.

[224] Gross CT, McGinnis W. DEAF-1, a novel protein that binds an essential region in a Deformed response element. *EMBO J.* 1996; 15(8):1961-70.

[225] Ansieau S, Leutz A. The conserved Mynd domain of BS69 binds cellular and oncoviral proteins through a common PXLXP motif. *J Biol Chem.* 2002; 277(7):4906-10.

[226] Gottlieb PD, Pierce SA, Sims RJ, Yamagishi H, Weihe EK, Harriss JV, Maika SD, Kuziel WA, King HL, Olson EN, Nakagawa O, Srivastava D. Bop encodes a muscle-

restricted protein containing MYND and SET domains and is essential for cardiac differentiation and morphogenesis. *Nat Genet.* 2002; 31(1):25-32.

[227] Gelmetti V, Zhang J, Fanelli M, Minucci S, Pelicci PG, Lazar MA. Aberrant recruitment of the nuclear receptor corepressor-histone deacetylase complex by the acute myeloid leukemia fusion partner ETO. *Mol Cell Biol.* 1998; 18(12):7185-91.

[228] Wang J, Hoshino T, Redner RL, Kajigaya S, Liu JM. ETO, fusion partner in t(8;21) acute myeloid leukemia, represses transcription by interaction with the human N-CoR/mSin3/HDAC1 complex. *Proc Natl Acad Sci U S A.* 1998; 95(18):10860-5.

[229] Amann JM, Nip J, Strom DK, Lutterbach B, Harada H, Lenny N, Downing JR, Meyers S, Hiebert SW. ETO, a target of t(8;21) in acute leukemia, makes distinct contacts with multiple histone deacetylases and binds mSin3A through its oligomerization domain. *Mol Cell Biol.* 2001; 21(19):6470-83.

[230] Lausen J, Cho S, Liu S, Werner MH. The nuclear receptor co-repressor (N-CoR) utilizes repression domains I and III for interaction and co-repression with ETO. *J Biol Chem.* 2004; 279(47):49281-8.

[231] Wang J, Wang M, Liu JM. Domains involved in ETO and human N-CoR interaction and ETO transcription repression. *Leuk Res.* 2004; 28(4):409-14.

[232] Hildebrand D, Tiefenbach J, Heinzel T, Grez M, Maurer AB. Multiple regions of ETO cooperate in transcriptional repression. *J Biol Chem.* 2001; 276(13):9889-95.

[233] Shaknovich R, Yeyati PL, Ivins S, Melnick A, Lempert C, Waxman S, Zelent A, Licht JD. The promyelocytic leukemia zinc finger protein affects myeloid cell growth, differentiation, and apoptosis. *Mol Cell Biol.* 1998; 18(9):5533-45.

[234] Melnick A, Carlile GW, McConnell MJ, Polinger A, Hiebert SW, Licht JD. AML-1/ETO fusion protein is a dominant negative inhibitor of transcriptional repression by the promyelocytic leukemia zinc finger protein. *Blood.* 2000; 96(12):3939-47.

[235] Melnick AM, Westendorf JJ, Polinger A, Carlile GW, Arai S, Ball HJ, Lutterbach B, Hiebert SW, Licht JD. The ETO protein disrupted in t(8;21)-associated acute myeloid leukemia is a corepressor for the promyelocytic leukemia zinc finger protein. *Mol Cell Biol.* 2000; 20(6):2075-86.

[236] McGhee L, Bryan J, Elliott L, Grimes HL, Kazanjian A, Davis JN, Meyers S. Gfi-1 attaches to the nuclear matrix, associates with ETO (MTG8) and histone deacetylase proteins, and represses transcription using a TSA-sensitive mechanism. *J Cell Biochem.* 2003; 89(5):1005-18.

[237] Gilks CB, Bear SE, Grimes HL, Tsichlis PN. Progression of interleukin-2 (IL-2)-dependent rat T cell lymphoma lines to IL-2-independent growth following activation of a gene (Gfi-1) encoding a novel zinc finger protein. *Mol Cell Biol.* 1993; 13(3): 1759-68.

[238] Tong B, Grimes HL, Yang TY, Bear SE, Qin Z, Du K, El-Deiry WS, Tsichlis PN. The Gfi-1B proto-oncoprotein represses p21WAF1 and inhibits myeloid cell differentiation. *Mol Cell Biol.* 1998; 18(5):2462-73.

[239] Karsunky H, Mende I, Schmidt T, Moroy T. High levels of the onco-protein Gfi-1 accelerate T-cell proliferation and inhibit activation induced T-cell death in Jurkat T-cells. *Oncogene.* 2002; 21(10):1571-9.

[240] Saleque S, Cameron S, Orkin SH. The zinc-finger proto-oncogene Gfi-1b is essential for development of the erythroid and megakaryocytic lineages. *Genes Dev.* 2002; 16(3):301-6.

[241] Wolford JK, Prochazka M. Structure and expression of the human MTG8/ETO gene. *Gene*. 1998; 212(1):103-9.

[242] Erickson PF, Dessev G, Lasher RS, Philips G, Robinson M, Drabkin HA. ETO and AML1 phosphoproteins are expressed in CD34+ hematopoietic progenitors: implications for t(8;21) leukemogenesis and monitoring residual disease. *Blood*. 1996; 88(5):1813-23.

[243] Calabi F, Pannell R, Pavloska G. Gene targeting reveals a crucial role for MTG8 in the gut. *Mol Cell Biol*. 2001; 21(16):5658-66.

[244] Erickson P, Gao J, Chang KS, Look T, Whisenant E, Raimondi S, Lasher R, Trujillo J, Rowley J, Drabkin H. Identification of breakpoints in t(8;21) acute myelogenous leukemia and isolation of a fusion transcript, AML1/ETO, with similarity to Drosophila segmentation gene, runt. *Blood*. 1992; 80(7):1825-31.

[245] Shimizu K, Miyoshi H, Kozu T, Nagata J, Enomoto K, Maseki N, Kaneko Y, Ohki M. Consistent disruption of the AML1 gene occurs within a single intron in the t(8;21) chromosomal translocation. *Cancer Res*. 1992; 52(24):6945-8.

[246] Miyoshi H, Kozu T, Shimizu K, Enomoto K, Maseki N, Kaneko Y, Kamada N, Ohki M. The t(8;21) translocation in acute myeloid leukemia results in production of an AML1-MTG8 fusion transcript. *EMBO J*. 1993; 12(7):2715-21.

[247] Minucci S, Maccarana M, Cioce M, De Luca P, Gelmetti V, Segalla S, Di Croce L, Giavara S, Matteucci C, Gobbi A, Bianchini A, Colombo E, Schiavoni I, Badaracco G, Hu X, Lazar MA, Landsberger N, Nervi C, Pelicci PG. Oligomerization of RAR and AML1 transcription factors as a novel mechanism of oncogenic activation. *Mol Cell*. 2000; 5(5):811-20.

[248] Zhang J, Hug BA, Huang EY, Chen CW, Gelmetti V, Maccarana M, Minucci S, Pelicci PG, Lazar MA. Oligomerization of ETO is obligatory for corepressor interaction. *Mol Cell Biol*. 2001; 21(1):156-63.

[249] Tanaka K, Tanaka T, Kurokawa M, Imai Y, Ogawa S, Mitani K, Yazaki Y, Hirai H. The AML1/ETO(MTG8) and AML1/Evi-1 leukemia-associated chimeric oncoproteins accumulate PEBP2beta(CBFbeta) in the nucleus more efficiently than wild-type AML1. *Blood*. 1998; 91(5):1688-99.

[250] Wang S, Wang Q, Crute BE, Melnikova IN, Keller SR, Speck NA. Cloning and characterization of subunits of the T-cell receptor and murine leukemia virus enhancer core-binding factor. *Mol Cell Biol*. 1993; 13(6):3324-39.

[251] Hiebert SW, Reed-Inderbitzin EF, Amann J, Irvin B, Durst K, Linggi B. The t(8;21) fusion protein contacts co-repressors and histone deacetylases to repress the transcription of the p14ARF tumor suppressor. *Blood Cells Mol Dis*. 2003; 30(2): 177-83.

[252] Klampfer L, Zhang J, Zelenetz AO, Uchida H, Nimer SD. The AML1/ETO fusion protein activates transcription of BCL-2. *Proc Natl Acad Sci U S A*. 1996; 93(24):14059-64.

[253] Banker DE, Radich J, Becker A, Kerkof K, Norwood T, Willman C, Appelbaum FR. The t(8;21) translocation is not consistently associated with high Bcl-2 expression in de novo acute myeloid leukemias of adults. *Clin Cancer Res*. 1998; 4(12):3051-62.

[254] Vangala RK, Heiss-Neumann MS, Rangatia JS, Singh SM, Schoch C, Tenen DG, Hiddemann W, Behre G. The myeloid master regulator transcription factor PU.1 is inactivated by AML1-ETO in t(8;21) myeloid leukemia. *Blood*. 2003; 101(1):270-7.

[255] Pabst T, Mueller BU, Harakawa N, Schoch C, Haferlach T, Behre G, Hiddemann W, Zhang DE, Tenen DG. AML1-ETO downregulates the granulocytic differentiation factor C/EBPalpha in t(8;21) myeloid leukemia. *Nat Med.* 2001; 7(4):444-51.

[256] Mao S, Frank RC, Zhang J, Miyazaki Y, Nimer SD. Functional and physical interactions between AML1 proteins and an ETS protein, MEF: implications for the pathogenesis of t(8;21)-positive leukemias. *Mol Cell Biol.* 1999; 19(5):3635-44.

[257] McNeil S, Zeng C, Harrington KS, Hiebert S, Lian JB, Stein JL, van Wijnen AJ, Stein GS. The t(8;21) chromosomal translocation in acute myelogenous leukemia modifies intranuclear targeting of the AML1/CBFalpha2 transcription factor. *Proc Natl Acad Sci U S A.* 1999; 96(26):14882-7.

[258] Barseguian K, Lutterbach B, Hiebert SW, Nickerson J, Lian JB, Stein JL, van Wijnen AJ, Stein GS. Multiple subnuclear targeting signals of the leukemia-related AML1/ETO and ETO repressor proteins. *Proc Natl Acad Sci U S A.* 2002; 99(24):15434-9.

[259] Nagel S, Hambach L, Krauter J, Venturini L, Heidenreich O, Ganser A, Heil G. Analysis of the nuclear distribution of the translocation t(8;21)-derived fusion protein AML1/ETO by confocal laser scanning microscopy. *J Hematother Stem Cell Res.* 2002; 11(2):401-8.

[260] Odaka Y, Mally A, Elliott LT, Meyers S. Nuclear import and subnuclear localization of the proto-oncoprotein ETO (MTG8) *Oncogene.* 2000; 19(32):3584-97.

[261] Le XF, Claxton D, Kornblau S, Fan YH, Mu ZM, Chang KS. Characterization of the ETO and AML1-ETO proteins involved in 8;21 translocation in acute myelogenous leukemia. *Eur J Haematol.* 1998; 60(4):217-25.

[262] Kohzaki H, Ito K, Huang G, Wee HJ, Murakami Y, Ito Y. Block of granulocytic differentiation of 32Dcl3 cells by AML1/ETO(MTG8) but not by highly expressed Bcl-2. *Oncogene.* 1999; 18(28):4055-62.

[263] Shimizu K, Kitabayashi I, Kamada N, Abe T, Maseki N, Suzukawa K, Ohki M. AML1-MTG8 leukemic protein induces the expression of granulocyte colony-stimulating factor (G-CSF) receptor through the up-regulation of CCAAT/enhancer binding protein epsilon. *Blood.* 2000; 96(1):288-96.

[264] Burel SA, Harakawa N, Zhou L, Pabst T, Tenen DG, Zhang DE. Dichotomy of AML1-ETO functions: growth arrest versus block of differentiation. *Mol Cell Biol.* 2001; 21(16):5577-90.

[265] Puccetti E, Obradovic D, Beissert T, Bianchini A, Washburn B, Chiaradonna F, Boehrer S, Hoelzer D, Ottmann OG, Pelicci PG, Nervi C, Ruthardt M. AML-associated translocation products block vitamin D(3)-induced differentiation by sequestering the vitamin D(3) receptor. *Cancer Res.* 2002; 62(23):7050-8.

[266] Mulloy JC, Cammenga J, MacKenzie KL, Berguido FJ, Moore MA, Nimer SD. The AML1-ETO fusion protein promotes the expansion of human hematopoietic stem cells. *Blood.* 2002; 99(1):15-23.

[267] Tonks A, Tonks AJ, Pearn L, Pearce L, Hoy T, Couzens S, Fisher J, Burnett AK, Darley RL. Expression of AML1-ETO in human myelomonocytic cells selectively inhibits granulocytic differentiation and promotes their self-renewal. *Leukemia.* 2004; 18(7):1238-45.

[268] Mulloy JC, Cammenga J, Berguido FJ, Wu K, Zhou P, Comenzo RL, Jhanwar S,Moore MA, Nimer SD. Maintaining the self-renewal and differentiation potential of human

CD34+ hematopoietic cells using a single genetic element. *Blood.* 2003; 102(13): 4369-76.

[269] Tonks A, Pearn L, Tonks AJ, Pearce L, Hoy T, Phillips S, Fisher J, Downing JR, Burnett AK, Darley RL. The AML1-ETO fusion gene promotes extensive self-renewal of human primary erythroid cells. *Blood.* 2003; 101(2):624-32.

[270] Yuan Y, Zhou L, Miyamoto T, Iwasaki H, Harakawa N, Hetherington CJ, Burel SA, Lagasse E, Weissman IL, Akashi K, Zhang DE. AML1-ETO expression is directly involved in the development of acute myeloid leukemia in the presence of additional mutations. *Proc Natl Acad Sci U S A.* 2001; 98(18):10398-403

[271] Grisolano JL, O'Neal J, Cain J, Tomasson MH. An activated receptor tyrosine kinase, TEL/PDGFbetaR, cooperates with AML1/ETO to induce acute myeloid leukemia in mice. *Proc Natl Acad Sci U S A.* 2003; 100(16):9506-11

[272] Nishida S, Hosen N, Shirakata T, Kanato K, Yanagihara M, Nakatsuka SI, Hoshida Y, Nakazawa T, Harada Y, Tatsumi N, Tsuboi A, Kawakami M, Oka Y, Oji Y, Aozasa K, Kawase I, Sugiyama H. AML1-ETO rapidly induces acute myeloblastic leukemia in cooperation with Wilms' tumor gene, WT1. *Blood.* 2006; 107(8):3303-12.

[273] Schessl C, Rawat VP, Cusan M, Deshpande A, Kohl TM, Rosten PM, Spiekermann K, Humphries RK, Schnittger S, Kern W, Hiddemann W, Quintanilla-Martinez L, Bohlander SK, Feuring-Buske M, Buske C. The AML1-ETO fusion gene and the FLT3 length mutation collaborate in inducing acute leukemia in mice. *J Clin Invest.* 2005; 115(8):2159-68.

[274] Schwieger M, Lohler J, Friel J, Scheller M, Horak I, Stocking C. AML1-ETO inhibits maturation of multiple lymphohematopoietic lineages and induces myeloblast transformation in synergy with ICSBP deficiency. *J Exp Med.* 2002; 196(9):1227-40.

[275] Berger R, Bernheim A, Daniel MT, Valensi F, Sigaux F, Flandrin G. Cytogenetic studies on acute myelomonocytic leukaemia (M4) with eosinophilia. *Leuk Res.* 1985; 9(2):279-88.

[276] Khoury H, Dalal BI, Barnett MJ, Nevill TJ, Horsman DE, Shepherd JD, Hogge DE, Toze CL, Conneally EA, Sutherland HJ, Nantel SH. "Quantitative Immunophenotype pattern of $CD34^{high}$, $CD13^{high}$ and $CD33^{low}$ Identifies Patients with Acute Myelogenous Leukemia (AML) With inv(16)". *Blood*, 2001; 98: 109a.

[277] Adriaansen HJ, te Boekhorst PA, Hagemeijer AM, van der Schoot CE, Delwel HR, van Dongen JJ. Acute myeloid leukemia M4 with bone marrow eosinophilia (M4Eo) and inv(16)(p13q22) exhibits a specific immunophenotype with CD2 expression. *Blood.* 1993; 81(11):3043-51.

[278] Bloomfield CD, Lawrence D, Byrd JC, Carroll A, Pettenati MJ, Tantravahi R, Patil SR, Davey FR, Berg DT, Schiffer CA, Arthur DC, Mayer RJ. Frequency of prolonged remission duration after high-dose cytarabine intensification in acute myeloid leukemia varies by cytogenetic subtype. *Cancer Res.* 1998; 58(18):4173-9.

[279] Ogawa E, Inuzuka M, Maruyama M, Satake M, Naito-Fujimoto M, Ito Y, Shigesada K. Molecular cloning and characterization of PEBP2 beta, the heterodimeric partner of a novel Drosophila runt-related DNA binding protein PEBP2 alpha. *Virology.* 1993; 194(1):314-31.

[280] Huang X, Crute BE, Sun C, Tang YY, Kelley JJ 3rd, Lewis AF, Hartman KL, Laue TM, Speck NA, Bushweller JH. Overexpression, purification, and biophysical

characterization of the heterodimerization domain of the core-binding factor beta subunit. *J Biol Chem.* 1998; 273(4):2480-7.

[281] Goger M, Gupta V, Kim WY, Shigesada K, Ito Y, Werner MH. Molecular insights into PEBP2/CBF beta-SMMHC associated acute leukemia revealed from the structure of PEBP2/CBF beta. *Nat Struct Biol.* 1999; 6(7):620-3.

[282] Warren AJ, Bravo J, Williams RL, Rabbitts TH. Structural basis for the heterodimeric interaction between the acute leukaemia-associated transcription factors AML1 and CBFbeta. *EMBO J.* 2000; 19(12):3004-15.

[283] Tahirov TH, Inoue-Bungo T, Sasaki M, Shiina M, Kimura K, Sato K, Kumasaka T, Yamamoto M, Kamiya N, Ogata K. Crystallization and preliminary X-ray analyses of quaternary, ternary and binary protein-DNA complexes with involvement of AML1/Runx-1/CBFalpha Runt domain, CBFbeta and the C/EBPbeta bZip region. *Acta Crystallogr D Biol Crystallogr.* 2001; 57(Pt 6):850-3.

[284] Tang YY, Crute BE, Kelley JJ, Huang X, Yan J, Shi J, Hartman KL, Laue TM, Speck NA, Bushweller JH. Biophysical characterization of interactions between the core binding factor alpha and beta subunits and DNA. *FEBS Lett.* 2000; 470(2):167-72.

[285] Tahirov TH, Inoue-Bungo T, Morii H, Fujikawa A, Sasaki M, Kimura K, Shiina M, Sato K, Kumasaka T, Yamamoto M, Ishii S, Ogata K. Structural analyses of DNA recognition by the AML1/Runx-1 Runt domain and its allosteric control by CBFbeta. *Cell.* 2001; 104(5):755-67.

[286] Kagoshima H, Akamatsu Y, Ito Y, Shigesada K. Functional dissection of the alpha and beta subunits of transcription factor PEBP2 and the redox susceptibility of its DNA binding activity. *J Biol Chem.* 1996; 271(51):33074-82.

[287] Golling G, Li L, Pepling M, Stebbins M, Gergen JP. Drosophila homologs of the proto-oncogene product PEBP2/CBF beta regulate the DNA-binding properties of Runt. *Mol Cell Biol.* 1996; 16(3):932-42.

[288] Adya N, Stacy T, Speck NA, Liu PP. The leukemic protein core binding factor beta (CBFbeta)-smooth-muscle myosin heavy chain sequesters CBFalpha2 into cytoskeletal filaments and aggregates. *Mol Cell Biol.* 1998; 18(12):7432-43.

[289] Huang G, Shigesada K, Ito K, Wee HJ, Yokomizo T, Ito Y. Dimerization with PEBP2beta protects RUNX1/AML1 from ubiquitin-proteasome-mediated degradation. *EMBO J.* 2001; 20(4):723-33.

[290] Kundu M, Chen A, Anderson S, Kirby M, Xu L, Castilla LH, Bodine D, Liu PP. Role of Cbfb in hematopoiesis and perturbations resulting from expression of the leukemogenic fusion gene Cbfb-MYH11. *Blood.* 2002; 100(7):2449-56.

[291] Lu J, Maruyama M, Satake M, Bae SC, Ogawa E, Kagoshima H, Shigesada K, Ito Y. Subcellular localization of the alpha and beta subunits of the acute myeloid leukemia-linked transcription factor PEBP2/CBF. *Mol Cell Biol.* 1995; 15(3):1651-61.

[292] Chiba N, Watanabe T, Nomura S, Tanaka Y, Minowa M, Niki M, Kanamaru R, Satake M. Differentiation dependent expression and distinct subcellular localization of the protooncogene product, PEBP2beta/CBFbeta, in muscle development. *Oncogene.* 1997; 14(21):2543-52.

[293] Tanaka Y, Watanabe T, Chiba N, Niki M, Kuroiwa Y, Nishihira T, Satomi S, Ito Y, Satake M. The protooncogene product, PEBP2beta/CBFbeta, is mainly located in the cytoplasm and has an affinity with cytoskeletal structures. *Oncogene.* 1997; 15(6): 677-83.

[294] Wang Q, Stacy T, Miller JD, Lewis AF, Gu TL, Huang X, Bushweller JH, Bories JC, Alt FW, Ryan G, Liu PP, Wynshaw-Boris A, Binder M, Marin-Padilla M, Sharpe AH, Speck NA. The CBFbeta subunit is essential for CBFalpha2 (AML1) function in vivo. *Cell.* 1996; 87(4):697-708.

[295] Sasaki K, Yagi H, Bronson RT, Tominaga K, Matsunashi T, Deguchi K, Tani Y, Kishimoto T, Komori T. Absence of fetal liver hematopoiesis in mice deficient in transcriptional coactivator core binding factor beta. *Proc Natl Acad Sci U S A.* 1996; 93(22):12359-63

[296] Wang Q, Stacy T, Binder M, Marin-Padilla M, Sharpe AH, Speck NA. Disruption of the Cbfa2 gene causes necrosis and hemorrhaging in the central nervous system and blocks definitive hematopoiesis. *Proc Natl Acad Sci U S A.* 1996; 93(8):3444-9.

[297] Miller JD, Stacy T, Liu PP, Speck NA. Core-binding factor beta (CBFbeta), but not CBFbeta-smooth muscle myosin heavy chain, rescues definitive hematopoiesis in CBFbeta-deficient embryonic stem cells. *Blood.* 2001; 97(8):2248-56.

[298] Nagai R, Kuro-o M, Babij P, Periasamy M. Identification of two types of smooth muscle myosin heavy chain isoforms by cDNA cloning and immunoblot analysis. *J Biol Chem.* 1989; 264(17):9734-7.

[299] Rayment I, Holden HM, Whittaker M, Yohn CB, Lorenz M, Holmes KC, Milligan RA. Structure of the actin-myosin complex and its implications for muscle contraction. *Science.* 1993; 261(5117):58-65.

[300] Parry DA. Structure of rabbit skeletal myosin. Analysis of the amino acid sequences of two fragments from the rod region. J Mol Biol. 1981; 153(2):459-64.

[301] Warrick HM, De Lozanne A, Leinwand LA, Spudich JA. Conserved protein domains in a myosin heavy chain gene from Dictyostelium discoideum. *Proc Natl Acad Sci U S A.* 1986; 83(24):9433-7.

[302] Strehler EE, Strehler-Page MA, Perriard JC, Periasamy M, Nadal-Ginard B. Complete nucleotide and encoded amino acid sequence of a mammalian myosin heavy chain gene. Evidence against intron-dependent evolution of the rod. *J Mol Biol.* 1986; 190(3):291-317.

[303] Rovner AS, Fagnant PM, Lowey S, Trybus KM. The carboxyl-terminal isoforms of smooth muscle myosin heavy chain determine thick filament assembly properties. *J Cell Biol.* 2002; 156(1):113-23.

[304] Liu PP, Hajra A, Wijmenga C, Collins FS. Molecular pathogenesis of the chromosome 16 inversion in the M4Eo subtype of acute myeloid *leukemia. Blood.* 1995; 85(9): 2289-302

[305] van der Reijden BA, Dauwerse HG, Giles RH, Jagmohan-Changur S, Wijmenga C, Liu PP, Smit B, Wessels HW, Beverstock GC, Jotterand-Bellomo M, Martinet D, Muhlematter D, Lafage-Pochitaloff M, Gabert J, Reiffers J, Bilhou-Nabera C, van Ommen GJ, Hagemeijer A, Breuning MH. Genomic acute myeloid leukemia-associated inv(16)(p13q22) breakpoints are tightly clustered. *Oncogene.* 1999; 18(2):543-50.

[306] Shurtleff SA, Meyers S, Hiebert SW, Raimondi SC, Head DR, Willman CL, Wolman S, Slovak ML, Carroll AJ, Behm F, et al. Heterogeneity in CBF beta/MYH11 fusion messages encoded by the inv(16)(p13q22) and the t(16;16)(p13;q22) in acute myelogenous leukemia. *Blood.* 1995; 85(12):3695-703.

[307] van der Reijden BA, Lombardo M, Dauwerse HG, Giles RH, Muhlematter D, Bellomo MJ, Wessels HW, Beverstock GC, van Ommen GJ, Hagemeijer A, et al. RT-PCR

diagnosis of patients with acute nonlymphocytic leukemia and inv(16)(p13q22) and identification of new alternative splicing in CBFB-MYH11 transcripts. *Blood.* 1995; 86(1):277-82.

[308] Lukasik SM, Zhang L, Corpora T, Tomanicek S, Li Y, Kundu M, Hartman K, Liu PP, Laue TM, Biltonen RL, Speck NA, Bushweller JH. Altered affinity of CBF beta-SMMHC for Runx1 explains its role in leukemogenesis. *Nat Struct Biol.* 2002; 9(9):674-9.

[309] Huang G, Shigesada K, Wee HJ, Liu PP, Osato M, Ito Y. Molecular basis for a dominant inactivation of RUNX1/AML1 by the leukemogenic inversion 16 chimera. *Blood.* 2004; 103(8):3200-7.

[310] Lutterbach B, Hou Y, Durst KL, Hiebert SW. The inv(16) encodes an acute myeloid leukemia 1 transcriptional corepressor. *Proc Natl Acad Sci U S A.* 1999; 96(22): 12822-7.

[311] Durst KL, Lutterbach B, Kummalue T, Friedman AD, Hiebert SW. The inv(16) fusion protein associates with corepressors via a smooth muscle myosin heavy-chain domain. *Mol Cell Biol.* 2003; 23(2):607-19.

[312] Liu P, Seidel N, Bodine D, Speck N, Tarle S, Collins FS. Acute myeloid leukemia with Inv (16) produces a chimeric transcription factor with a myosin heavy chain tail. *Cold Spring Harb Symp Quant Biol.* 1994; 59:547-53.

[313] Liu PP, Wijmenga C, Hajra A, Blake TB, Kelley CA, Adelstein RS, Bagg A, Rector J, Cotelingam J, Willman CL, Collins FS. Identification of the chimeric protein product of the CBFB-MYH11 fusion gene in inv(16) leukemia cells. *Genes Chromosomes Cancer.* 1996; 16(2):77-87.

[314] Kanno Y, Kanno T, Sakakura C, Bae SC, Ito Y. Cytoplasmic sequestration of the polyomavirus enhancer binding protein 2 (PEBP2)/core binding factor alpha (CBFalpha) subunit by the leukemia-related PEBP2/CBFbeta-SMMHC fusion protein inhibits PEBP2/CBF-mediated transactivation. *Mol Cell Biol.* 1998; 18(7):4252-61.

[315] Kanto S, Chiba N, Tanaka Y, Fujita S, Endo M, Kamada N, Yoshikawa K, Fukuzaki A, Orikasa S, Watanabe T, Satake M. The PEBP2beta/CBF beta-SMMHC chimeric protein is localized both in the cell membrane and nuclear subfractions of leukemic cells carrying chromosomal inversion 16. *Leukemia.* 2000; 14(7):1253-9.

[316] Wijmenga C, Gregory PE, Hajra A, Schrock E, Ried T, Eils R, Liu PP, Collins FS. Core binding factor beta-smooth muscle myosin heavy chain chimeric protein involved in acute myeloid leukemia forms unusual nuclear rod-like structures in transformed NIH 3T3 cells. *Proc Natl Acad Sci U S A.* 1996; 93(4):1630-5.

[317] Cao W, Adya N, Britos-Bray M, Liu PP, Friedman AD. The core binding factor (CBF) alpha interaction domain and the smooth muscle myosin heavy chain (SMMHC) segment of CBFbeta-SMMHC are both required to slow cell proliferation. *J Biol Chem.* 1998; 273(47):31534-40.

[318] Lou J, Cao W, Bernardin F, Ayyanathan K, RauscherIII FJ, Friedman AD. Exogenous cdk4 overcomes reduced cdk4 RNA and inhibition of G1 progression in hematopoietic cells expressing a dominant-negative CBF - a model for overcoming inhibition of proliferation by CBF oncoproteins. *Oncogene.* 2000; 19(22):2695-703.

[319] Bernardin F, Friedman AD. AML1 stimulates G1 to S progression via its transactivation domain. *Oncogene.* 2002; 21(20):3247-52.

[320] Bernardin-Fried F, Kummalue T, Leijen S, Collector MI, Ravid K, Friedman AD. AML1/RUNX1 increases during G1 to S cell cycle progression independent of cytokine-dependent phosphorylation and induces cyclin D3 gene expression. *J Biol Chem.* 2004; 279(15):15678-87.

[321] D'Costa J, Chaudhuri S, Civin CI, Friedman AD. CBFbeta-SMMHC slows proliferation of primary murine and human myeloid progenitors. *Leukemia.* 2005; 19(6):921-9.

[322] Cilloni D, Carturan S, Gottardi E, Messa F, Messa E, Fava M, Diverio D, Guerrasio A, Lo-Coco F, Saglio G. Down-modulation of the C/EBPalpha transcription factor in core binding factor acute myeloid leukemias. *Blood.* 2003; 102(7):2705-6.

[323] Castilla LH, Wijmenga C, Wang Q, Stacy T, Speck NA, Eckhaus M, Marin-Padilla M, Collins FS, Wynshaw-Boris A, Liu PP. Failure of embryonic hematopoiesis and lethal hemorrhages in mouse embryos heterozygous for a knocked-in leukemia gene CBFB-MYH11. *Cell.* 1996; 87(4):687-96.

[324] Britos-Bray M, Ramirez M, Cao W, Wang X, Liu PP, Civin CI, Friedman AD. CBFbeta-SMMHC, expressed in M4eo acute myeloid leukemia, reduces p53 induction and slows apoptosis in hematopoietic cells exposed to DNA-damaging agents. *Blood.* 1998; 92(11):4344-52.

[325] Cao W, Britos-Bray M, Claxton DF, Kelley CA, Speck NA, Liu PP, Friedman AD. CBF beta-SMMHC, expressed in M4Eo AML, reduced CBF DNA-binding and inhibited the G1 to S cell cycle transition at the restriction point in myeloid and lymphoid cells. *Oncogene.* 1997; 15(11):1315-27.

[326] Castilla LH, Garrett L, Adya N, Orlic D, Dutra A, Anderson S, Owens J,Eckhaus M, Bodine D, Liu PP. The fusion gene Cbfb-MYH11 blocks myeloid differentiation and predisposes mice to acute myelomonocytic leukaemia. *Nat Genet.* 1999; 23(2):144-6. No

[327] Kogan SC, Lagasse E, Atwater S, Bae SC, Weissman I, Ito Y, Bishop JM. The PEBP2betaMYH11 fusion created by inv(16)(p13;q22) in myeloid leukemia impairs neutrophil maturation and contributes to granulocytic dysplasia. *Proc Natl Acad Sci U S A.* 1998; 95(20):11863-8.

[328] Kuo YH, Landrette SF, Heilman SA, Perrat PN, Garrett L, Liu PP, Le Beau MM, Kogan SC, Castilla LH. Cbf beta-SMMHC induces distinct abnormal myeloid progenitors able to develop acute myeloid leukemia. Cancer *Cell.* 2006; 9(1):57-68.

[329] Castilla LH, Perrat P, Martinez NJ, Landrette SF, Keys R, Oikemus S, Flanegan J, Heilman S, Garrett L, Dutra A, Anderson S, Pihan GA, Wolff L, Liu PP. Identification of genes that synergize with Cbfb-MYH11 in the pathogenesis of acute myeloid *leukemia. Proc Natl Acad Sci U S A.* 2004; 101(14):4924-9.

[330] Aplan PD. Chromosomal translocations involving the MLL gene: molecular mechanisms. DNA Repair (Amst). 2006 Sep 8;5(9-10):1265-72.

[331] Basecke J, Whelan JT, Griesinger F, Bertrand FE. The MLL partial tandem duplication in acute myeloid leukaemia. *Br J Haematol.* 2006; 135(4):438-49.

[332] Gu Y, Nakamura T, Alder H, Prasad R, Canaani O, Cimino G, Croce CM, Canaani E. The t(4;11) chromosome translocation of human acute leukemias fuses the ALL-1 gene, related to Drosophila trithorax, to the AF-4 gene. *Cell.* 1992; 71(4):701-8.

[333] Tkachuk DC, Kohler S, Cleary ML. Involvement of a homolog of Drosophila trithorax by 11q23 chromosomal translocations in acute leukemias. *Cell.* 1992; 71(4):691-700

[334] Yokoyama A, Kitabayashi I, Ayton PM, Cleary ML, Ohki M. Leukemia proto-oncoprotein MLL is proteolytically processed into 2 fragments with opposite transcriptional properties. *Blood.* 2002; 100(10):3710-8.

[335] Hsieh JJ, Cheng EH, Korsmeyer SJ. Taspase1: a threonine aspartase required for cleavage of MLL and proper HOX gene expression. *Cell.* 2003; 115(3):293-303

[336] Hsieh JJ, Ernst P, Erdjument-Bromage H, Tempst P, Korsmeyer SJ. Proteolytic cleavage of MLL generates a complex of N- and C-terminal fragments that confers protein stability and subnuclear localization. *Mol Cell Biol.* 2003; 23(1):186-94.

[337] Yokoyama A, Wang Z, Wysocka J, Sanyal M, Aufiero DJ, Kitabayashi I, Herr W, Cleary ML. Leukemia proto-oncoprotein MLL forms a SET1-like histone methyltransferase complex with menin to regulate Hox gene expression. *Mol Cell Biol.* 2004; 24(13):5639-49.

[338] Yokoyama A, Somervaille TC, Smith KS, Rozenblatt-Rosen O, Meyerson M, Cleary ML. The menin tumor suppressor protein is an essential oncogenic cofactor for MLL-associated leukemogenesis. *Cell.* 2005; 123(2):207-18.

[339] Milne TA, Hughes CM, Lloyd R, Yang Z, Rozenblatt-Rosen O, Dou Y, Schnepp RW, Krankel C, Livolsi VA, Gibbs D, Hua X, Roeder RG, Meyerson M, Hess JL. Menin and MLL cooperatively regulate expression of cyclin-dependent kinase inhibitors. *Proc Natl Acad Sci U S A.* 2005; 102(3):749-54

[340] Zeleznik-Le NJ, Harden AM, Rowley JD. 11q23 translocations split the "AT-hook" cruciform DNA-binding region and the transcriptional repression domain from the activation domain of the mixed-lineage leukemia (MLL) gene. *Proc Natl Acad Sci U S A.* 1994; 91(22):10610-4.

[341] Aravind L, Landsman D. AT-hook motifs identified in a wide variety of DNA-binding proteins. *Nucleic Acids Res.* 1998; 26(19):4413-21

[342] Yano T, Nakamura T, Blechman J, Sorio C, Dang CV, Geiger B, Canaani E. Nuclear punctate distribution of ALL-1 is conferred by distinct elements at the N terminus of the protein. *Proc Natl Acad Sci U S A.* 1997; 94(14):7286-91.

[343] Caslini C, Alarcon AS, Hess JL, Tanaka R, Murti KG, Biondi A. The amino terminus targets the mixed lineage leukemia (MLL) protein to the nucleolus, nuclear matrix and mitotic chromosomal scaffolds. *Leukemia.* 2000; 14(11):1898-908

[344] Cross SH, Meehan RR, Nan X, Bird A. A component of the transcriptional repressor MeCP1 shares a motif with DNA methyltransferase and HRX proteins. *Nat Genet.* 1997; 16(3):256-9.

[345] Birke M, Schreiner S, Garcia-Cuellar MP, Mahr K, Titgemeyer F, Slany RK. The MT domain of the proto-oncoprotein MLL binds to CpG-containing DNA and discriminates against methylation. *Nucleic Acids Res.* 2002; 30(4):958-65.

[346] Xia ZB, Anderson M, Diaz MO, Zeleznik-Le NJ. MLL repression domain interacts with histone deacetylases, the polycomb group proteins HPC2 and BMI-1, and the corepressor C-terminal-binding protein. *Proc Natl Acad Sci U S A.* 2003; 100(14): 8342-7.

[347] Aasland R, Gibson TJ, Stewart AF. The PHD finger: implications for chromatin-mediated transcriptional regulation. *Trends Biochem Sci.* 1995; 20(2):56-9.

[348] Anderson M, Fair K, Amero S, Nelson S, Harte PJ, Diaz MO. A new family of cyclophilins with an RNA recognition motif that interact with members of the trx/MLL protein family in Drosophila and human cells. *Dev Genes Evol.* 2002; 212(3):107-13.

[349] Fair K, Anderson M, Bulanova E, Mi H, Tropschug M, Diaz MO. Protein interactions of the MLL PHD fingers modulate MLL target gene regulation in human cells. *Mol Cell Biol.* 2001; 21(10):3589-97.

[350] Wysocka J, Myers MP, Laherty CD, Eisenman RN, Herr W. Human Sin3 deacetylase and trithorax-related Set1/Ash2 histone H3-K4 methyltransferase are tethered together selectively by the cell proliferation factor HCF-1. *Genes Dev.* 2003; 17(7):896-911.

[351] Ernst P, Wang J, Huang M, Goodman RH, Korsmeyer SJ. MLL and CREB bind cooperatively to the nuclear coactivator CREB-binding protein. *Mol Cell Biol.* 2001; 21(7):2249-58.

[352] Dou Y, Milne TA, Tackett AJ, Smith ER, Fukuda A, Wysocka J, Allis CD, Chait BT, Hess JL, Roeder RG. Physical association and coordinate function of the H3 K4 methyltransferase MLL1 and the H4 K16 acetyltransferase MOF. *Cell.* 2005; 121(6):873-85.

[353] Milne TA, Briggs SD, Brock HW, Martin ME, Gibbs D, Allis CD, Hess JL. MLL targets SET domain methyltransferase activity to Hox gene promoters. *Mol Cell.* 2002; 10(5):1107-17.

[354] Rozenblatt-Rosen O, Rozovskaia T, Burakov D, Sedkov Y, Tillib S, Blechman J, Nakamura T, Croce CM, Mazo A, Canaani E. The C-terminal SET domains of ALL-1 and TRITHORAX interact with the INI1 and SNR1 proteins, components of the SWI/SNF complex. *Proc Natl Acad Sci U S A.* 1998; 95(8):4152-7.

[355] Rozovskaia T, Rozenblatt-Rosen O, Sedkov Y, Burakov D, Yano T, Nakamura T, Petruck S, Ben-Simchon L, Croce CM, Mazo A, Canaani E. Self association of the SET domains of human ALL-1 and of Drosophila TRITHORAX and ASH1 proteins. *Oncogene.* 2000; 19(3):351-7.

[356] Nakamura T, Mori T, Tada S, Krajewski W, Rozovskaia T, Wassell R, Dubois G,Mazo A, Croce CM, Canaani E. ALL-1 is a histone methyltransferase that assembles a supercomplex of proteins involved in transcriptional regulation. *Mol Cell.* 2002; 10(5):1119-28.

[357] Nakamura T, Blechman J, Tada S, Rozovskaia T, Itoyama T, Bullrich F, Mazo A, Croce CM, Geiger B, Canaani E. huASH1 protein, a putative transcription factor encoded by a human homologue of the Drosophila ash1 gene, localizes to both nuclei and cell-cell tight junctions. *Proc Natl Acad Sci U S A.* 2000; 97(13):7284.

[358] Pray-Grant MG, Schieltz D, McMahon SJ, Wood JM, Kennedy EL, Cook RG, Workman JL, Yates JR 3rd, Grant PA. The novel SLIK histone acetyltransferase complex functions in the yeast retrograde response pathway. *Mol Cell Biol.* 2002; 22(24):8774-86.

[359] Yu BD, Hess JL, Horning SE, Brown GA, Korsmeyer SJ. Altered Hox expression and segmental identity in Mll-mutant mice. *Nature.* 1995; 378(6556):505-8.

[360] Schumacher A, Magnuson T. Murine Polycomb- and trithorax-group genes regulate homeotic pathways and beyond. *Trends Genet.* 1997; 13(5):167-70.

[361] Hanson RD, Hess JL, Yu BD, Ernst P, van Lohuizen M, Berns A, van der Lugt NM, Shashikant CS, Ruddle FH, Seto M, Korsmeyer SJ. Mammalian Trithorax and polycomb-group homologues are antagonistic regulators of homeotic development. *Proc Natl Acad Sci U S A.* 1999; 96(25):14372-7.

[362] Ernst P, Mabon M, Davidson AJ, Zon LI, Korsmeyer SJ. An Mll-dependent Hox program drives hematopoietic progenitor expansion. *Curr Biol.* 2004; 14(22):2063-9.

[363] Yu BD, Hanson RD, Hess JL, Horning SE, Korsmeyer SJ. MLL, a mammalian trithorax-group gene, functions as a transcriptional maintenance factor in morphogenesis. *Proc Natl Acad Sci U S A.* 1998; 95(18):10632-6.

[364] Arakawa H, Nakamura T, Zhadanov AB, Fidanza V, Yano T, Bullrich F, Shimizu M, Blechman J, Mazo A, Canaani E, Croce CM. Identification and characterization of the ARP1 gene, a target for the human acute leukemia ALL1 gene. *Proc Natl Acad Sci U S A.* 1998; 95(8):4573-8.

[365] Ferrari N, Palmisano GL, Paleari L, Basso G, Mangioni M, Fidanza V, Albini A, Croce CM, Levi G, Brigati C. DLX genes as targets of ALL-1: DLX 2,3,4 down-regulation in t(4;11) acute lymphoblastic leukemias. *J Leukoc Biol.* 2003; 74(2):302-5.

[366] Schraets D, Lehmann T, Dingermann T, Marschalek R. MLL-mediated transcriptional gene regulation investigated by gene expression profiling. *Oncogene.* 2003; 22(23):3655-68.

[367] Hess JL, Yu BD, Li B, Hanson R, Korsmeyer SJ. Defects in yolk sac hematopoiesis in Mll-null embryos. *Blood.* 1997; 90(5):1799-806.

[368] Yagi H, Deguchi K, Aono A, Tani Y, Kishimoto T, Komori T. Growth disturbance in fetal liver hematopoiesis of Mll-mutant mice. *Blood.* 1998; 92(1):108-17.

[369] Corral J, Lavenir I, Impey H, Warren AJ, Forster A, Larson TA, Bell S, McKenzie AN, King G, Rabbitts TH. An Mll-AF9 fusion gene made by homologous recombination causes acute leukemia in chimeric mice: a method to create fusion oncogenes. *Cell.* 1996; 85(6):853-61.

[370] Dobson CL, Warren AJ, Pannell R, Forster A, Lavenir I, Corral J, Smith AJ,Rabbitts TH. The mll-AF9 gene fusion in mice controls myeloproliferation and specifies acute myeloid leukaemogenesis. *EMBO J.* 1999; 18(13):3564-74.

[371] Johnson JJ, Chen W, Hudson W, Yao Q, Taylor M, Rabbitts TH, Kersey JH. Prenatal and postnatal myeloid cells demonstrate stepwise progression in the pathogenesis of MLL fusion gene *leukemia. Blood.* 2003; 101(8):3229-35.

[372] DiMartino JF, Ayton PM, Chen EH, Naftzger CC, Young BD, Cleary ML. The AF10 leucine zipper is required for leukemic transformation of myeloid progenitors by MLL-AF10. *Blood.* 2002; 99(10):3780-5.

[373] DiMartino JF, Miller T, Ayton PM, Landewe T, Hess JL, Cleary ML, Shilatifard A. A carboxy-terminal domain of ELL is required and sufficient for immortalization of myeloid progenitors by MLL-ELL. *Blood.* 2000; 96(12):3887-93.

[374] Lavau C, Luo RT, Du C, Thirman MJ. Retrovirus-mediated gene transfer of MLL-ELL transforms primary myeloid progenitors and causes acute myeloid leukemias in mice. *Proc Natl Acad Sci U S A.* 2000; 97(20):10984

[375] Slany RK, Lavau C, Cleary ML. The oncogenic capacity of HRX-ENL requires the transcriptional transactivation activity of ENL and the DNA binding motifs of HRX. *Mol Cell Biol.* 1998; 18(1):122-9.

[376] So CW, Cleary ML. Common mechanism for oncogenic activation of MLL by forkhead family proteins. *Blood.* 2003; 101(2):633-9.

[377] Prasad R, Yano T, Sorio C, Nakamura T, Rallapalli R, Gu Y, Leshkowitz D, Croce CM, Canaani E. Domains with transcriptional regulatory activity within the ALL1 and AF4 proteins involved in acute *leukemia. Proc Natl Acad Sci U S A.* 1995; 92(26):12160-4.

[378] Lavau C, Du C, Thirman M, Zeleznik-Le N. Chromatin-related properties of CBP fused to MLL generate a myelodysplastic-like syndrome that evolves into myeloid *leukemia. EMBO J.* 2000; 19(17):4655-64.

[379] Wiederschain D, Kawai H, Gu J, Shilatifard A, Yuan ZM. Molecular basis of p53 functional inactivation by the leukemic protein MLL-ELL. *Mol Cell Biol.* 2003; 23(12):4230-46.

[380] Martin ME, Milne TA, Bloyer S, Galoian K, Shen W, Gibbs D, Brock HW, Slany R, Hess JL. Dimerization of MLL fusion proteins immortalizes hematopoietic cells. *Cancer Cell.* 2003; 4(3):197-207.

[381] Milne TA, Dou Y, Martin ME, Brock HW, Roeder RG, Hess JL. MLL associates specifically with a subset of transcriptionally active target genes. *Proc Natl Acad Sci U S A.* 2005; 102(41):14765-70.

[382] Milne TA, Martin ME, Brock HW, Slany RK, Hess JL. Leukemogenic MLL fusion proteins bind across a broad region of the Hox a9 locus, promoting transcription and multiple histone modifications. *Cancer Res.* 2005; 65(24):11367-74.

[383] Nie Z, Yan Z, Chen EH, Sechi S, Ling C, Zhou S, Xue Y, Yang D, Murray D, Kanakubo E, Cleary ML, Wang W. Novel SWI/SNF chromatin-remodeling complexes contain a mixed-lineage leukemia chromosomal translocation partner. *Mol Cell Biol.* 2003; 23(8):2942-52.

[384] Debernardi S, Bassini A, Jones LK, Chaplin T, Linder B, de Bruijn DR, Meese E, Young BD. The MLL fusion partner AF10 binds GAS41, a protein that interacts with the human SWI/SNF complex. *Blood.* 2002; 99(1):275-81.

[385] Armstrong SA, Staunton JE, Silverman LB, Pieters R, den Boer ML, Minden MD, Sallan SE, Lander ES, Golub TR, Korsmeyer SJ. MLL translocations specify a distinct gene expression profile that distinguishes a unique leukemia. *Nat Genet.* 2002; 30(1):41-7.

[386] Yeoh EJ, Ross ME, Shurtleff SA, Williams WK, Patel D, Mahfouz R, Behm FG, Raimondi SC, Relling MV, Patel A, Cheng C, Campana D, Wilkins D, Zhou X, Li J, Liu H, Pui CH, Evans WE, Naeve C, Wong L, Downing JR. Classification, subtype discovery, and prediction of outcome in pediatric acute lymphoblastic leukemia by gene expression profiling. *Cancer Cell.* 2002; 1(2):133-43.

[387] Ferrando AA, Armstrong SA, Neuberg DS, Sallan SE, Silverman LB, Korsmeyer SJ, Look AT. Gene expression signatures in MLL-rearranged T-lineage and B-precursor acute leukemias: dominance of HOX dysregulation. *Blood.* 2003; 102(1):262-8

[388] Zeisig BB, Milne T, Garcia-Cuellar MP, Schreiner S, Martin ME, Fuchs U, Borkhardt A, Chanda SK, Walker J, Soden R, Hess JL, Slany RK. Hoxa9 and Meis1 are key targets for MLL-ENL-mediated cellular immortalization. *Mol Cell Biol.* 2004; 24(2):617-28.

[389] Kumar AR, Hudson WA, Chen W, Nishiuchi R, Yao Q, Kersey JH. Hoxa9 influences the phenotype but not the incidence of Mll-AF9 fusion gene leukemia. *Blood.* 2004; 103(5):1823-8.

[390] Collins EC, Pannell R, Simpson EM, Forster A, Rabbitts TH. Inter-chromosomal recombination of Mll and Af9 genes mediated by cre-loxP in mouse development. *EMBO Rep.* 2000; 1(2):127-32.

[391] Tabe Y, Jin L, Contractor R, Gold D, Ruvolo P, Radke S, Xu Y, Tsutusmi-Ishii Y,Miyake K, Miyake N, Kondo S, Ohsaka A, Nagaoka I, Andreeff M, Konopleva M.

Novel role of HDAC inhibitors in AML1/ETO AML cells: activation of apoptosis and phagocytosis through induction of annexin A1. *Cell Death Differ*. 2007; 14(8):1443-56.
[392] Liu S, Klisovic RB, Vukosavljevic T, Yu J, Paschka P, Huynh L, Pang J, Neviani P, Liu Z, Blum W, Chan KK, Perrotti D, Marcucci G. Targeting AML1/ETO-histone deacetylase repressor complex: a novel mechanism for valproic acid-mediated gene expression and cellular differentiation in AML1/ETO-positive acute myeloid leukemia cells. *J Pharmacol Exp Ther*. 2007; 321(3):953-60.
[393] Moe-Behrens GH, Pandolfi PP. Targeting aberrant transcriptional repression in acute myeloid leukemia. *Rev Clin Exp Hematol*. 2003; 7(2):139-59.
[394] Cheng Y, Paz K.Tandutinib, an oral, small-molecule inhibitor of FLT3 for the treatment of AML and other cancer indications. *IDrugs*. 2008;11(1):46-56.
[395] Lee BD, Sevcikova S, Kogan SC. Dual treatment with FLT3 inhibitor SU11657 and doxorubicin increases survival of leukemic mice. *Leuk Res*. 2007; 31(8):1131-4.
[396] Knapper S, Burnett AK, Littlewood T, Kell WJ, Agrawal S, Chopra R, Clark R, Levis MJ, Small D. A phase 2 trial of the FLT3 inhibitor lestaurtinib (CEP701) as first-line treatment for older patients with acute myeloid leukemia not considered fit for intensive chemotherapy. *Blood*. 2006; 108(10):3262-70.

In: Genetic Translocations and Other Chromosome Aberrations
Editor: Guy T. Leyden, pp. 65-116
ISBN: 978-1-60456-344-3
© 2008 Nova Science Publishers, Inc.

Chapter 2

Chromosomal Translocations and Fusion Genes in Human Soft Tissue Sarcomas

Masanori Hisaoka*
Department of Pathology and Oncology, School of Medicine,
University of Occupational and Environmental Health, Japan

Abstract

Soft tissue tumors are rare and a heterogeneous group of neoplastic lesions that mostly display variable morphologic features of non-epithelial or mesenchymal tissues. Because of the rarity and morphologic diversity, the diagnoses of soft tissue tumors are often challenging and difficult to make. Besides, pathogenetic mechanisms underlying the development of soft tissue tumors still remain largely unknown. In the past decades, great strides in our understanding of molecular or genetic aberrations of various neoplastic diseases have been made. One of the most profound impacts in this field is an identification of tumor-associated chromosomal translocations that result in formation of unique fusion (or chimeric) genes. It has been also shown that such aberrations are specifically found in certain morphologic types of neoplasms, many of which are belonging to soft tissue tumors with malignant biological potential, namely sarcomas. Currently more than ten types of soft tissue sarcomas are known to be translocation-associated. Many tumor-associated chromosomal translocations or fusion genes have been repeatedly and increasingly detected by standard karyotyping in combination with recent molecular techniques such as fluorescence *in situ* hybridization and reverse transcription-polymerase chain reaction, which can be also applied to archival formalin-fixed, paraffin-embedded tumor tissues. In addition to an insight into their oncogenic roles, molecular detection of such cytogenetic aberrations can provide valuable information with respect to tumor diagnosis, because such alterations are most often tumor type-specific. Notably, most translocation-associated tumors have some variations in the chromosomal regions or genes involved by the translocations or in resultant fusion gene transcripts. This phenomenon implies that morphologic features of the tumors and their biological behavior or

* Address for correspondence to M. Hisaoka, Department of Pathology and Oncology, School of Medicine, University of Occupational and Environmental Health, 1-1 Iseigaoka Yahatanishi-ku, Kitakyushu 807-8555, Japan

therapeutic susceptibility may also vary according to such cytogenetic or molecular variations. In this chapter, summarized and updated molecular genetic and clinicopathologic features of translocation-associated soft tissue sarcomas are described.

Introduction

Soft tissue tumors are uncommon and encompass a wide variety of neoplasms or tumor-like conditions classified into more than 50 histologic types based on variable lines of differentiation towards mesenchymal tissues such as adipose tissue, skeletal and smooth muscles, blood vessels, and so forth [1,2]. Generally, soft tissue tumors can affect individuals at any age and develop anywhere in the body, although each type of tumor has some predilections for particular patient's ages and anatomical locations. According to their biologic behaviors, the tumors are categorized into benign lesions (tumors usually do not recur or metastasize and are extirpated by complete local excision), those of low-grade or intermediate malignancy (tumors often recur in a locally aggressive growth pattern or rarely metastasize), and malignant tumors (tumors grow in a locally destructive manner with significant risk of local recurrence and distant metastasis) [1]. A clinical treatment as well as patient's prognosis generally depends on the histologic type and grade of a given tumor. However, the rarity and morphologic diversity of the soft tissue tumors often create diagnostic challenges. Moreover, it is not uncommon that distinct types of tumor show similar or overlapping histologic appearances that often make the pathologic diagnosis (or histologic typing) difficult. In this case, various ancillary techniques such as immunohistochemistry, electron microscopy, and molecular genetic analyses are used as diagnostic adjuncts.

The etiology of soft tissue tumors remains still largely unknown. In some cases, genetic and environmental factors, irradiation, viral infection, and immune deficiency have been suggested to be associated with the tumor development [1,2]. Recent identifications of unique chromosomal translocations and resultant fusion (or chimeric) genes in certain types of tumors, particularly soft tissue sarcomas, have provided a valuable insight into their pathogenetic mechanisms [3,4]. Such translocation-derived fusion genes can generate in-frame fusion transcripts, encoding novel fusion proteins that are considered to function as aberrant biological molecules, such as transcription factors and growth factors, and to play a pivotal role in tumorigenesis. In addition, such chromosomal translocations or fusion genes are essentially tumor type-specific, leading to a notion that the detection of tumor type-specific cytogenetic changes can be a useful diagnostic aid [5,6]. Indeed, the cytogenetic and molecular identifications of such chromosomal translocations or fusion genes have been applied to confirmation of the tumor diagnoses [7,8]. Currently, more than 10 types of soft tissue sarcoma are considered to be translocation- or fusion gene-associated (Table 1) [4,6,8].

Table 1. Sarcomas with Characteristic Chromosomal Translocations and Fusion Genes

Tumor	Representative translocation	Fusion gene (estimated frequency)
Synovial sarcoma	t(X;18)(p11.2;q11.2)	SS18-SSX1 (65-75%)
		SS18-SSX2 (25-35%)
		SS18-SSX4 (rare)
	t(18;20)(q11.2;q13.3)	SS18L1-SSX1 (rare)

Table 1. Continued

Tumor	Representative translocation	Fusion gene (estimated frequency)
Ewing family of tumors	t(11;22)(q24;q12)	EWSR1-FLI-1 (≈85%)
	t(21;22)(q22;q12)	EWSR1-ERG (5-10%)
	t(7;22)(p22;q12)	EWSR1-ETV1 (rare)
	t(2;22)(q33;q12)	EWSR1-FEV (rare)
	t(17;22)(q21;q12)	EWSR1-ETV4 (rare)
	t(16;21)(p11;q22)	FUS-ERG (rare)
Alveolar rhabdomyosarcoma	t(2;13)(q35;q14)	PAX3-FOXO1A (80-85%)
	t(1;13)(q36;q14)	PAX7-FOXO1A (15-20%)
	t(2;2)(q35;p23)	PAX3-NCOA1 (rare)
Myxoid liposarcoma	t(12;16)(q13;p11)	FUS-DDIT3 (≈90%)
	t(12;22)(q13;q12)	EWSR1-DDIT3 (≈10%)
Desmoplastic small round cell Tumor	t(11;22)(p13;q12)	EWSR1-WT1
Extraskeletal myxoid Chondrosarcoma	t(9;22)(q22;q12)	EWSR1-NR4A3 (≈85%)
	t(9;17)(q22;q11)	TAF15-NR4A3 (≈15%)
	t(9;15)(q22;q21)	TCF12-NR4A3 (rare)
	t(3;9)(q21;q22)*	TFG-NR4A3 (rare)
Dermatofibrosarcoma Protuberans	t(17;22)(q22;q13) or r(17;22)	COL1A1-PDGFB
Infantile fibrosarcoma	t(12;15)(p13;q25)	ETV6-NTRK3
Inflammatory myofibroblastic tumor	t(2;2)(p23;p35)	ATIC-ALK (?%)
	t(2;2)(p23;q13)	RANBP2-ALK (?%)
	t(1;2)(q21;p23)	TMP3-ALK (?%)
	t(2;19)(p23;p13)	TMP4-ALK (?%)
	t(2;11)(p23;p15)	CARS-ALK (?%)
	t(2;17)(p23;q23)	CLTC-ALK (?%)
	t(2;4)(p23;q21)	SEC31L1-ALK (?%)
Clear cell sarcoma	t(12;22)(q13;p12)	EWSR1-ATF1 (> 90%)
	t(2;22)(q32;q12)*	EWSR1-CREB1 (rare)
Alveolar soft part sarcoma	der(17)t(X;17)(p11.2;q25)	ASPSCR1-TFE3
Low grade fibromyxoid sarcoma	t(7;16)(q32;p11)	FUS-CREB3L2 (> 90%)
	t(11;16)(p11;p11)*	FUS-CREB3L1 (rare)
Angiomatoid fibrous histiocytoma	t(12;16)(q13;p11)*	FUS-ATF1
	t(12;22)(q13;p11)*	EWSR1-ATF1

* putative but yet undetected translocations

In this chapter, summarized methodological aspects of cytogenetic and other recent molecular techniques to detect the chromosomal translocations or fusion genes are described. Subsequently, representative soft tissue sarcomas with characteristic chromosomal translocations and fusion genes are reviewed with special reference to their updated clinicopathologic and molecular genetic profiles.

Cytogenetic and Molecular Genetic Approaches

Traditionally, many chromosomal aberrations are identified by conventional karyotyping of short-term cultured tumor cells. After semi-confluent growths in flasks, metaphase spreads of the tumor cells are prepared and usually stained by the trypsin-Giemsa method [9]. The stained or banded chromosomes are examined microscopically and assessed according to the guidelines of An International System for Human Cytogenetic Nomenclature [10]. Despite the comprehensive analysis, this approach requires a labor-intensive, time-consuming procedure that visualizes only coarse chromosomal changes and is less applicable in many institutions than other molecular techniques such as fluorescent *in situ* hybridization (FISH) and reverse transcription-polymerase chain reaction (RT-PCR). However, the conventional karyotyping is considered to be still valuable as an initial step especially to identify novel cytogenetic aberrations such as translocations.

FISH can provide information regarding genetic and cytogenetic changes in metaphase or interphase cells by means of a microscopic detection of fluorescent signals of gene- or chromosome-specific DNA probes hybridized specifically onto target genes or chromosomal regions. This approach can be applied to a touch-smear preparation of tumor cells, isolated or 'free' nuclei after an enzymatic treatment, and a frozen or fixed tumor tissue such as an archival formalin-fixed, paraffin-embedded specimen (FFPE). FISH is a more feasible method with higher resolution than ordinary karyotyping. DNA probes labeled with fluorescent dyes are commercially available or can be generated in each institution using BAC- or YAC-derived DNA. However, optimal conditions for the hybridization procedure may be variable according to the probes or samples used. In addition, the fluorescent signal intensity becomes gradually reduced in a time-dependent manner. FISH usually requires prior knowledge of the chromosomal region(s) of interest. Detectable cytogenetic aberrations by FISH are mostly simple and include aneuploidy, deletions, translocations, duplications, inversions, marker chromosome identification, polymorphisms, mosaicism, and cellular localization of aneuploidy [9]. Complex numerical or structural changes involving multiple chromosomes or chromosomal regions are difficult to be identified by ordinary FISH. Recently developed molecular techniques using multiple fluorescent dyes such as multicolor FISH (m-FISH) and spectral karyotyping (SKY) can expand the ability of FISH and overcome such a problem [11,12].

Nucleic acids (DNA and RNA) extracted from tumor cells or tissues have been widely used to be analyzed for variable genomic alternations such as a loss of heterozygosity (LOH), genomic or DNA polymorphisms, gene mutations and amplifications. Specifically PCR-based approaches are most frequently employed as a molecular detection of fusion genes resulting from chromosomal translocations. PCR using primers flanking each fusion partner gene can amplify only a small amount of template DNA derived from a targeted fusion gene. Because of divergent fusion points, a size of the amplified fragments of fusion genes is usually variable or unpredictable. However, transcripts of a given fusion gene are often consistent in size and detectable by RT-PCR that is a sensitive molecular technique to amplify synthesized DNA complementary to mRNA in samples. This method can be also applied to FFPE, allowing us to assess cases without fresh or snap-frozen tumor tissues or cultured tumor cells and to analyze a large series of archival samples [13,14].

Southern blotting is also a molecular approach that can be used to detect genomic rearrangements caused by translocations. However, this method requires a substantial amount of fresh or snap-frozen tumor tissue for DNA extraction and complicated procedures. Hence, the current application of this method seems to be limited [6,8].

Soft Tissue Sarcomas with Characteristic Chromosomal Translocations

1. Synovial Sarcoma

Synovial sarcoma accounts for 5 to 10% of all soft tissue sarcomas and affects predominantly adolescents and young adults with ages mostly ranging from 15 to 40 years and a slight male preponderance [1,2,15,16]. Although the tumors can occur in almost any part of the human body, they are often found in juxta-articular locations of the upper and lower extremities as deeply seated, slowly growing masses. Histologically, synovial sarcomas are classified into three subtypes: monophasic, biphasic, or poorly differentiated synovial sarcoma [1,2]. Biphasic synovial sarcoma is characterized by glandular epithelial and spindle cell components in varying proportions. Monophasic synovial sarcoma is composed exclusively of either the epithelial or spindle cell component, although monophasic epithelial tumors are exceptionally rare [17,18]. Poorly differentiated synovial sarcoma is a hypercellular tumor of short spindle or oval or rounded small primitive cells, posing a diagnostic difficulty in distinguishing from other small round cell tumors such as the Ewing family of tumors [19,20]. Distant metastasis such as in the lung is common, and 5-year and 10-year survival rates of patients are estimated to be 36-76% and 20-60%, respectively [1,2,15,16].

t(X;18)(p11.2;q11.2) is a cytogenetic hallmark of synovial sarcoma (Fig. 1), and was initially described in 1986 [21,22]. This chromosomal aberration is identified in more than 90% of the cases, irrespective of their histologic variations, and generates so far three types of the fusion gene formed by parts of the *SS18* (formerly known as *SYT*) gene located at 18q11.2 and of the *SSX1*, *SSX2*, or *SSX4* gene at Xq11.2 [23-25]. Complex chromosomal changes involving one or more chromosomes in addition to t(X;18) have been described in less than 10% of the reported cases [24]. Recently, a variant of the fusion gene made up of parts of the *SS18L1* gene at 20q13.3 and of the *SSX1* has been recently described in a single case [26]. Approximately two thirds of the reported cases are associated with the *SS18-SSX1* fusion gene, and the remaining cases are mostly associated with the *SS18-SSX2* gene [23,24]. The *SS18-SSX4* fusion gene or its transcript has been identified only in a minor population of synovial sarcomas [27-29].

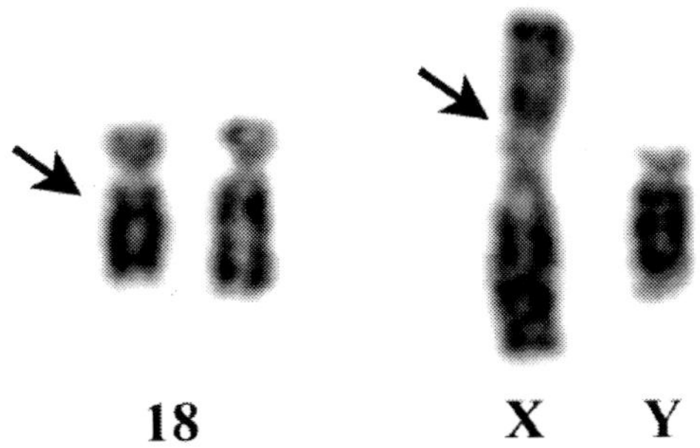

Figure 1. t(18;X)(p11.2;q11.2) in synovial sarcoma. Arrows indicate chromosomal breakpoints.

Notably, the *SS18-SSX* and *SS18L1-SSX* fusion genes are present exclusively in synovial sarcoma and virtually never found in any other types of tumors including histologically mimicking spindle cell sarcomas, such as fibrosarcoma, leiomyosarcoma, malignant peripheral nerve sheath tumor, and hemangiopericytoma [30-32]. Besides, there seems to be close correlation between the *SS18-SSX* fusion type and the histologic subtype of synovial sarcoma; most of the tumors with the *SS18-SSX2* display monophasic histology, and biphasic tumors almost invariably harbor the *SS18-SSX1* fusion gene [23,24,33]. The patients' prognosis appears to also differ according to the fusion type; five-year overall survivals for the *SS18-SSX1* and *SS18-SSX2* groups have been reported to be 53% and 73%, respectively, based on an analysis of a series of more than 200 cases [33].

The *SS18* gene encodes a protein composed of 387 amino acids and ubiquitously expressed in the early stages of embryonic development or in several organs from the adult mouse such as the heart, kidney, and the testis [34]. The protein is localized in the nucleus in a punctuated pattern [35], and contains three potential SH2-binding and one potential SH3-binding domains [36], which have been implicated in protein-protein interactions that occur in signal transduction pathways [37]. This protein has the QPGY domain (rich in glutamine, praline, glycine, and tyrosine) at its C-terminal region, which can activate transcription when coupled to a DNA-binding domain [38]. However, SS18 itself lacks any obvious DNA-binding domain, which suggests that the protein is a transcriptional co-activator. It has also been suggested that the N-terminal of SS18 acts as an inhibitor of the activation domain by a deletion mutagenesis study [39]. SS18 has been shown to interact with a variety of proteins such as a chromatin remodeling factor hBRM/hSNF2 alpha protein [39], a transcriptional factor AF10 [40], a putative transcriptional co-activator p300 protein [41], and mSin3A which is a component of histone deacetylase complex [42]. Recently, it has been shown that the *SS18* gene encodes a nuclear receptor co-activator protein that interacts with steroid receptor coactivator-1-containing histone acetyltransferase complex [43]. In contrast to SS18, the expression pattern and function of the product of the *SS18L1* gene are poorly understood [44]. The SS18L1 protein possesses an SS18 N-terminal homology domain and also the QPGY domain at its C terminal [26].

The *SSX* genes comprise a gene family of 9 members (*SSX1* to *SSX9*) mapped to the short arm of chromosome X [45]. The members share sequence homology ranging from 73-92% and 87-96% in protein and cDNA levels, respectively, and the expressions of SSX6, SSX8, and SSX9 have not been identified in normal tissues [45]. The other SSXs have been found to be expressed in normal adult testis and faintly in the thyroid tissue. Proteins of the *SSX* genes are localized diffusely in the nucleus [35], and likely act as transcriptional repressors. The N-terminal ends of SSXs show extensive homology to the so-called Krüppel-associated box (KRAB), which is recognized in a subgroup of zinc finger proteins functioning as transcriptional repressors [46,47]. The acidic C-terminal end of SSX, designated as SSXRD, has been found to be also an active repressor of transcription that can co-localize with polycomb-group repressor proteins such as HPC2, BMI1, and RING1 [48,49]. Like SS18, no DNA-binding domain is identified in SSXs.

In the majority of SS18-SSX fusion proteins, C-terminal eight amino acids of SS18 are replaced by C-terminal 78 amino acids of SSX. As a result, the KRAB domain of SSX is deleted, while its SSXRD domain is preserved in SS18-SSX fusion proteins (Fig. 2) [23]. Some minor variants of SS18-SSX fusions with more N-terminally located fusion points in either SS18 or SSX have been described in sporadic cases of synovial sarcoma [50-54]. Two splicing isoforms of SS18 with or without an in-frame insertion of 93 bp sequence derived from a genomic DNA of chromosome 18 can be expressed by the *SS18-SSX* genes as well as the native *SS18* gene [55]. This insertion has been shown to increase transcriptional activity of the native SS18 or SS18-SSX proteins and to accelerate cell proliferation [56]. By binding to an as yet unknown DNA-binding protein or recruiting the other transcriptional co-activator or co-repressor, SS18-SSX or SS18L1-SSX may alter the transcriptional machinery that is tightly regulated under a normal condition [23]. Recently, it has been shown that SS18-SSX1 can transform rat fibroblasts by binding to a chromatin remodeling factor hBRM/hSNF2 alpha and by potential down-regulation of a tumor suppressor DCC [57]. Genome-wide analyses of gene-expression profiles of synovial sarcomas have demonstrated a variety of up-regulated genes such as *XRCC4*, *IGFBP2*, *ERBB2*, *IGF2*, and *FZD10*, which are also possible downstream targets of SS18-SSX [58-61].

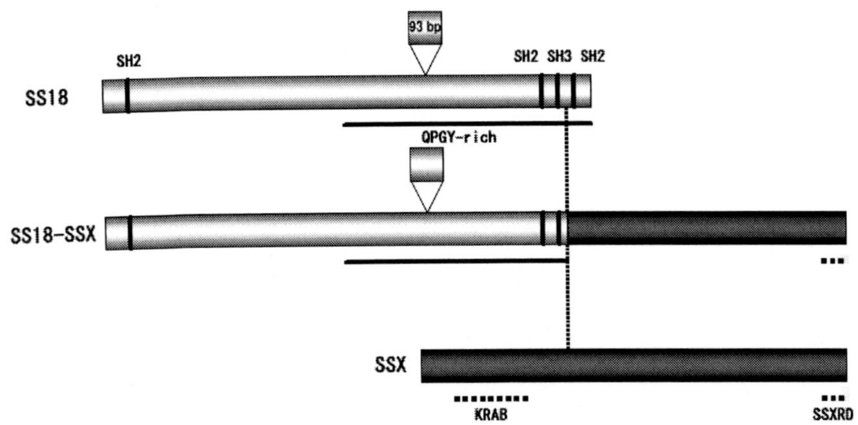

Figure 2. *SS18-SSX* fusion transcript in synovial sarcoma (middle) consisting of parts of SS18 (upper) and SSX (lower) transcripts. In SS18-SSX, the N-terminal region of SSX containing Krüppel-associated box (KRAB) is deleted, whereas SSX repressor domain (SSXRD) is preserved.

Although the SS18-SSX fusion proteins are considered to play a crucial role in the development of synovial sarcoma, the etiological and molecular mechanisms of the formation of t(X;18) or the *SS18-SSX* genes are still largely unknown. Generally, unrepaired DNA double-strand breaks and/or genomic instability can lead to chromosomal rearrangements including translocations, in which telomere dysfunction and unique promoting sequences susceptible to homologous recombination have been suggested to be involved [62-64]. Such promoting sequences include heptamer/nonamer sequences, Translin consensus sequences, topoisomerase I or II cleavage sequences, chi-consensus sequences, *Alu* repeats, palindrome sequences, alternating purine/pyrimidine and polypurine or polypyrimidine sequences, which are often identified in the vicinity of the breakpoints. Indeed, sequences homologous to consensus topoisomerase II cleavage sites are consistently present at or near the breakpoints in both the *SS18* and *SSX* genes [65]. Moreover, Translin-binding sequences, an *Alu-Alu* fusion, and an insertion of a 206 bp LINE-1 have been found at the breakpoints in some of the cases examined [64]. By contrast, telomere lengths of translocation-associated sarcomas including synovial sarcoma have been reported to be similar to or reduced compared to surrounding non-neoplastic tissues [66].

2. The Ewing Family of Tumors

Both Ewing's sarcoma and primitive neuroectodermal tumor (PNET) are morphologically characterized by densely packed atypical or primitive small round cells. Although Ewing's sarcoma was previously considered to be a distinct bone tumor, its extraosseous counterpart has been repeatedly described and thus referred to as extraskeletal Ewing's sarcoma [2,67]. In contrast to the least differentiated morphology of Ewing's sarcoma or extraskeletal Ewing's sarcoma, PNET usually shows variable degrees of neural or neuroectodermal differentiation. However, skeletal or extraskeletal Ewing's sarcoma and PNET are now regarded as members of a tumor family based on their significant clinicopathologic overlaps and the recent identification of common cytogenetic aberrations described below. This tumor family commonly affects adolescents or young adults and rarely occurs in those older than 30 years. A mild predilection for male patients is observed. The tumors can arise at any body sites including visceral organs such as the lung, heart, kidney, and ovary, and the extremities and trunk are most frequently involved by the tumor [1,2]. This tumor family commonly behaves aggressively and has a dismal outcome. However, the prognosis of the Ewing family has steadily improved, and current 5-year survival rates are estimated to be roughly 40-55% [1,68].

Most of the tumors in this family are cytogenetically characterized by chromosomal translocations, such as t(11;22)(q24;q12) and t(21;22)(q22;q12) (Fig. 3). These characteristic chromosomal aberrations are detectable in approximately 90 to 95% of the cases examined, and t(11;22)(q24;q12), resulting in a fusion of the *EWSR1* (formerly *EWS*) gene at 22q12 and the *FLI-1* gene at 11q24, is most commonly found (nearly in 85% of cases) [1,2]. t(21;22)(q22;q12) is identified less commonly (in 5-10%) and generates a fusion gene composed of the *EWSR1* gene and the *ERG* gene at 21q22. Other minor variants of the chromosomal translocation involving the 22q12, such as t(7;22)(p22;q12), t(2;22)(q33;q12), and t(17;22)(q21;q12), have been described in less than 1% of cases, and the *ETV1*, *FEV*, and *E1AF* genes are fused to the *EWSR1* in these variant translocations, respectively. Recently, t(16;21)(p11;q22) resulting in a gene fusion of the *FUS* (also known as *TLS*) gene at chromosome 16p11 and the *ERG* gene, has been reported in a small subset of the family [69]. Incidentally, t(16;21)(p11;q22) or the *FUS-ERG* has been also identified in rare cases of acute myeloid leukemia [70,71].

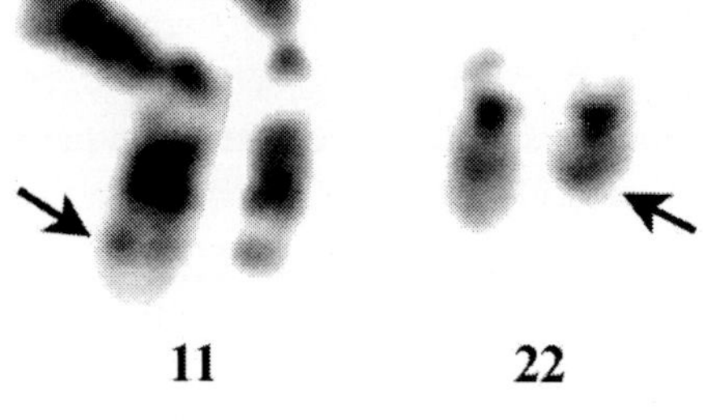

Figure 3. t(11;22)(q24;q12) in the Ewing family of tumors. Arrows indicate chromosomal breakpoints.

The *EWSR1* gene encodes a ubiquitously expressed RNA-binding nuclear protein that potentially is involved in the basal RNA transcription machinery and as yet unknown signal transduction pathways. EWSR1 is a member of the TET family of proteins, which includes TAF15 (also called TAF2N or TAFII68 or RBP56) and FUS and is characterized by a distinctive 87-amino-acid RRM/RNP-CS domain that is thought to mediate protein-RNA binding [72]. A C-terminal of EWSR1 contains three RGG-rich regions and an RNA recognition motif (otherwise called the RNP or RBP motif), which functions as an RNA-binding domain [73,74]. An N-terminal region of EWSR1 is considered an activation domain and contains a repeated degenerated polypeptide of 7 to 12 residues rich in thyrosine, serine, glycine, and glutamine [75]. The nuclear localization of EWSR1 is dependent on the N-terminal activation domain [76]. In the Ewing family of tumors, EWSR1-associated fusion genes encode fusion proteins composed of the N-terminal domain of EWSR1 fused to a C-terminal part of a member of the ETS family of proteins such as FLI-1, ERG, ETV1, ETV4 (also known as E1AF), and FEV, thus resulting in deletion of the RNA-binding domain of the EWSR1 (Fig. 4).

The ETS family comprises proteins that share a unique DNA-binding domain designated as the ETS domain, which shows a highly conserved structure containing three α-helices and four β-sheets [77]. The ETS domain is also a target for protein-protein interactions that are mediated either intramolecularly or *in trans* by co-regulatory proteins such as other DNA-binding or non-DNA-biding transcription factors and co-activator-co-repressor complexes [77]. Consequently, the ETS proteins are considered to be transcription factors that are assumed to be linked with diverse biologic processes, although their functions in normal cells still remain to be defined. EWSR1-ETS fusion proteins have been suggested to act as also sequence-specific transcriptional activators, which are capable of transforming cells [78-80].

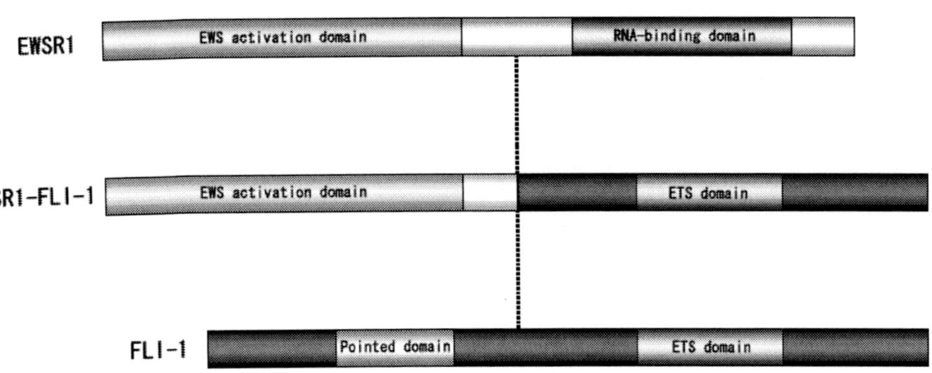

Figure 4. Schematic representations of EWSR1 (upper), EWSR1-FLI-1 (middle), and FLI-1 (lower). EWSR1-FLI-1 contains an N-terminally located EWS activation domain and an ETS domain of the FLI-1.

The *EWSR1-FLI-1* fusion gene creates at least 12 variant transcripts according to distinct combinations of exons derived from the *EWSR1* and *FLI-1* genes [81,82]. The most common form of the transcript, designated type I, has an in-frame fusion between exon 7 of the *EWSR1* and exon 6 of the *FLI-1* and accounts for approximately 60% of cases with the *EWSR1-FLI1* fusion genes. The product of this variant has been shown to be a stronger transcriptional activator than native FLI-1, although EWSR1-FLI-1 and FLI-1 display similar DNA-binding specificity and affinity [78,83]. A less common variant of the transcript, designated type 2 and accounting for approximately 25% of the cases, is composed of a fusion between exon 7 of the *EWSR1* and exon 5 of the *FLI-1*.

Interestingly, differential transactivation by alternative *EWSR1-FLI-1* fusion gene products has been suggested, and the type 1 transcript appears to generate a functionally weaker transactivator compared to other EWSR1-FLI-1 fusion transcripts [84]. Moreover, tumors with the type 1 fusion show lower proliferative activity and lower expression of a receptor for insulin-like growth factor-1 (IGF-1R) than those with non-type 1 fusions [82]. These findings seem to conform to a better clinical outcome of the patients with tumors bearing the type 1 EWSR-FLI-1 fusion than those with other fusion types [85,86]. A microarray-based analysis of gene expression has demonstrated that over-expressions of genes regulating the cell cycle (*CDK2*, *E2F*, *RAF*, and *MAPKs*) and of invasion- or metastasis-associated genes (*cadherin-11* and *MTA1*), down-regulations of tumor suppressor genes (*FHIT*, *NEURL*, and *LLGL1*), and those of apoptosis-inducers (*IL-2*, *IL1RL1*, and *HLA-DOB*) are potential predictors of the poor patient's prognosis [87]. However, a recent chromatin immunoprecipitation experiment combined with RNAi-mediated suppression of EWSR1-FLI-1 could not support that these genes are putative target genes of EWSR1-FLI-1 [88]. Although a wide variety of other candidate genes have been indicated as direct molecular targets of EWSR1-ETS fusion proteins, their authenticity still remains to be attested [88-95].

ERG shares significant sequence homology (98%) in the DNA-binding domain with FLI-1, suggesting that ERG and FLI-1 may interact with the same molecular targets. The EWSR1-ERG fusion protein has been shown to have more potent transcriptional activation properties compared to native ERG [74]. At least 4 types of EWSR1-ERG fusion transcripts have been described [81,82]. ERG has been suggested to be involved in cartilaginous development from mesenchyme, and EWSR1-ERG, but not ERG, transactivates COL11A2 that encodes a type of constituent collagens of hyaline cartilage [96]. By contrast, ETV1, which is a rare partner of the EWSR1 in EWSR1-ETS fusions, has only 58% and 63% amino acid identity in its ETS domain with those of FLI-1 and ERG, respectively [97]. However, EWSR1-ETV1 has sequence-specific DNA-binding activity [98]. ETV4, also referred to as E1AF, is an ETS protein that can bind to adenovirus E1A enhancer sites and is involved in the EWSR1-ETS fusions in limited cases of the Ewing family of tumors [98,99]. FEV is also a member of the ETS family that constitutes the EWSR1-associated fusion partner proteins, and can repress both basal transcription and, even more strongly, ectopically ETS-activated transcription through its ETS and alanine-rich C-terminal domains [100, 101].

Although each EWSR1-ETS fusion protein seems to bind to its own molecular target, targeted and transcriptionally deranged genes are considered to share a similar or common biological role in the development of the Ewing family of tumors. Thus, suppression of the oncogenic EWS-ETS fusion proteins using antisense oligonucleotides or small interfering RNAs have recently attracted special attention as potential therapeutic applications in this pediatric sarcoma group [102-104].

3. Alveolar Rhabdomyosarcoma

Rhabdomyosarcomas are malignant mesenchymal tumors with variable skeletal muscle differentiation, accounting for about 10-20% of all malignant soft tissue tumors [2]. Histologically, they are classified into embryonal, alveolar, and pleomorphic subtypes that differ in the degree of differentiation, clinical presentation, and prognosis of the patients [1,2]. Alveolar rhabdomyosarcoma (ARMS) occurs less frequently (about 20-30% of all rhabdomyosarcomas) than the embryonal subtype (ERMS) and mostly affects adolescents and young adults [1,2,105]. AMRS often arises in deep soft tissues of the extremities and trunk and behaves aggressively with an estimated 5-year survival rate of 53%, which is in contrast with ERMS that usually affects much younger children and has a better prognosis (a 5-year survival rate of 64%) [105]. ARMS is morphologically characterized by ill-defined aggregates of poorly differentiated small round or oval cells with frequent central losses of cellular cohesion, displaying irregular alveolar spaces surrounded by a framework of fibrous septa. The tumor may rarely form densely packed sheets or groups of tumor cells, lacking a distinct alveolar pattern, which is designated as a solid variant of ARMS [106].

The cytogenetic hallmarks of ARMS are chromosomal translocations, t(2;13)(q35;q14) and t(1;13)(p36;q14), resulting in the *PAX3-FOXO1A* (formerly *FKHR*) and *PAX7-FOXO1A* fusion genes, respectively. These fusion genes are assumed to be detectable in more than 80-85% of ARMS with the *PAX3-FOXO1A* being 4.5-6 fold more prevalent than the *PAX7-FOXO1A* [2,107]. However, these fusion genes have been identified also in small fractions of embryonal and pleomorphic subtypes, which might be related to an unsampled alveolar component in these tumors or due to possibly different classification systems used [2,105,108]. Other than these exceptional cases and another of unusual biphenotypic small cell sarcoma with neuroectodermal and muscle differentiation [109], the chromosomal translocations, t(2;13)(q35;q14) and t(1;13)(p36;q14), or their resultant fusion genes have not been described in any other type of tumors. Thus, the cytogenetic or molecular identification of such translocations or fusion genes could provide valuable information in the diagnosis of ARMS. Moreover, patients with the *PAX3-FOXO1A* pursuit a poorer clinical outcome than those bearing the *PAX7-FOXO1A* (an estimated 4-year overall survival rate of 77% for cases with *PAX7-FOXO1A* vs. 52% for those with *PAX3-FOXO1A*) [110]. A novel fusion gene, *PAX3-NCOA1* resulting from t(2;2)(q35;p23), has been identified in a single case of alveolar rhabdomyosarcoma lacking the *PAX3/7-FOXO1A* [111].

The *PAX3* and *PAX7* genes encode highly related members of the paired box transcription factor family that are organized with N-terminal DNA-binding domains and C-terminal transcriptional activation domains [112]. The N-terminal DNA-binding domain is also known as the paired domain consisting of two subdomains, each of which structurally resembles a helix-turn-helix motif, an ancient structure similar to that found in homeodomains. Besides, several Pax proteins including Pax3 and Pax7 possess an additional DNA-binding homeodomain [113]. Pax3 and Pax7 play pivotal roles in skeletal muscle development and are expressed with spatially and temporally distinct patterns in mesodermal compartments that give rise to skeletal muscle progenitors [112]. Although Pax3 is not expressed in adult muscle, Pax7 is expressed by adult myogenic satellite cells, which can proliferate and differentiate into functional myocytes [114]. In addition, both Pax proteins are expressed in the neural crest and neural tube, suggesting their essential roles in the development of neural tissues. A recent gene targeting experiment has demonstrated that

Pax7 can substitute for Pax3 function in dorsal neural tube, neural crest cell, and somite development [115].

FOXO1A is a member of the forkhead transcription factor (FOX) family that mediates molecular pathways regulating diverse cellular functions such as metabolism, differentiation, proliferation, and survival [116]. In addition, FOXO1A controls the rate of myotube fusion during murine myogenic differentiation [117]. The FOX family consists of approximately 90 members that are evolutionarily conserved, and their conserved DNA-binding domains are characterized by three α-helices and two characteristic large loops or wings [118]. FOXO1A and other members (FKHR-L1/FOXO3A and AFX/FOXO4) constitute a FOXO subfamily based on their consensus DNA-binding sequence (TTGTTTAC) [119], and are expressed in a variety of tissues or cells such as the liver, prostate, heart, endometrium, lymphocyte, adipocyte, and endothelial cell. FOXO1A has an N-terminal forkhead DNA-binding domain, a C-terminal transcriptional activation domain, and three consensus PKB (also known as AKT) phosphorylation sites, which are responsible for phosphoinositide 3-kinase mediated signal transduction that also regulates subcellular localization of the FOXO1A [116, 120]. It has recently been demonstrated that FOXO1A is not expressed in primary ARMS or ARMS-derived tumor cell lines, and that restoration of FOXO1A expression not in ERMS but ARMS cells is sufficient to induce cell cycle arrest and apoptosis, suggesting that FOXO1A is a potent and specific tumor suppressor in ARMS [121].

In PAX3-FOXO1A or PAX7-FOXO1A, an intact DNA-binding domain of the PAX3 or PAX7 in the N-terminal region is fused in frame with a C-terminal FOXO1A segment containing the transactivation domain (Fig. 5).

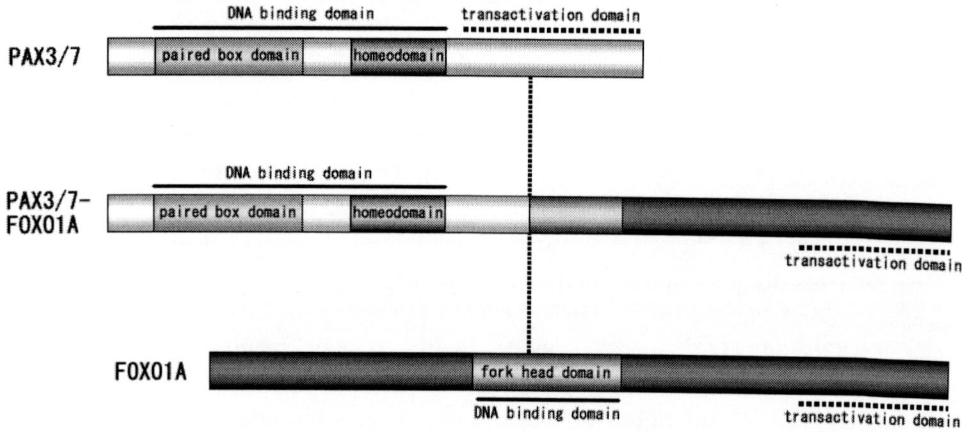

Figure 5. PAX3/7-FOXO1A in alveolar rhabdomyosarcoma (middle) consisting of a fusion between PAX3/7 (upper) and FOXO1A (lower).

The fusion protein is constitutively localized in nuclei and has more potent transcriptional activity than wild-type PAX3 or PAX7 [120,122,123]. The fusion genes are believed to act as oncogenes, probably by inappropriate activation of PAX3- or PAX7-target genes such as the *MET* and *MyoD1* or by forcing somatic cells down an aberrant embryonic myogenic developmental pathway [124,125]. Expressed PAX3-FOXO1A induces cellular transformation, anchorage-independent growth, and focus formation in vitro [126]. In addition, PAX3-FOXO1A expressed in ERMS cells can enhance cell proliferative activity, promote cell growth in the absence of added growth factors, and induce a more locally invasive and pleomorphic phenotype in xenografts than controls lacking the *PAX3-FOXO1A* [127]. A recent study using conditional knock-in and knock-out techniques has suggested differentiating skeletal muscle rather than embryonic or postnatal muscle stem cells is the cellular origin of ARMS, and that the tumor can develop artificially in mice with knock-out alleles of either TP53 or CDKN2A with a PAX3-FOXO1A knock-in allele [128,129].

4. Myxoid Liposarcomas

Liposarcoma is one of the most common soft tissue sarcomas of adults and classified into five histologic groups including well differentiated (or synonymously atypical lipomatous tumor), myxoid (including a round cell variant), pleomorphic, dedifferentiated, and mixed subtypes [1,2]. Myxoid liposarcoma, accounting for about 10% of all adult soft tissue sarcomas, is the second most common subtype of liposarcoma. Unlike other subtypes, myxoid liposarcoma occurs in younger age groups (a peak incidence in the 4th and 5th decades of life) with a predilection for deep soft tissues of the extremities, particularly the medial thigh and popliteal area [1,2]. Grossly, the tumor tends to exhibit a glistening, gelatinous appearance. Histologically, purely myxoid tumors are composed of a hypocellular bland-looking spindle cell proliferation with a variable number of immature-appearing fat cells or lipoblasts embedded in an abundant myxoid background rich in delicate capillary vessels. The lesion may display hypercellular areas of primitive or undifferentiated round cells together with the hypocellular myxoid portions, and the tumor with a predominantly hypercellular round cell area (more than 75%) is also designated round cell liposarcoma, the term of which was formerly used as a distinct variant of liposarcoma [130]. However, overlapping or mostly identical clinicopathologic and cytogenetic features of myxoid and round cell liposarcomas indicate that both represent a morphologic continuum. The biological behavior of the tumor differs significantly between lesions with or without hypercellular round cell areas; the 70% of 5-year survival rate of myxoid liposarcoma decreases to 20% when associated with a round cell component [131,132].

Cytogenetically, myxoid liposarcomas including those with round cells are characterized by chromosomal translocations, t(12;16)(q13;p11) and t(12;22)(q13;q12) (Fig. 6), resulting in gene fusions between the *DDIT3* (formerly *CHOP* or *GADD153*) gene located at 12q13 and either the *FUS* (*TLS*) at 16p11 or *EWSR1* at 22q12. t(12;16)(q13;p11) is detectable cytogenetically in more than 90% of myxoid liposarcoma cases, and the remaining cases are assumed to harbor t(12;22)(q13;q12) [133]. The myxoid morphology in myxoid liposarcoma often creates diagnostic challenges especially in distinction from other myxoid soft tissue tumors such as well differentiated liposarcoma with prominent myxoid change and myxofibrosarcoma. Except for an unusual case of liposarcoma with mixed myxoid and well-differentiated lipoma-like morphology [134], no *FUS-DDIT3* gene rearrangement or fusion gene transcript has been detected in such myxoid soft tissue tumors other than myxoid liposarcoma, suggesting a diagnostic value of the cytogenetic or molecular alterations [135,136].

According to the distinct combinations of exons in the gene fusions, at least 5 and 2 variant transcripts have been described in the *FUS-DDIT3* and *EWS-DDIT3* fusion genes, respectively [137,138]. The most prevalent FUS-DDIT3 fusion transcript is composed of exons 1 to 5 of the *FUS* gene and exons 2 to 4 of the *DDIT3* gene (designated type II), accounting for 67% of the cases examined, followed by that consisting of exons 1 to 7 of the *FUS* and exons 2 to 4 of the *DDIT3* (type I, 20% of cases) [139]. However, no significant correlation between types of the fusion transcripts and histological grade or patient's survival has been suggested [139].

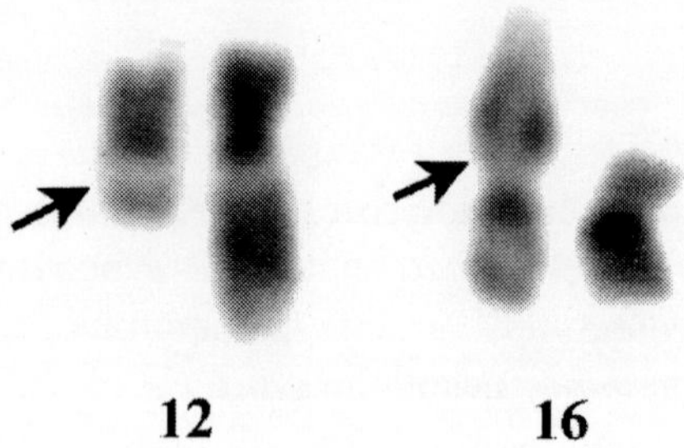

Figure 6. t(12;16)(q13;p11) in myxoid liposarcoma. Arrows indicate chromosomal breakpoints.

The *FUS* gene encodes an RNA-binding protein that is belonging to the TET protein family and has extensive sequence homology to the *EWSR1* gene (55.6% identity) [72,140]. Like EWSR1, an N-terminal domain of FUS is rich in glutamine, threonine, serine, proline, tyrosine, and glycine, and its C-terminal domain contains a conserved 80-amino acid RNP core sequence [141]. FUS can bind to RNA in a poly(G)-specific manner and shuttles between the nucleus and cytoplasm [142,143]. FUS has been shown to interact with an ETS protein, Spi-1/PU.1, and modulate transcriptional functions of the Spi-1/PU.1 [144]. The *FUS* gene is also known as a fusion partner of the *ERG* gene in acute myeloid leukemia with t(16;21)(p11;q22) [145].

DDIT3 is a member of the CCAAT/enhancer binding protein (C/EBP) family, which is involved through heterodimer formation with other C/EBP family proteins in the regulation of adipocytic differentiation and growth arrest [140,146]. The *DDIT3* gene is ubiquitously expressed at a relatively low level in several types of cells, and its expression is induced by a variety of stresses or growth arrest signals [147]. DDIT3 has an N-terminal transactivation domain and a C-terminal basic-leucine zipper (bZIP) domain consisting of a basic amino-acid-rich DNA-binding region followed by a leucine zipper dimerization motif, the latter of which is important for DDIT3-induced apoptosis [148].

In the fusion gene products in myxoid liposarcoma, the RNA-binding domain of FUS or EWSR1 is replaced by DDIT3 containing the DNA-binding and leucine zipper dimerization domains, which seems to act as an aberrant transcription factor (Fig. 7). The expression of the *FUS-DDIT3* gene can transform NIH3T3 cells, induce tumor development in nude mice, and inhibit terminal differentiation of preadipocytic cells [149,150]. Moreover, the *FUS-DDIT3* gene product does not induce arrest at the G1/S checkpoint, although wild-type DDIT3 interferes with the G1/S cell cycle progression [151]. These abnormal functions of FUS-DDIT3 are considered to play a crucial role in the development of myxoid liposarcoma. Although biological functions of EWSR1-DDIT3 remain to be determined, this protein seems to be functionally equivalent to that of the *FUS-DDIT3* fusion gene.

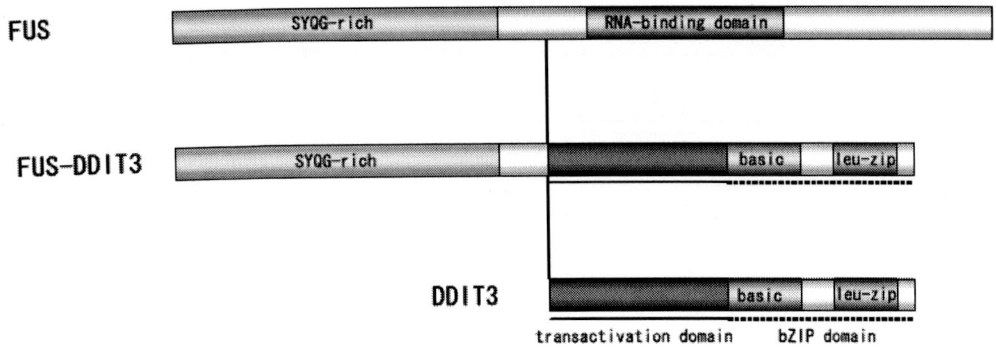

Figure 7. Schematic representations of FUS (upper), FUS-DDIT3 (middle), and DDIT3 (lower) proteins, the latter of which is composed of an N-terminal transactivation domain and a C-terminal basic-leucine zipper (bZIP) domain consisting of a basic amino-acid-rich DNA-binding region followed by a leucine zipper dimerization motif.

5. Desmoplastic Small Round Cell Tumor

Desmoplastic small round cell tumor (DSRCT) is a recently recognized, rare malignant tumor with a striking predilection for young male patients (a peak incidence in the 3rd decade of life) [1]. The tumor usually develops in and extends along mostly the peritoneum or occasionally the pleura or a paratesticular region. DSRCT is characterized histologically by a proliferation of small round cells arranged in nests or strands embedded in a dense desmoplastic stroma and immunohistochemically by polyphenotypic differentiation such as towards epithelial, myogenic, and neural or neuroendocrine cells. The histogenetic origin of DSRCT remains uncertain, although its mesothelial or submesothelial mesenchymal cell origin has been suggested [2]. DSRCT is a highly aggressive tumor that often develops multiple local recurrences and rarely distant metastases. The patient's prognosis is generally dismal, and the average survival is less than 2 or 3 years [152].

A reciprocal chromosomal translocation, t(11;22)(p13;q12), is a consistent cytogenetic feature of DSRCT. This chromosomal aberration creates a fusion of the *EWSR1* gene at 22q12 and the *WT1* gene at 11p13, and the *EWSR1-WT1* fusion gene transcript is considered to be detectable more than 90% of cases by RT-PCR [1,2]. The major EWSR1-WT1 transcript is composed of an in-frame fusion of the first seven exons of the *EWSR1* gene, encoding its putative transactivation domain, and exon 8 through exon 10 of the *WT1* gene, encoding the last three zinc-fingers of the DNA-binding domain (Fig. 8) [153,154]. Based on the heterogeneity in exonic combinations, at least other nine variant transcripts of the *EWSR1-WT1* have been described in a minor subset of DSRCTs [155]. However, clinical or biological significance of such variant transcripts remains to be determined.

The *WT1* gene encodes a transcription factor with four C_2H_2 zinc fingers in its C-terminal region. WT1 was initially identified in Wilms' tumor, a pediatric malignant renal tumor, as an inactive form. Subsequently, it was found to play an essential role in the normal development of a variety of organs or tissues such as the kidney, gonads, spleen, retina, and mesodermal tissues [156]. There are multiple isoforms of WT1, which are produced by a combination of alternative splicing, RNA editing, and alternative translation start sites. Among the isoforms,

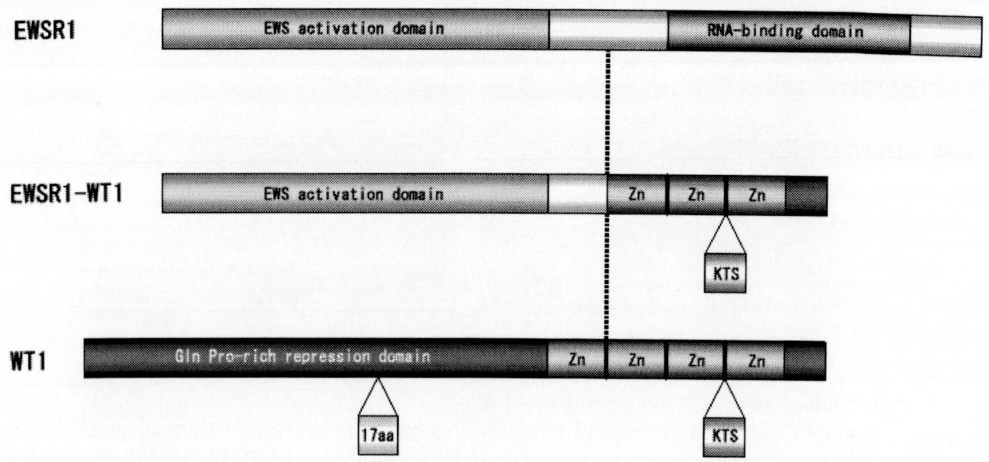

Figure 8. EWSR1-WT1 in desmoplastic small round cell tumor (middle) is composed of an N-terminal part of the EWSR1 (upper) fused to the C-terminal of WT1 (lower) containing three of the four C_2H_2 zinc fingers with or without insertion of three amino acids (lysine, threonine, and serine (KTS)).

those with or without insertion of three amino acids (lysine, threonine, and serine, KTS) between the third and fourth zinc finger are of particular interest. The uninterrupted zinc finger domain of the WT1(-KTS) can bind to the consensus DNA sequences, such as 5'-GCGGGGGCG-3' and 5'-GCGTGGGAGT-3', leading to the transactivation of a number of target genes including the epidermal growth factor family member amphiregulin, $p21^{Cip1}$, a major structural membrane protein of glomerular podocytes podocalyxin, and Vitamin D receptor [157]. However, the insertion of KTS in WT1 abrogates its binding to such consensus sequences. Since products of the WT1(+KTS) co-localize with nuclear $poly(A)^+$ ribonucleoprotein and co-immunoprecipitate with the splicing factor U2AF65, this variant might play a role in RNA processing [158]. Another alternative splicing event occurs in exon 5 of the *WT1*, resulting in the presence or absence of 17 amino acids positioned between its transcriptional regulatory and DNA-bind domains. Because the alternatively spliced exon 5 is only found in placental mammals, this variant isoform may be important for mammal-specific functions, such as embryonic implantation or lactation [158].

In DSRCT, EWSR1-WT1(+KTS) is the most abundant isoform, which is expressed at a ratio of ~2:1 with respect to the (-KTS) variant [154,159]. EWSR1-WT1(-KTS) can bind to the consensus sequence 5'-(G/C)(C/G)(G/C)TGGGGG-3' and induce expressions of target genes including *PDGFA*, *MLF1*, *IL-2/15Rβ*, *BAIAP3*, and *TALLA-1*, whereas they are not induced by WT1(-KTS) [160]. By contrast, only a gene encoding a leucine-rich transmembrane protein LRRC15 has recently been shown to be dramatically induced by EWSR1-WT1(+KTS), which seems to recognize the different consensus sequence 5'-GGAGG(A/G)-3' frequently located in an up-stream of the *LRRC15* gene [159]. The dysregulated expressions of such target genes of EWSR1-WT1 are assumed to play a crucial role in the development and progression of DSRCT.

6. Extraskeletal Myxoid Chondrosarcoma

Extraskeletal myxoid chondrosarcoma (EMC) is a rare soft tissue sarcoma, and its estimated incidence is less than 2.5% of all soft tissue sarcomas [2]. EMC mostly arises in deep soft parts of the proximal extremities and limb girdles in middle-age adults. EMC is histologically characterized by multilobular structures of a proliferation of short spindle or oval cells with an abundance of a chondromyxoid matrix, somewhat resembling chondroblastic tissue in the early stages of cartilaginous development or other chondroid tumors [1,2,161]. However, its cartilaginous differentiation has been a matter of great controversy, and the tumor is provisionally classified into a group of tumors of uncertain differentiation [1]. Moreover, neural or neuroendocrine differentiation of the tumor cells has been reported in a fraction of EMC [162]. EMC tends to behave aggressively with frequent local recurrences and/or distant metastases, and an estimated survival rate of 70% at 10 years has been described [163].

Reciprocal chromosomal translocations, t(9;22)(q22;q12) and t(9;17)(q22;q11), have been reported as representative cytogenetic features of EMC. t(9;22)(q22;q12) results in a fusion of the *EWSR1* gene to the *NR4A3* (formerly *CHN* or *NOR1*) gene located at 9q22, and this translocation or fusion gene is assumed to be present in approximately 85% of EMC [164]. t(9;17)(q22;q11) resulting in gene fusion of the *TAF15* (also known as *TAF2N*, *RBP56*, or *hTAFII68*) at 17q11 with the *NR4A3* is less frequently found in EMC (approximately 15%) [164]. Minor fusion genes involving the *NR4A3*, such as the *TCF12-NR4A3* resulting from t(9;15)(q22;q21) and the *TFG-NR4A3* probably derived from t(3;9)(q21;q22), have been also described in a few examples of EMC [165,166].

The *EWSR1-NR4A3* gene generates a fusion transcript consisting of a sequence encoding an N-terminal transactivation domain of EWSR1 fused to 5' untranslated region of NR4A3 followed by its entire coding sequence (Fig. 9). The breakpoints in the *EWSR1* and *NR4A3* genes may vary, and at least nine variants of the fusion transcripts have been described [164]. The most common *EWSR1-NR4A3* fusion transcript contains exon 12 of the *EWSR1*, which is fused to exon 3 of the *NR4A3*. In the *TAF15-NR4A3* fusion transcript, exon 6 of the *TAF15* is fused exclusively to exon 3 of the *NR4A3*. Information regarding functional differences among the distinct *NR4A3* fusion genes or their variant transcripts is so far unavailable.

The *NR4A3* gene encodes an orphan nuclear receptor protein belonging to the Nur77/NGFI-B subfamily of the nuclear steroid/thyroid/retinoid receptor superfamily. NR4A3 has structural features of a ligand-activated transcription factor; a centrally located zinc-finger type DNA-binding domain close to its carboxyl end. NR4A3 is normally expressed ubiquitously but most significantly in the central nervous system, skeletal and cardiac muscles, and lung [167]. Like Nurr77 and NGFI-B, NR4A3 has been implicated in cell proliferation as an immediate to early response protein induced by various stimuli including liver regeneration or in T-cell receptor-mediated apoptosis of immature thymocytes and T-cell hybridomas [168-170]. In addition, it has been recently shown that NR4A3 is essential for the development of the semicircular canals of the inner ear as well as early embryogenesis of mice and mediates the genotoxic stress response [171-173].

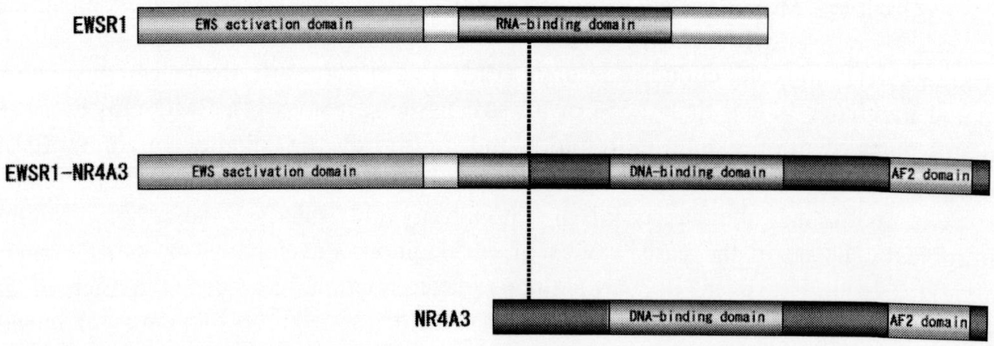

Figure 9. A representative fusion transcript, EWSR1-NR4A3 (middle), in extraskeletal myxoid chondrosarcoma is composed of an in-frame fusion between transcripts of the EWSR1 (upper) and of NR4A3 (lower).

TAF15 is also a ubiquitously expressed protein that contains a putative RNA-binding RNP-1 motif and a possible DNA-binding zinc-finger motif. Because of extensive sequence homology, TAF15 is also considered a member of the TET protein family including EWSR1 and FUS [72]. The *TCF12* gene encodes a basic helix-loop-helix transcription factor with a serine-rich N-terminal domain [165]. TFG is also ubiquitously expressed and has a putative N-terminal coiled-coil domain and SPYGQ-rich regions that are similar to the N-terminal parts of EWSR1 and FUS [174].

EWSR1-NR4A3 functions as a highly potent transcriptional factor and modulates pre-mRNA splicing probably through the zinc-finger motif of NR4A3, leading to instability of RNA splicing [175]. NR4A3 fusion proteins may disrupt and modulate the normally regulated NR4A3 interaction with its hitherto unknown target genes. This view is suggested by the fact that there is no obvious sequence homology between the variable N-terminal regions of NR4A3 fusions, which implies that the N-terminal part of each fusion protein is likely to be functionally equivalent. Consequently, common functional or oncogenic properties of the NR4A3 fusion proteins are probably owing to NR4A3 itself, which is expressed aberrantly in cells of EMC under the control of a promoter region of the various fusion partner gene.

7. Dermatofibrosarcoma Protuberans

Dermatofibrosarcoma protuberans (DFSP) is an uncommon tumor that primarily arises in the dermis and often involves the subcutis predominantly on the trunk or in the proximal extremities of young or middle-aged adults. Clinically, DFSP is characterized by an indolent growth and a relatively high rate of local recurrence, but distant metastases (less than 5%) or tumor-related deaths are extremely rare [2,176]. Although DFSP has been considered a prototype of fibrohistiocytic neoplasms, DFSP usually lacks discernible histiocytic differentiation. There seems to be a general consensus that such a tentative classification should be reassessed in future [1]. Histologically, ordinary DFSP is composed of a monomorphic population of bland-appearing spindle cells arranged in a distinct storiform or cartwheel pattern. There are several unusual histological variants including pigmented DFSP (also referred to as Bednar tumor), giant cell fibroblastoma, and fibrosarcomatous DFSP [2].

Fibrosarcomatous DFSP is regarded as a form of high-grade tumor progression of DFSP and seems to be associated with an adverse clinical outcome [176,177].

Consistent cytogenetic features of DFSP are a reciprocal or unbalanced chromosomal translocation, t(17;22)(q22;q13) or der(22)t(17;22)(q22;q13), and a supernumerary ring chromosome derived from the translocation, r(17;22), the latter of which is more often detected in DFSP (Fig. 10) [178-180]. These chromosomal aberrations generate a fusion of the *COL1A1* gene at 17q21-22 and the *PDGFB* gene at 22q13 [181]. The *COL1A1-PDGFB* fusion genes have been so far identified exclusively in DFSP and its histologic variants including giant cell fibroblastoma, pigmented DFSP, and fibrosarcomatous DFSP, except for some cases of superficial adult fibrosarcoma, suggesting a genetic and clinicopathologic rink among these spindle cell mesenchymal tumors [181-184].

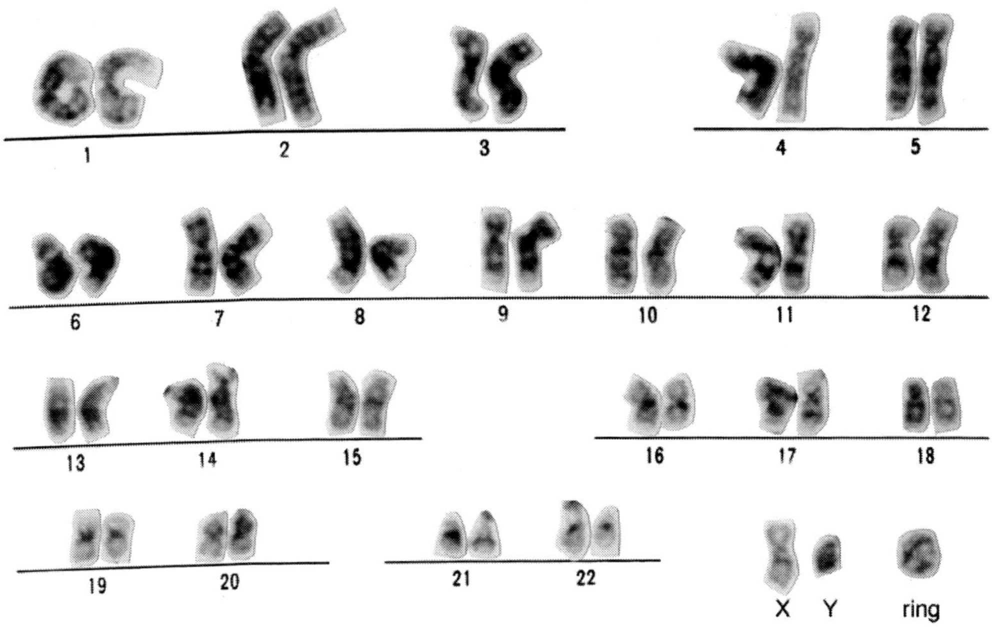

Figure 10. Karyotypic profile of dermatofibrosarcoma protuberans showing a supernumerary ring chromosome as a sole cytogenetic abnormality.

The *COL1A1* gene encodes the α1(1) chain of type 1 collagen, a member of a highly characteristic fibrous protein family of the extracellular matrix. Type 1 collagen is a heterotrimer composed of two α1(1) and one α2(1) chains, the latter of which is encoded by a distinct gene, *COL1A2*. The *COL1A1* gene contains 52 exons, and an α-helical domain consisting of 338 repeats of a Gly-X-Y triplet (X and Y are often proline) is assigned to the region between exon 6 and exon 49 [185].

The *PDGFB* gene is the cellular homologue of the v-sis oncogene and encodes the B chain of PDGF, which forms a homodimer, PDGF-BB, or a heterodimer, PDGF-AB, with another B chain or the PDGF A chain by disulfide linking. PDGF-BB is a potent growth factor that acts as a mitogen and chemoattractant for a variety of connective tissue cells through binding to its specific receptor tyrosine kinase on cellular surfaces [186]. The co-expression of PDGFB and its receptors has been described in several types of tumors, including DFSP, and PDGFB autocrine signaling has been suggested to be involved in the development of such tumors [187].

In *COL1A1-PDGFB* fusion transcripts, the sequences, located upstream of the *PDGFB* exon 2 and known to contain repressor elements for PDGFB transcription and translation, are replaced by COL1A1 elements, in which fusion points are located in variable positions (ranging from exon 7 to exon 47) within it's α-helical coding region [188]. In contrast to a probable subordinate role of the COL1A1 elements, mature PDGFB protein generated from exons 3 to 6 of the preserved *PDGFB* sequence in the *COL1A1-PDGFB* fusion gene seems to play an essential oncogenic role in DFSP. Indeed, the *COL1A1-PDGFB* fusion cDNA-transfected cells can transform and secrete PDGFB-like molecules processed from COL1A1-PDGFB, which stimulate a cell growth in an autocrine or paracrine fashion through the activation of PDGF β-receptors [189,190]. Such PDGFB-mediated oncogenic signaling is inhibited by a chemical compound STI571, a specific inhibitor of tyrosine kinases [191,192]. This finding implicates a potential therapeutic application of STI571 in DFSP cases. Clinical response to imatinib mesylate (STI571) has been recently described in cases with locally advanced or metastatic DFSP [193-195].

8. Infantile (Congenital) Fibrosarcoma

Infantile or congenital fibrosarcoma represents a rare soft tissue tumor with essentially identical histologic features to classic fibrosarcoma of adults [1]. The vast majority of the tumors is congenital or occurs in infants aged less than 2 years. The superficial or deep soft tissues of the extremities are the major anatomical sites involved by the tumor, followed by the trunk and head and neck [1,2]. Histologically, the tumor is hypercellular and composed of a proliferation of almost uniform oval or spindle-shaped cells arranged in fascicles or whorls, often displaying a herringbone appearance, with a variable amount of stromal collagen and scattered chronic inflammatory cells. Despite the histologic similarities, infantile fibrosarcoma is much less aggressive than adult-type fibrosarcoma; the five-year survival rate of infantile fibrosarcoma has been reported to be 84% [196].

Infantile fibrosarcoma is cytogenetically characterized by a chromosomal translocation, t(12;15)(p13;q25), resulting in a fusion of the *ETV6* (also known as *TEL*) gene at 12p13 and the *NTRK3* (also referred to as *TRKC*) at 15q25 [197]. This type of chromosomal translocation or fusion gene has been also described in congenital mesoblastic nephroma, acute myeloid leukemia, and secretory carcinoma of the breast [198-201]. Although a close relationship between infantile fibrosarcoma and congenital mesoblastic nephroma may be also supported by their overlapping morphologic features in addition to the common cytogenetic change [199,200], any other histogenetic link among the tumors harboring t(12;15)(p13;q25) or the *ETV6-NTRK3* fusion gene is difficult to recognize.

NTRK3 is a member of the neurotrophic tyrosine receptor kinase (NTRK) family and plays an important role primarily in the development and maintenance of neural tissues [202]. NTRK3 is activated through interaction with its specific ligand, neurotrophin-3, at cellular surfaces, resulting in oligomerization and subsequent autophosphorylation of the tyrosine kinase [203]. Since NTRK3 is also expressed in a variety of cells or tissues including hematopoietic cells, glandular epithelium, epidermis of the skin, and vascular smooth muscle cells, pleiotropic functions in multiple organ or tissue systems are suggested [204-208]. The

NTRK3 gene encodes a transmembrane protein containing an extracellular domain in its N-terminal, a transmembrane domain, and a C-terminally located tyrosine kinase domain. There are alternatively spliced variants of the *NTRK3* transcripts with a 24-nucleotide insert in the extracellular domain and a 42-nucleotide insert in the tyrosine kinase domain. In addition, at least two types of truncated receptors lacking entirely or a part of the tyrosine kinase domain have been described [205,209].

ETV6 is a member of the ETS family of transcription factors and known to be frequently involved in gene rearrangements resulting from chromosomal translocations identified in myeloid or lymphoid leukemia [210]. Like some other ETS transcription factors, ETV6 has an N-terminal helix-loop-helix domain (also known as sterile alpha motif (SAM) or pointed domain), responsible for oligomerization, and an ETS DNA-binding domain in its C-terminal [211]. ETV6 has been shown to be essential for angiogenesis in the yolk sac, survival for selected neural and mesenchymal cells in embryos, and hematopoiesis in the bone marrow [212].

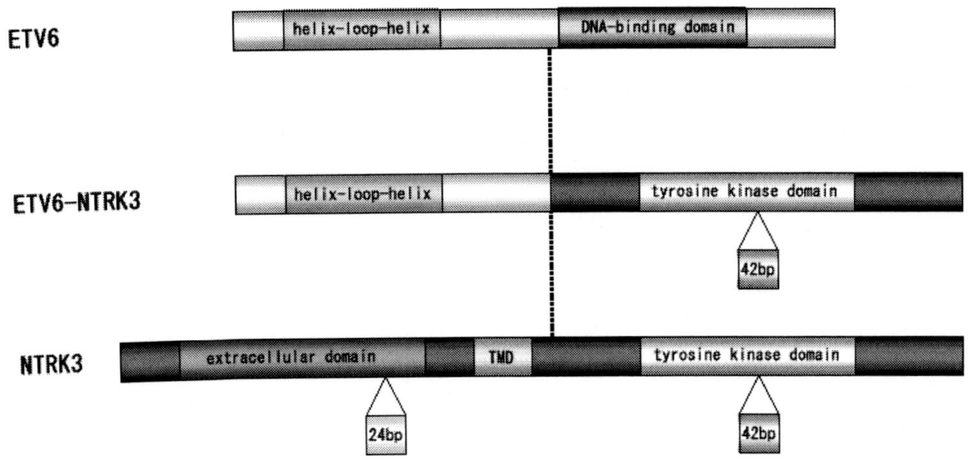

Figure 11. Schematic representations of ETV6 (upper), ETV6-NTRK3 (middle), and NTRK3 (lower). ETV6-NTRK3 contains an N-terminally located helix-loop-helix domain of the ETV6 and a tyrosine kinase domain of the NTRK3 in the C-terminal.

ETV6-NTRK3 is composed of the helix-loop-helix domain of ETV6 fused to an intracellular portion containing the tyrosine kinase domain of NTRK3 (Fig. 11). The helix-loop-helix domain is considered to enable the fusion protein to undergo oligomerization leading to ligand-independent activation of the tyrosine kinase domain [213]. The fusion protein transforms NIH3T3 cells through constitutive activation of both the Ras-mitogen-activated protein kinase (MAPK) and the phosphatidylinositol-3'kinase (PI3K)-Akt pathway [213,214]. Moreover, an intact insulin-like growth factor (IGF) signaling axis has been suggested to be essential for the ETV6-NTRK3-mediated transformation of NIH3T3 cells [215]. The *IGF2* gene is localized to 11p15.5, and trisomy 11 is often identified in infantile fibrosarcoma and congenital mesoblastic nephroma, implicating that the extra copy of chromosome 11 might lead to over-expression of IGF2 and activation of the IGF pathway in these tumors [215]. Interestingly, it has been shown that the expression of the *ETV6-NTRK3* also induces transformation of mammary epithelial cells, supporting its oncogenetic role in the development of secretory breast carcinoma, although the IGF pathway has not been examined in such an epithelial tumor [216]. Thus, the fusion gene is not always specific for a given tumor type, and some additional or alternative mechanisms seem to be required to define the histologic phenotype of tumors bearing the *ETV6-NTRK3*. Recently, ETV6-NTRK3 has been demonstrated to suppress the TGF-β tumor suppressor pathway by inactivating TGF-β type II receptor, providing a novel mechanism in ETV-NTRK3-induced tumorigenesis [217].

9. Inflammatory Myofibroblastic Tumor

Inflammatory myofibroblastic tumor (IMT) is an uncommon mass-forming lesion that primarily affects young individuals and occurs at virtually any anatomical sites with a predisposition to the lung and abdominal cavity. IMT has been described synonymously as a variety of lesions such as plasma cell granuloma, plasma cell pseudotumor, inflammatory myofibroblastic proliferation, omental mesenteric myxoid hamartoma, inflammatory pseudotumor, and inflammatory fibrosarcoma [1]. As this situation implies, fundamental issues, such as whether a single or multiple entities have been referred to as IMT and whether such a lesion represents a true neoplasm or a tumefactive reactive process (namely 'pseudotumor'), still exist. Some investigators consider that IMT is a disease spectrum encompassing multiple lesions ranging from a non-neoplastic inflammatory change mostly due to infection to an overtly malignant tumor that was previously described as inflammatory fibrosarcoma [218].

Despite such an enigma, IMT shows a fairly consistent histologic appearance of a mixed population of chronic inflammatory cells including lymphocytes, plasma cells, and histiocytes in a variable fibro-cellular background of spindle-shaped cells displaying features of fibroblasts and myofibroblasts. A minority (5-25%) of patients with IMT manifests a symptomatic complex including fever of unknown origin, weight loss, and a variety of laboratory abnormalities such as anemia, polyclonal hypergammaglobulinemia, thrombocytosis, and plasmacytosis [219]. Patients with IMT often pursue a favorable or uneventful clinical course. Approximately 25% of cases with extrapulmonary IMT have been reported to recur locally, and metastases occur in only a minor population (less than 5%) [1,219].

Cytogenetic studies have demonstrated that a subset of IMTs, most of which occur in children or young adults, harbors clonal chromosomal rearrangements or translocations involving a region 2p23, in which the *ALK* gene is located (Fig. 12). These clonal and recurrent cytogenetic changes reinforce the notion that at least such a subset of IMTs is truly neoplastic. The rearranged *ALK* gene creates multiple fusion genes with various fusion partners including the *ATIC* located at 2p35, *CARS* at 11p15, *CLTC* at 17q23, *RANBP2* at 2q13, *TMP3* at 1p23, *TMP4* at 19p13, and *SEC31L1* at 4q21 [220-225]. Some identical or different *ALK* rearrangements have been also described in fractions of lymphoid malignancies such as anaplastic large cell lymphoma and diffuse large B-cell lymphoma [226-228]. The *CARS-ALK*, *RANBP2-ALK*, and *SEC31L1-ALK* fusion genes have been so far identified exclusively in IMT. Curiously, a few *ALK* fusion genes such as the *ATIC-ALK* have been reported to be expressed at low levels in non-neoplastic cells [229].

The *ALK* gene encodes a receptor tyrosine kinase with a large extracellular domain, lipophilic transmembrane segment, and a cytoplasmic tyrosine kinase domain. Based on the similarity in the extracellular domain, ALK is belonging to the insulin receptor subfamily of tyrosine kinases including leukocyte tyrosine kinase (LTK) [230]. ALK is expressed specifically in the central nervous system such as the neonatal brain, suggesting a role for the receptor in brain development [231]. Pleiotrophin, a polypeptide growth factor that induces proliferation in a wide range of cells including epithelial, endothelial and mesenchymal cell lineages, and midkine, a secreted growth factor homologous to pleiotrophin, can activate ALK and are considered its potential ligands [232,233], although other receptors mediating multiple functions of pleiotrophin may be also present [230].

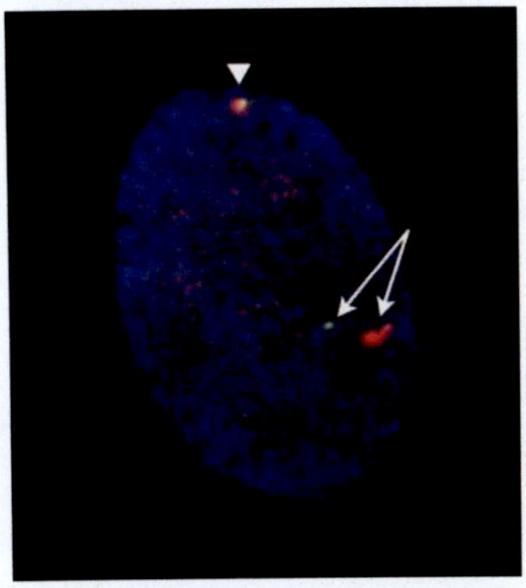

Figure 12. *ALK* gene rearrangement in inflammatory myofibroblastic tumor demonstrated by interphase FISH. The gene rearrangement is shown by split signals (arrows) of an ALK probe, whereas merged signals represent the non-rearranged *ALK* gene (arrow head).(permission to use this copyrighted illustration has been granted by Bunkodo Co., Ltd. Tokyo, Japan).

The *TPM3* and *TPM4* genes encode actin filament-binding filamentous proteins, tropomyosins 3 and 4, belonging to the tropomyosin family comprising of more than 15 isoforms present in muscle or non-muscle cells and primarily mediating cell contraction or stabilization of cytoskeleton [234]. The product of the *ATIC* gene is an enzyme, 5-aminoimidazole-4- carboxaminde ribonucleotide formyltransferase/IMP cyclohydolase, which is required for DNA synthesis [235]. The CARS gene encodes the cysteinyl-tRNA synthetase, which is also an enzyme catalyzing the formation of cysteinyl-tRNA that is a key step in protein biosynthesis [236]. Clathrin heavy chain, encoded by the *CLTC* gene, is a component of a three-legged molecule of clathrin, which is the major protein constituent of the coat that surrounds the cytoplasmic face of the organelles (coated vesicles) mediating selective protein transport [237]. SEC31L1 has been shown to localize to cytoplasmic vesicular structures, and is a component of COPII coat complex, which mediates transport from the endoplasmic reticulum to the Golgi machinery [238]. The *RANMP2* gene encodes a 358-kDa multi-domain nuclear pore protein that functions probably in nuclear import and export [223]. Despite the divergent molecular structures and functions, these partners of the ALK-fusion proteins have common biological and structural features; they are constitutively expressed and contain a known or putative N-terminal oligomerization motif [225].

The *ALK* fusion genes encode aberrant proteins composed of an N-terminal part of each partner protein containing the oligomerization motif fused to the last 563 amino acids of the ALK kinase including its catalytic domain (Fig. 13). The oligomerization of the constitutively expressed ALK-fusion proteins results in ligand-independent autoactivation of the ALK kinase underlying their oncogenic properties. Indeed, several types of ALK fusions such as TPM3-ALK, CLTC-ALK, and ATIC-ALK have been shown to transform NIH3T3 cells through the PI3-kinase/AKT signaling pathway and to confer tumorigenic capacity to the cells in nude mice [226]. Although the functional differences between the fusion variants may be due to differential activation of various signaling pathways as previously speculated [226], a potential relationship between the fusion type and clinicopathologic features of IMT remains to be addressed. Like other tumor-associated fusion proteins involving the tyrosine kinase signal pathways such as COL1A1-PDGFB and NTRK3-ETV6, ALK fusions are also potential molecular targets for the tyrosine kinase inhibitor, imatinib mesylate (STI571).

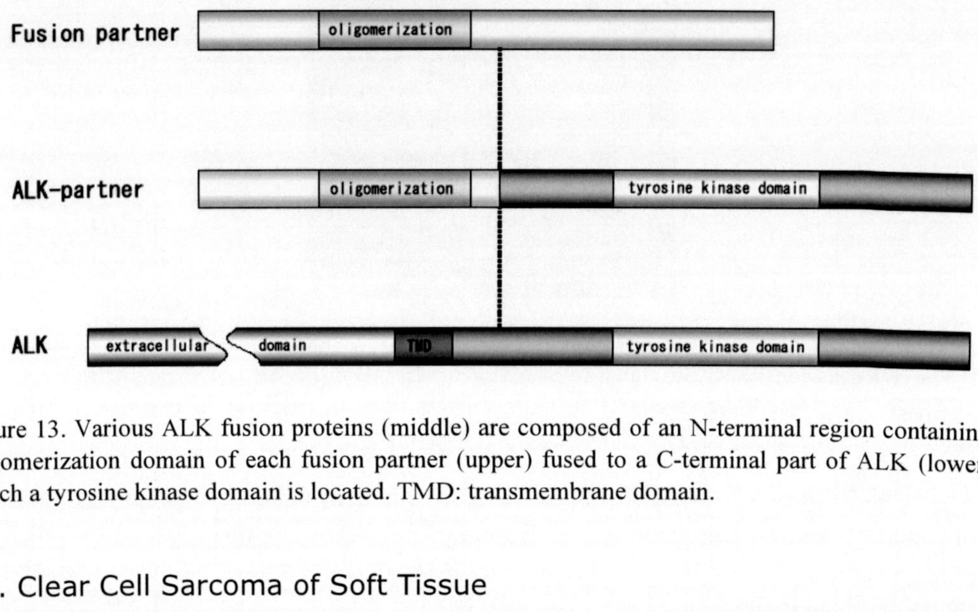

Figure 13. Various ALK fusion proteins (middle) are composed of an N-terminal region containing an oligomerization domain of each fusion partner (upper) fused to a C-terminal part of ALK (lower) in which a tyrosine kinase domain is located. TMD: transmembrane domain.

10. Clear Cell Sarcoma of Soft Tissue

Clear cell sarcoma of soft tissue (formerly referred to as malignant melanoma of soft parts) is a rare soft tissue tumor that most commonly affects adolescents and young adults. It typically afflicts deep soft tissues or is attached to tendons and aponeuroses particularly in the distal extremities [1,2]. The head and neck, trunk, visceral organs, retroperitoneum, bone, penis, and spinal nerve roots have been described as rare anatomical locations in which the tumor may arise [1,2]. Clear cell sarcoma is histologically characterized by a uniformly nested or fascicular growth pattern of epithelioid or spindle-shaped cells possessing clear or granular eosinophilic cytoplasm, sometimes containing melanin pigments, and frequently admixed with wreath-like multinucleated giant cells. Because of similar morphologic features including melanin synthesis and immunohistochemical expressions of melanoma-associated proteins such as S-100 protein and HMB45, kinship between clear cell sarcoma of soft tissue and malignant melanoma has been suggested. However, both tumors are currently considered distinct entities based on their unique cytogenetic profiles described below [2,239]. Like malignant melanoma, clear cell sarcoma is, however, regarded as a tumor of neuroectodermal derivation. The prognosis of this tumor is generally poor, and many patients develop recurrences or distant metastases. Five-year overall and disease-free survival rates of 68% and 50%, respectively, have been recently reported [240].

The chromosomal translocation t(12;22)(q13;p12), resulting in a gene fusion between the *EWSR1* gene at 22p12 and the *ATF1* gene at 12q13, is a cytogenetic hallmark of clear cell sarcoma of soft tissue. By means of various cytogenetic or molecular approaches, between 70% and over 90% of clear cell sarcomas have been shown to have t(12;22)(q13;p12) or the resultant *EWSR1-ATF1* fusion gene [241]. This genetic aberration has so far described exclusively in clear cell sarcoma of soft tissue, except for each case of angiomatoid fibrous histiocytoma, which is a distinct soft tissue tumor with putative fibrohistiocytic differentiation, and of intraosseous round cell sarcoma with polyphenotypic differentiation [242,243]. Cutaneous malignant melanoma demonstrates different and much divergent cytogenetic changes often involving chromosomes 1, 5, and 6q, and t(12;22) has not yet been identified in such a melanocytic malignancy [239]. Another type of the fusion gene, the *EWSR1-CREB1* probably resulting from t(2;22)(q32;q12), has been very recently identified in a minor fraction of the tumors arising in the gastrointestinal tract [244].

ATF1 is a member of a subgroup of basic leucine-zipper (bZIP) transcription factors that includes the cAMP-response-element-binding protein (CREB) and activator forms of the cAMP-response-element-modulator (CREM) and regulates gene expression [245]. ATF1 homodimerizes and forms heterodimers with CREB, binds directly to cAMP-inducible promoters, and activates transcription on phosphorylation by the cAMP-dependent protein kinase (PKA) [245]. The consensus recognition site for phosphorylation by PKA is located in the N-terminal of ATF1, whereas the bZIP domain is present in its C-terminal end.

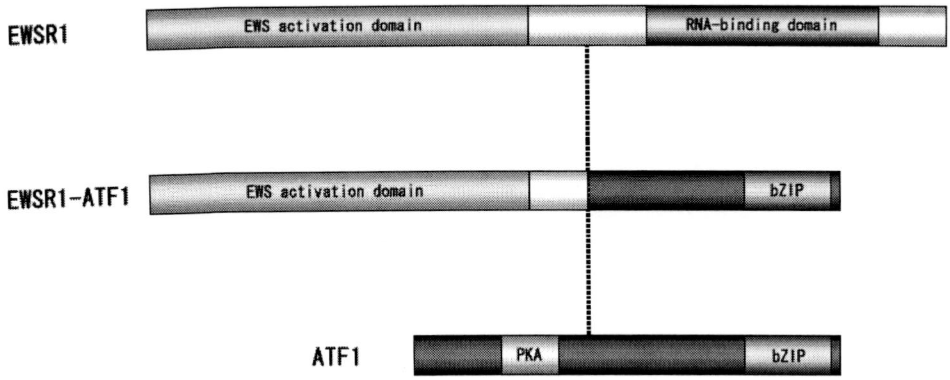

Figure 14. In clear cell sarcoma of soft tissue, EWSR1-ATF1 fusion protein (middle) composed of an N-terminal region of EWSR1 (uppe) and a C-terminal part of ATF1 (lower) containing a basic leucine-zipper domain (bZIP) is expressed. PKA: cAMP-dependent protein kinase.

In EWSR1-ATF1, the C-terminal region of EWSR1 containing the RNA-binding domain is replaced by the C-terminal part of ATF1 containing the bZIP domain (Fig. 14). Therefore, the N-terminal kinase-inducible domain is lacked in this fusion protein. At least 4 types of the *EWSR1-ATF1* fusion transcript have been described, and type 1 transcript showing a fusion between *EWSR1* exon 8 and *ATF1* exon 4 seems to be most prevalent [246]. EWSR1-ATF1 binds to ATF sites in cAMP-responsive promoters independently of cAMP induction through the bZIP domain derived from ATF1 and functions as a constitutive and potent transcriptional activator [247-249].

Thus, EWSR1-ATF1 may exert its oncogenic potential via unscheduled or aberrant induction of target genes. It has been suggested that EWSR1-ATF1 represses TP53 function and decreases endogenous expression of POSH, a putative inducer of apoptosis in clear cell sarcoma, further supporting the oncogenic potential of EWSR1-ATF1 [250,251]. In addition, the *ERBB3* gene encoding a member of the epidermal growth factor receptor family has been shown to be significantly overexpressed in clear cell sarcoma at both the mRNA and protein levels [252]. However, whether the *ERBB3* gene is also involved in the oncogenic pathway of EWSR1-ATF1 or not still remains to be determined.

11. Alveolar Soft Part Sarcoma

Alveolar soft part sarcoma is also a rare soft tissue sarcoma that mostly affects adolescents and young adults. The tumor usually develops as a slowly growing mass in deep soft tissues of the thigh, buttock, and abdominal wall as well as in the head and neck region, or especially in the tongue and orbit [1,2]. Alveolar soft part sarcoma shows highly characteristic histology of an organoid or nesting pattern of polygonal large tumor cells having vesicular nuclei and ample eosinophilic granular cytoplasm, which are compartmentalized by a delicate connective tissue stroma often containing sinusoidal vascular channels. Its line of differentiation remains yet undetermined, although myogenic and neurogenic origins have been proposed [253,254]. Despite such a slowly growing tumor, alveolar soft part sarcoma frequently metastasizes into the lung, bone, and brain, and 5- and 10-years metastasis-free survivals have been reported to be 60% and 38%, respectively [255].

Among the limited cases examined cytogenetically, chromosomal aberrations involving chromosome 17 have been recurrently and frequently identified in alveolar soft part sarcoma [256]. Especially, an unbalanced translocation, der(17)t(X;17)(p11.2;q25), is considered to be a specific cytogenetic change for this sarcoma, although a balanced form of the same translocation has been detected in some cases with pediatric renal cell carcinomas [257]. The balanced t(X;17)(p11.2;q25) has been also found in a minority of alveolar soft part sarcomas [258]. The different chromosomal representations consisting of the same structural change in these tumors are difficult to explain. In contrast to the general trend in cancers with somatic translocations involving the X chromosome in male patients, there is a predilection of alveolar soft part sarcoma with a der(17)t(X;17) and a normal X in men [258]. This finding suggests that the der(17)t(X;17) forms in the G2 phase of cell cycle [258]. The chromosomal translocation t(X;17)(p11.2;q25) creates a fusion between the *ASPSCR1* (also known as *ASPS*) gene at 17q25 and the *TFE3* gene at Xp11.2 [259].

The *ASPSCR1* is a recently identified gene that encodes a deduced 476-amino acid protein containing a UBX-like domain in its C-terminal and is ubiquitously expressed [259]. UBX is one of the ubiquitin-like domains that have a close structural relationship with ubiquitin [260], which plays an essential role in intracytoplasmic degradation of a wide variety of proteins. Recently, ASPSCR1 has been shown to be the probable human homologue of murine TUG that functions as a putative regulator of trafficking of the glucose transporter GLUT4 [261].

The *TFE3* gene encodes a member of the microphthalmia-TFE subfamily of basic helix-loop-helix leucine zipper (bHLH-LZ) transcription factors that include TFEB, TFEC, and MITF. Although diverse biological roles of TFE3 are suggested basedon its ubiquitous expression, it has been implicated in cell type-specific expressions of target genes in the melanocytic and peripheral neuroendocrine cells and osteoclasts in bone [262-264]. Besides, synergistic or cooperative functions of TFE3 and Smad proteins have been suggested in the TFG β-activated transduction pathway [265]. The TFE3 gene is known to be frequently rearranged in renal carcinomas particularly affecting children and young adults and fused to a variety of genes including the *PRCC* at 1q21, *SFPQ* (or *PSF*) at 1q34, and *NONO* (or *NRB54*) at Xq13.1 as well as to the *ASPSCR1* [266]. In such tumors with *TFE3* gene fusions, a portion of the *TFE3* gene is placed under the control of a novel promoter that may result in over-expression of TFE3 fusion protein, which can be readily detectable immunohistochemically using an anti-TFE3 antibody [267].

The *ASPSCR1-TFE3* fusion gene is transcribed into at least two mutually exclusive types of fusion transcripts, which are composed of the first 4 exons of the *ASPSCR1* fused in-frame to either exon 4 (type 1) or exon 3 (type 2) of the *TFE3* followed by its C-terminal part including a sequence coding the bHLH-LZ domain (Fig. 15) [259]. ASPSCR1-TFE3 is localized to the nucleus and presumed to act as an aberrant transcription factor in a gain-of-function or a dominant negative manner [266]. Detailed biological functions of ASPSCR1-TFE3, however, are still largely unknown.

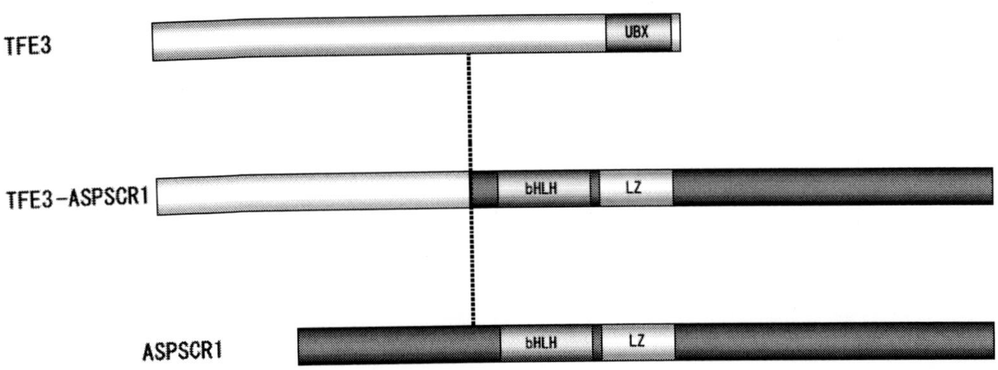

Figure 15. TFE3-ASPSCR1 fusion protein (middle) is composed of an N-terminal part of TFE3 and a C-terminal region of ASPSCR1 containing its basic helix-loop-helix leucine zipper domain (bHLH-LZ). UBX: a UBX (ultrabithorax homeo domain)-like domain.

12. Low Grade Fibromyxoid Sarcoma

Low grade fibromyxoid sarcoma is a recently recognized, rare variant of fibrosarcoma characterized by a bland-looking histologic appearance and a paradoxically aggressive biological behavior [268]. The sarcoma commonly affects young adults and typically develops as a deeply located painless soft tissue mass in the trunk or proximal extremities such as the thigh [1,2,269]. Histologically, the lesion is essentially hypocellular and composed of a proliferation of fibroblast-like spindle cells lacking significant cellular atypia in a fascicular or whorl-like growth pattern embedded in an abundant and alternating fibrocollagenous and myxoid background. Such a deceptively bland histologic appearance may create diagnostic challenges, and the distinction from other benign or low grade fibromyxoid tumors such as desmoid-type fibromatoses and low grade myxofibrosarcoma is often difficult. Another fibromyxoid tumor with hyalinized collagenous nodules surrounded by cuffs of epithelioid tumor cells ('collagen rosettes') has been referred to as hyalinizing spindle cell tumor with giant rosettes, which is currently regarded as a histologic variant of low grade fibromyxoid sarcoma because these tumors have similar clinicopathologic features and an identical cytogenetic aberration [1,269]. Although earlier reports described high rates of local recurrences (68%), distant metastases (41%), and deaths of disease (18%), a recent study based on a large series of low grade fibromyxoid sarcoma has shown recurrences, metastases, and deaths in 9%, 6%, and 2% of patients, respectively [269].

Recently both low grade fibromyxoid sarcoma and hyalinized spindle cell tumor with giant rosettes have been shown to share a common chromosomal translocation, t(7;16)(q33;p11), resulting in a fusion of the *FUS* gene at 16p11 and the *CREB3L2* (also known as *BBF2H7*) gene at 7q33 [270-272]. Subsequently, a variant fusion gene, *FUS-CREB3L1*, presumably resulting from an yet undetected chromosomal translocation t(11;16)(p11;p11), has been identified in a single case of low grade fibromyxoid sarcoma [273]. Specifically, t(7;16)(q33;p11) and the *FUS-CREB3L2/1* fusion genes are so far confined to low grade fibromyxoid sarcoma and its morphologic variant, indicating that these genetic aberrations are specific for this type of soft tissue sarcoma [272,274]. Thus, the detection of such chromosomal or genetic changes would greatly facilitate the diagnosis of low grade fibromyxoid sarcoma.

The *CREB3L2* and *CREB3L1* genes encode protein members belonging to the OASIS basic-leucine zipper (bZIP) family of transcription factors together with CREB-H, CREB3, and CREB4. These transcription factors can dimerize with each other through their bZIP domain, can bind CRE DNA sequences, can activate transcription through box-B and ATF6 elements, and contain a transcriptional activation domain at their N-terminal parts [270,275]. CREB3L2 and CREB3L1 are significantly homologous proteins with 80% identity in their bZIP domains [273]. CREB3L2 is widely expressed in various fetal and adult tissues including the placenta, lung, spleen, and small intestine [270]. Like other OASIS family members, CREB3L2 also has a transmembrane domain and is probably localized to endoplasmic reticulum (ER), which may function as an ER-stress transducer [270, 276].

The fusion gene generated by t(7;16)(q33;p11) is composed of the 5' part of the *FUS* gene and the 3' part of the *CREB3L2* gene. The breakpoints of these fusion partners, however, mostly vary from case to case, and distribute in the regions ranging from exon 5 to exon 7 of the *FUS* gene and from exon 5 to intron 6 of the *CREB3L2* gene [270,272-274]. Based on the variable genomic sites involved in the gene fusion, at least 38 patters of fusion transcripts, all of which are formed in an in-frame manner, have been identified [270,272-274]. In FUS-CREB3L2 fusion proteins, the bZIP domain encoded by exons 6 and 7 of the *CREB3L2* is placed under the control of the *FUS* promoter, which may derange the expression of its target genes that are normally regulated by native CREB3L2 (Fig. 16).

Although the *FUS-CREB3L2/1* fusion genes are considered to be responsible for the development of low grade fibromyxoid sarcoma, their detailed oncogenic mechanisms remain to be elucidated.

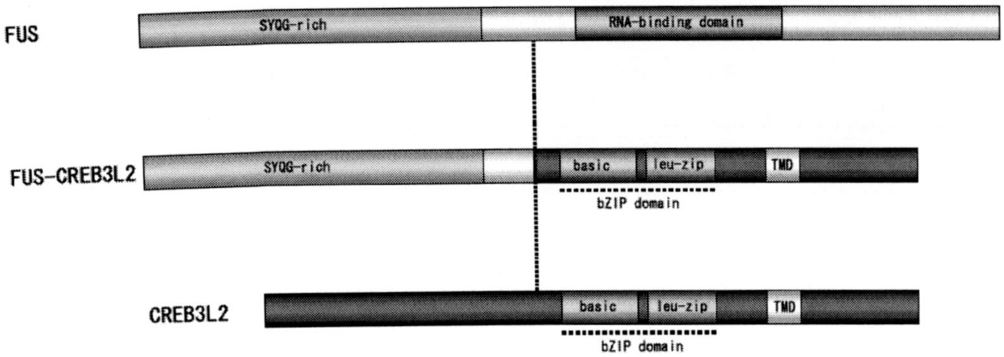

Figure 16. Schematic representations of FUS (upper), FUS-CREB3L2 (middle), and CREB3L2 (lower) proteins, the latter of which contain a basic leucine-zipper (bZIP) domain and a transmembrane domain (TMD) in its C-terminal region.

13. Angiomatoid Fibrinous Histiocytoma

Angiomatoid fibrous histiocytoma is a slow-growing, rare soft tissue tumor that most commonly affects children and young adults. The tumor typically occurs in the deep dermis or subcutis of the extremities followed by the trunk and the head and neck regions. The lesion is often multinodular with blood-filled cystic spaces, displaying an angiomatoid appearance or simulating a hematoma. Histologically, the lesion is composed of lobules or sheets of a proliferation of almost uniform short spindle or oval or epithelioid cells with vesicular nuclei and eosinophilic cytoplasm, surrounding the blood-filled pseudoangiomatous spaces. The lobules are often accompanied by peripheral aggregates or cuffs of lymphoplasmacytic cells, which are another histologic feature of angiomatoid fibrous histiocytoma. Although the lesion was originally described under the rubric of angiomatoid malignant fibrous histiocytoma [277], most of the patients pursuit an excellent clinical course; the rates of local recurrence and metastasis have been described to be 2-11% and less than 1% respectively, leading to the revised and currently used designation of angiomatoid fibrous histiocytoma [2,278,279]. As its name implies, angiomatoid fibrous histiocytoma is generally considered as a so-called fibrohistiocytic neoplasm although its potential myoid or myofibroblastic differentiation has been also suggested [279].

In angiomatoid fibrous histiocytoma, only two cases with the *FUS-ATF1* fusion gene and one with the *EWSR1-ATF1* fusion gene have been described, together with complex chromosomal rearrangements involving 12q13 and 16p11 in one of the former cases [242,280,281]. Both reported cases with the *FUS-ATF1* have been shown to share an identical, in-frame fusion transcript having interruptions at codon 175 of the *FUS* gene and codon 110 of the *ATF1* gene with a glycine to valine transition at the junction [280,281]. A paucity of cytogenetically well-characterized cases warrants further analyses on cytogenetic or molecular features of angiomatoid fibrous histiocytoma. The *ATF1* fusion genes are suggested to play an important role in the development of angiomatoid fibrous histiocytoma as well as clear cell sarcoma of soft tissue, although there is little clinicopathologic overlap between both tumors.

Special Consideration of Chromosomal Translocation and Fusion Gene in Soft Tissue Sarcoma

Despite the diverse clinicopathologic features in soft tissue sarcomas, they can be genetically divided into two categories; one group of sarcomas is characterized by tumor-specific translocations or fusion genes, which are highlighted in this chapter, and another group without such recurrent and characteristic cytogenetic aberrations but with random and rather complex karyotypic changes [4]. In the former group, there is a close association between distinct chromosomal translocations or fusion genes and certain histologic types of sarcoma, which has offered us special opportunities to investigate novel molecular mechanisms in sarcoma genesis, refined the tumor classification, and facilitated tumor diagnoses particularly in challenging cases. Moreover, prognostic and therapeutic applications of such fusion genes have been implicated in some of the translocation-associated sarcomas [282].

There have been limited studies to elucidate the molecular mechanisms underlying the formation of the nonrandom genomic rearrangements that cause such recurrent chromosomal translocations. However, several mechanisms have been proposed to explain these chromosomal rearrangements in hematopoietic neoplasms and sarcomas; the high densities of repetitive DNA, such as *Alu* elements, at chromosomal breakpoint regions, translin activity, cleavage at sites of Z-DNA structure, topoisomerase II subunit exchange, repair of DNA breaks with nonhomologous chromosomes by nonhomologous end joining, and fragile sites may mediate illegitimate or homologous recombinations [62-65,283]. In addition, a spatial proximity or juxtaposition of two chromosomes or chromosomal regions in an interphase nucleus can influence the probability of translocations [284]. Incidentally, a close topographical association between territories of chromosomes 12 and 16, which are known to be involved in the translocation in myxoid liposarcoma, is seen in differentiated adipocytes [285].

The translocations that generate fusion proteins might transform several types of mesenchymal progenitor or stem cells with their distinct differentiation programs, leading to the development of certain sarcomas. Indeed, some of the fusion gene-derived cDNAs have been shown to be capable of transforming mammalian mesenchymal cells such as NIH3T3 cells and preadipocytes *in vitro* and tumor formation in nude mice [149,150,213,214,226]. In addition, fusion gene products are structurally or functionally regarded as aberrant transcription factors or growth factors or receptors for growth factors. Even though such translocations also occur in any cell type, only cells at specific points in certain differentiation pathways might be susceptible to and tolerant of the effects of a given fusion gene [4]. This view may explain the tissue-type specificity of translocation-associated tumors.

Some interesting but somewhat paradoxical features are noted with regard to the association between tumor types and characteristic chromosomal translocations or fusion genes; the tumor type-specificity of the translocations or fusion genes is not entirely perfect, and several identical translocations or fusion genes are shared by distinct tumor types of different cellular lineages (Table 2). Moreover, variant fusion genes exist in many translocation-associated sarcomas, and their fusion partner genes, which are mostly those encoding transcription factors, are also involved in gene fusions of tumors other than soft tissue sarcomas.

The first point is most exemplified by t(12;15)(p13;q25) or ETV6-NTRK3 in infantile fibrosarcoma, congenital mesoblastic nephroma, acute myeloid leukemia, and secretory breast carcinoma. This phenomenon suggests that such fusion proteins may activate transformation pathways that function independently of lineage constraints and do not directly affect differentiation programs, and that modulate common signal transduction cascades present in diverse cell types that potentiate downstream proliferative effects [201]. Additional genetic changes or types of progenitor cells in which the translocations are initiated may contribute to defining a tumor phenotype.

By contrast, the other fusion proteins seem to potently derange or affect cellular differentiation pathways, which may be reflected by morphologic phenotypes of tumor cells. A recently established tetracycline-inducible EWSR1-FLI-1 expression system in a rhabdomyosarcoma cell line has demonstrated that EWSR1-FLI-1 can induce an Ewing-like neural crest phenotype in multiple cellular contexts and diminish inherent myogenic differentiation, suggesting that the *EWSR1-FLI-1* acts as a cell lineage determinator rather than a pure oncogene [289].

Table 2. Fusion Genes Shared by Tumors of Multiple Cellular Lineages

Fusion gene	Tumor	Reference
FUS-ERG	Ewing family of tumors	[69]
	Acute myeloid leukemia	[70,71]
ETV6-NTRK3	Infantile fibrosarcoma	[197]
	Congenital mesoblastic nephroma	[198,199]
	Acute myeloid leukemia	[200]
	Secretory mammary carcinoma	[201]
ATIC/TMP3/TMP4-ALK	Inflammatory myofibroblastic tumor	[220,224]
	Anaplastic large cell lymphoma	[226,286,287]
CLTC-ALK	Inflammatory myofibroblastic tumor	[221]
	Anaplastic large cell lymphoma	[288]
	Diffuse large B cell lymphoma	[228]
EWSR1-ATF1	Clear cell sarcoma of soft tissue	[246]
	Angiomatoid fibrous histiocytoma	[242]
ASPSCR1-TFE3	Alveolar soft part sarcoma	[259]
	Pediatric renal cell carcinoma	[257]

Furthermore, in a mouse model of alveolar rhabdomyosarcoma generated by conditional knock-in and knock-out techniques, the expression of PAX3-FOXO1A in muscle stellate cells, which are assumed to be postnatal myogenic precursors, did not produce tumors, but it did in differentiating myofibers, implicating that such a pediatric sarcoma may not arise primarily from dormant precursor or stem cells but instead may arise by dedifferentiation of rather mature cells [128,290]. Thus, the timing as well as the cellular origin seems to be important for the fusion gene-induced tumorigenesis at least in this type of sarcoma.

In spite of the divergent genetic involvements in translocations, some variant fusion genes occurring in the same tumor type are considered to exert common oncogenic properties because of their homologous structural and functional features. These variant fusion genes include those of the *SS18-SSXs* in synovial sarcoma, the *PAX3/PAX7-FOXO1A* in alveolar rhabdomyosarcoma, and the *FUS-CREB3L2/CREB3L1* in low grade fibromyxoid sarcoma. Such fusion variants are so far mutually exclusive in each case, and any case with multiple or entirely different fusion genes has not been described to date.

In many tumors with fusion genes, fusion partner genes that are expressed ubiquitously seem to contribute to aberrant or ectopic expression of fusion proteins by exchanging promoter regions of host genes with those of the partners ('promoter swapping') [291]. Fusion genes in myxoid liposarcoma, extraskeletal myxoid chondrosarcoma, and angimatoid fibrous histiocytoma are examples with this molecular mechanism, and, thus, their fusion partners are interchangeable, explaining multiple fusion variants existing in such sarcomas.

Specifically in a variety of EWSR1-ETS fusions, an intact ETS DNA-binding domain derived from diverse genes of ETS transcription factors is consistently present despite their structural heterogeneity, and seems to play a common essential role in transcriptional modulation of their target genes [292]. A recent identification of the *FUS-ERG* fusion gene in the Ewing family of tumors has provided evidence that the transactivation domains of the TET family of RNA-binding proteins such as EWSR1 and FUS are also interchangeable in this tumor family [69]. In various ALK fusions identified in inflammatory myofibroblastic tumors and lymphomas, each of the fusion partners containing similar oligomerization motives is juxtaposed to the N-terminal of ALK, resulting in constitutive, ligand-independent activation of tyrosine kinase of ALK. Based on these phenomena, hitherto undescribed but theoretically possible fusion variants, such as FUS-WT1 in desmoplastic small round cell tumor and FUS-NR4A3 in extraskeletal myxoid chondrosarcoma, may be present.

However, ample unresolved issues on the variant fusion genes or transcripts still remain. For example, the significantly different detection frequency between fusion variants in a given sarcoma type has not been well explained. In addition, it is unknown why such limited and specialized genes are recurrently involved in the rearrangements creating multiple fusion variants or shared by different tumor types, despite a wide variety of genes with similar structural or functional features. Although differential biological functions of fusion variants have been suggested in some types of sarcomas [292-294], there is still a paucity of comparative studies with respect to functional properties or oncogenic mechanisms between variant chromosomal translocations or fusion genes.

Conclusion

Together with summarized clinicopathologic features, molecular and cellular biology of sarcoma-associated fusion genes resulting from chromosomal translocations are reviewed. These molecular genetic aberrations have provided novel and valuable insights into mostly elusive sarcoma genesis. Such genetic changes are not entirely but most often tumor type-specific, leading to their diagnostic applications and some refinements of tumor classification. Several fusion gene products such as receptor tyrosine kinases have been suggested to be promising molecular targets in therapeutic strategies. Further detailed and extensive analyses would expand our knowledge of molecular biology of sarcoma-associated chromosomal translocations and fusion genes.

References

[1] Fletcher CDM, Unni KK, Mertens F. World Health Organization Classification of Tumours. *Pathology and Genetics. Tumours of Soft Tissue and Bone.* Lyon: *IARC* press; 2002.
[2] Weiss SW, Goldblum JR. *Soft Tissue Tumors*. 4th ed. St. Louis: Mosby; 2001.
[3] Bennicelli JL, Barr FG. Genetics and the biologic basis of sarcomas. *Curr Opin Oncol.* 1999; 11:267-274.
[4] Helman LJ, Meltzer P. Mechanisms of sarcoma development. *Nat Rev Cancer* 2003; 3: 685-694.
[5] Ladanyi M. The emerging molecular genetics of sarcoma translocations. *Diagn Mol Pathol.* 1995; 4: 162-173.
[6] Chang CC, Shidham VB. Molecular genetics of pediatric soft tissue tumors. *J Mol Diagn*. 2003; 5: 143-154.
[7] Barr FG, Chatten J, D'Cruz CM, Wilson AE, Nauta LE, Nycum LM, Biegel JA, Womer RB. Molecular assays for chromosomal translocations in the diagnosis of pediatric soft tissue sarcomas. *JAMA.* 1995; 273: 553-557.
[8] Ladanyi M, Bridge JA. Contribution of molecular genetic data to the classification of sarcomas. *Hum Pathol*. 2000; 31: 532-538.
[9] Sandberg AA, Bridge JA. *The Cytogenetics of Bone and Soft Tissue Tumors*: R.G. Landes Company: Austin; 1994.
[10] Shaffer LG, Tommerup N. *An International System for Human Cytogenetic Nomenclature* (2005): Karger: Basel; 2005.
[11] Schrock E, du Manoir S, Veldman T, Schoell B, Weinberg J, Ferguson-Smith MA, Ning Y, Ledbetter DH, Bar-Am I, Soenksen D, Garini Y, Ried T. Multicolor spectral karyotyping of human chromosomes. *Science*. 1996; 273:494-497.
[12] Jalal SM, Law ME. Utility of multicolor fluorescent in situ hybridization in clinical cytogenetics. *Genet Med*. 1999; 1:181-186.
[13] Williams GH, Williams ED. Identification of tumor-specific translocations in archival material. *J Pathol*. 1995; 175:279-281.
[14] Hill DA, O'Sullivan MJ, Zhu X, Vollmer RT, Humphrey PA, Dehner LP, Pfeifer JD. Practical application of molecular genetic testing as an aid to the surgical pathologic diagnosis of sarcomas. A prospective study. *Am J Surg Pathol*. 2002; 26:965-977.
[15] Spillane AJ, A'Hern R, Judson IR, Fisher C, Thomas JM. Synovial sarcoma: a clinicopathologic, staging, and prognostic assessment. *J Clin Oncol*. 2000; 18:3794-3803.
[16] Ferrari A, Gronchi A, Casanova M, Meazza C, Gandola L, Collini P, Lozza L, Bertulli R, Olmi P, Casali PG. Synovial sarcoma: a retrospective analysis of 271 patients of all ages treated at a single institution. *Cancer*. 2004; 101:627-634.
[17] Farris KB, Reed RJ. Monophasic, glandular, synovial sarcomas and carcinomas of the soft tissues. *Arch Pathol Lab Med*. 1982; 106:129-132.
[18] Majeste RM, Beckman EN. Synovial sarcoma with an overwhelming epithelial component. *Cancer*. 1988; 61:2527-2531.
[19] Dei Tos AP, Wadden C, Calonje E, Sciot R, Pauwels P, Knight JC, Dal Cin P, Fletcher CD. Immunohistochemical demonstration of glycoprotein p30/32MIC2 (CD99) in

synovial sarcoma. A potential cause of diagnostic confusion. *Appl Immunohistochem*. 1995; 3:168-173.

[20] Folpe AL, Schmidt RA, Chapman D, Gown AM. Poorly differentiated synovial sarcoma. Immunohistochemical distinction from primitive neuroectodermal tumors and high-grade malignant peripheral nerve sheath tumors. *Am J Surg Pathol*. 1998; 22: 673-682.

[21] Limon J, Dal Cin P, Sandberg AA. Translocations involving the X chromosome in solid tumors: presentation of two sarcomas with t(X;18)(q13;p11). *Cancer Genet Cytogenet*. 1986; 23:87-91.

[22] Turc-Carel C, Dal Cin P, Limon J, Li F, Sandberg AA. Translocation X; 18 in synovial sarcoma. *Cancer Genet Cytogenet*. 1986; 23:93.

[23] dos Santos NR, de Bruijn DRH, van Kessel AG. Molecular mechanisms underlying human synovial sarcoma development. *Genes Chromosomes Cancer*. 2001; 30:1-14.

[24] Sandberg AA, Bridge JA. Updates on the cytogenetics and molecular genetics of bone and soft tissue tumors. Synovial sarcoma. *Cancer Genet Cytogenet*. 2002; 133:1-23.

[25] Mitelman F, Johansson B, Mertens F. *Mitelman Database of Chromosome Aberrations in Cancer. 2004*. Available form: http://cgap.nci.nih.gov/Chromosomes/Mitelman.

[26] Storlazzi CT, Mertens F, Mandahl N, Gisselsson D, Isaksson M, Gustafson P, Domanski A, Panagopoulos I. A novel fusion gene, SS18L1/SSX1, in synovial sarcoma. *Genes Chromosomes Cancer*. 2003; 37:195-200.

[27] Skytting B, Nilsson G, Brodin B, Xie Y, Landeberg J, Uhlen M, Larsson O. A novel fusion gene, SYT-SSX4, in synovial sarcoma. *J Natl Cancer Inst*. 1999; 91:974-975.

[28] Mezzelani A, Dagrada GP, Sozzi G, Pierotti MA, Pilotti S. SYT-SSX fusion transcripts and epithelial differentiation in synovial sarcoma. *Diagn Mol Pathol*. 2000; 9:234.

[29] Törnkvist M, Brodin B, Bartolazzi A, Larsson O. A novel type of SYT/SSX fusion: methodological and biological implications. *Mod Pathol*. 2002; 15:679-785.

[30] Guillou L, Coindre JM, Gallagher G, Terrier P, Gebhard S, de Saint Aubain Somerhausen N, Michels JJ, Jundt G, Vince DR, Collin F, Trassard M, Le Doussal V, Benhattar J. Detection of the synovial sarcoma translocation t(X;18) (SYT;SSX) in paraffin-embedded tissues using reverse transcriptase-polymerase chain reaction: a reliable and powerful diagnostic tool for pathologists. A molecular analysis of 221 mesenchymal tumors fixed in different fixatives. *Hum Pathol*. 2001; 32:105-112.

[31] Tamborini E, Agus V, Perrone F, Papini D, Romanò R, Pasini B, Gronchi A, Colecchia M, Rosai J, Pierotti MA, Pilotti S. Lack of SYT-SSX fusion transcripts in malignant peripheral nerve sheath tumors on RT-PCR analysis of 34 archival cases. *Lab Invest*. 2002; 82:609-618.

[32] Coindre JM, Hostein I, Benhattar J, Lussan C, Rivel J, Guillou L. Malignant peripheral nerve sheath tumors are t(X;18)-negative sarcomas. Molecular analysis of 25 cases occurring in neurofibromatosis type 1 patients, using two different RT-PCR-based methods of detection. *Mod Pathol*. 2002; 15:589-592.

[33] Ladanyi M, Antonescu CR, Leung DH, Woodruff JM, Kawai A, Healey JH, Brennan MF, Bridge JA, Neff JR, Barr FG, Goldsmith JD, Brooks JSJ, Goldblum JR, Ali SZ, Shipley J, Cooper CS, Fisher C, Skytting B, Larsson O. Impact on SYT-SSX fusion type on the clinical behavior of synovial sarcoma: a multi-institutional retrospective study of 243 patients. *Cancer Res*. 2002; 62:135-140.

[34] de Bruijn DR, Baats E, Zechner U, de Leeuw B, Balemans M, Olde Weghuis D, Hirning-Folz U, van Kessel AG. Isolation and characterization of the mouse homolog of SYT, a gene implicated in the development of human synovial sarcomas. *Oncogene.* 1996; 13:643-648.
[35] dos Santos NR, de Bruijn DRH, Balemans M, Janssen B, Gärtner F, Lopes JM, de Leeuw B, van Kessel AG. Nuclear localization of SYT, SSX and the synovial sarcoma-associated SYT-SSX fusion proteins. *Hum Mol Genet.* 1997; 6:1549-1558.
[36] Clark J, Rocques PJ, Crew AJ, Gill S, Shipley J, Chan AML, Gusterson BA, Cooper CS. Identification of novel genes, SYT and SSX, involved in the t(X;18)(p11.2;q11.2) translocation found in human synovial sarcoma. *Nat Genet.* 1994; 7:502-508.
[37] Pawson T, Schlessinger J. SH2 and SH3 domains. Curr Biol. 1993; 3:434-442.
[38] Brett D, Whitehouse S, Antonson P, Shipley J, Cooper C, Goodwin G. The SYT protein involved in the t(X;18) synovial sarcoma translocation is a transcriptional activator localised in nuclear bodies. *Hum Mol Genet.* 1997; 6:1559-1564.
[39] Thaete C, Brett D, Monaghan P, Whitehouse S, Rennie G, Rayner E, Cooper CS, Goodwin G. Functional domains of the SYT and SYT-SSX synovial sarcoma translocation proteins and co-localization with the SNF protein BRM in the nucleus. *Hum Mol Genet.* 1999; 8:585-591.
[40] de Bruijn DR, dos Santos NR, Thijssen J, Balemans M, Debernardi S, Linder B, Young BD, van Kessel AG. The synovial sarcoma associated protein SYT interacts with the acute leukemia associated protein AF10. *Oncogene.* 2001; 20:3281-3289.
[41] Eid JE, Kung AL, Scully R, Livingston DM. p300 interacts with the nuclear proto-oncoprotein SYT as part of the active control of cell adhesion. *Cell.* 2000; 102:839-848.
[42] Ito T, Ouchida M, Ito S, Jitsumori Y, Morimoto Y, Ozaki T, Kawai A, Inoue H, Shimizu K. SYT, a partner of SYT-SSX oncoprotein in synovial sarcomas, interacts with mSin3A, a component of histone deacetylase complex. *Lab Invest.* 2004; 84:1484-1490.
[43] Iwasaki T, Koibuchi N, Chin WW. Synovial sarcoma translocation (SYT) encodes a nuclear receptor coactivator. *Endocrinology.* 2005; 146:3892-3899.
[44] de Bruijn DR, Kater-Baats E, Eleveld M, Merkx G, van Kessel AG. Mapping and characterization of the mouse and human SS18 genes, two human SS18-like genes and a mouse Ss18 pseudogene. *Cytogenet Cell Genet.* 2001; 92:310-319.
[45] Güre AO, Wei IJ, Old LJ, Chen YT. The SSX gene family: characterization of 9 complete genes. *Int J Cancer.* 2002; 101:448-453.
[46] Margolin JF, Friedman JR, Meyer WKH, Vissing H, Thiesen HJ, Rauscher FJ. Krüppel-associated boxes are potent transcriptional repression domains. *Proc Natl Acad Sci USA.* 1994; 91:4509-4513.
[47] Witzgall R, O'Leary E, Leaf A, Onaldi D, Bonventre JV. The Krüppel-associated box-A (KRAB-A) domain of zinc-finger proteins mediates transcriptional repression. *Proc Natl Acad Sci USA.* 1994; 91:4514-4518.
[48] Soulez M, Saurin AJ, Freemont PS, Knight JC. SSX and the synovial-sarcoma-specific chimaeric protein SYT-SSX co-localize with the human Polycomb group complex. *Oncogene.* 1999; 18:2739-2746.
[49] dos Santos NR, de Bruijn DRH, Kater-Baats E, Otte AP, van Kasssel AG. Delineation of the protein domains responsible for SYT, SSX, and SYT-SSX nuclear localization. *Exp Cell Res.* 2000; 256:192-202.

[50] Crew AJ, Clark J, Fisher C, Gill S, Grimer R, Chand A, Shipley J, Guestersson BA, Cooper CS. Fusion of SYT to two genes, SSX1 and SSX2, encoding proteins with homology to the Kruppel-associated box in human synovial sarcoma. *EMBO J.* 1995; 14:2333-2340.

[51] Fligman I, Lonardo F, Jhanwar SC, Gerald WL, Woodruff J, Ladanyi M. Molecular diagnosis of synovial sarcoma and characterization of a variant SYT-SSX2 fusion transcript. *Am J Pathol.* 1995; 147:1592-1599.

[52] Safar A, Wickert R, Nelson M, Neff M, Bridge JA. Characterization of a variant SYT-SSX1 synovial sarcoma fusion transcript. *Diagn Mol Pathol.* 1998; 7:283-287.

[53] Sanders ME, van de Rijn M, Barr FG. Detection of a variant SYT-SSX1 fusion in a case of preominantly epithelioid synovial sarcoma. *J Mol Diagn.* 1999; 4:65-70.

[54] Sonobe H, Takeuchi T, Liag SB, Taguchi T, Yuri K, Shimizu K, Iwata J, Furihata M, Ohtsuki Y, Testa JR. A new human synovial sarcoma cell line, HS-SY-3, with a truncated form of hybrid SYT/SSX1 gene. *Int J Cancer.* 1999; 82:459-464.

[55] Tamborini E, Agus V, Mezzelani A, Riva C, Sozzi G, Azzarelli A, Pierotti MA, Pilotti S. Identification of a novel spliced variant of the SYT gene expressed in normal tissues and in synovial sarcoma. *Br J Cancer.* 2001; 84:1087-1094.

[56] Morimoto Y, Ouchida M, Ozaki T, Kawai A, Ito T, Yoshida A, Inoue H, Shimizu K. Splicing isoform of SYT-SSX fusion protein accelerates transcriptional activity and cell proliferation. *Cancer Lett.* 2003; 199:35-43.

[57] Nagai M, Tanaka S, Tsuda M, Endo S, Kato H, Sonobe H, Minami A, Hiraga H, Nishihara H, Sawa H, Nagashima K. Analysis of transforming activity of human synovial sarcoma-associated chimeric protein SYT-SSX1 bound to chromatin remodeling factor hBRM/hSNF2α. *Proc Natl Acad Sci USA.* 2001; 98:3843-3848.

[58] Nagayama S, Katagiri T, Tsunoda T, Hosaka T, Nakashima Y, Araki N, Kusuzaki K, Nakayama T, Tsuboyama T, Nakamura T, Imamura M, Nakayama Y, Toguchida J. Genome-wide analysis of gene expression in synovial sarcomas using a cDNA microarray. *Cancer Res.* 2002; 62:5859-5866.

[59] Allander SV, Illei PB, Chen Y, Antonescu, C, Bittner M, Ladanyi M, Meltzer PS. Expression profiling of synovial sarcoma by cDNA microarrays. Association of ERBB2, IGFBP2, and ELF3 with epithelial differentiation. *Am J Pathol.* 2002; 161:1587-1595.

[60] Xie Y, Törnkvist M, Aalto Y, Nilsson G, Girnita L, Nagy B, Kunuutila S, Larsson O. Gene expression profile by blocking the SYT-SSX fusion gene in synovial sarcoma cells. Identification of XRCC4 as a putative SYT-SSX target gene. *Oncogene.* 2003; 22:7628-7631.

[61] Nagayama S, Fukukawa C, Katagiri T, Okamoto T, Aoyama T, Oyaizu N, Imamura M, Toguchida J, Nakamura Y. Therapeutic potential of antibodies against FZD10, a cell-surface protein, for synovial sarcomas. *Oncogene.* 2005; 24:6201-6212.

[62] Ferguson DO, Alt FW. DNA double strand break repair and chromosomal translocation: lessons from animal models. *Oncogene.* 2001; 20:5572-5579.

[63] Kolomietz E, Meyn MS, Pandita A, Squire JA. The role of Alu repeat clusters as mediators of recurrent chromosomal aberrations in tumors. *Genes Chromosomes Cancer.* 2002; 35:97-112.

[64] Qi L, Strong MA, Karim BO, Huso DL, Greider CW. Telomere fusion to chromosome breaks reduces oncogenic translocations and tumour formation. *Nat Cell Biol.* 2005; 7:706-711.

[65] Wei Y, Sun M, Nilsson G, Dwight T, Xie Y, Wang J, Hou Y, Larsson O, Larsson C, Zhu X. Characteristic sequence motifs located at the genomic breakpoints of the translocation t(X;18) in synovial sarcomas. *Oncogene*. 2003; 22:2215-2222.

[66] Montogomery E, Argani P, Hicks JL, DeMarzo AM, Meeker AK. Telomere lengths of translocation-associated and nontranslocation-associated sarcomas differ dramatically. *Am J Pathol*. 2004; 164:1523-1529.

[67] Fechner RE, Mills SE. Small cell sarcomas. In: *Tumors of the Bones and Joints, Atlas of Tumor Pathology,* 3rd series, Fascicles 8. Washington DC: Armed Forces Institute of Pathology; 1993, pp187-201.

[68] Cotterill SJ, Ahrens S, Paulussen M, Jürgens HF, Voûte H, Gadner PA, Craft AW. Prognostic factors in Ewing's tumor of bone: analysis of 975 patients from the European Intergroup Cooperative Ewing's Sarcoma Study Group. *J Clin Oncol.* 2000; 18:3108-3114.

[69] Shing DC, McMullan DC, Roberts P, Smith K, Chin SF, Nicholson J, Tillman RM, Ramila P, Cullinane C, Coleman N. FUS/ERG gene fusions in Ewing's tumors. *Cancer* Res. 2003; 63:4568-4576.

[70] Panagopoulos I, Åman P, Fioretos T, Hoglund M, Johansson B, Mandahl N, Heim S, Behrendtz M, Mitelman F. Fusion of the FUS gene with ERG in acute myeloid leukemia with t(16;21)(p11;q22). *Genes Chromosomes Cancer*. 1994; 11:256-262.

[71] Ichikawa H, Shimizu K, Hayashi Y, Ohki M. An RNA-binding protein gene, TLS/FUS, is fused to ERG in human myeloid leukemia with t(16;21) chromosomal translocation. *Cancer Res.* 1994; 54:2865-2868.

[72] Bertolotti A, Lutz Y, Heard DJ, Chambon P, Tora L. hTAF(II)68, a novel RNA/ssDNA-binding protein with homology to the pro-oncoproteins TLS/FUS and EWS is associated with both TFIID and RNA polymerase II. *EMBO J.* 1996; 15:5022-5031.

[73] Haynes SR. The RNP motif protein family. *New Biol*. 1992; 4:421-429.

[74] Ohno T, Ouchida M, Lee L, Gatalica Z, Rao VN, Reddy ESP. The EWS gene, involved in Ewing family of tumors, malignant melanoma of soft parts and desmoplastic small round cell tumors, codes for an RNA binding protein with novel regulatory domains. *Oncogene*. 1994; 2:3717-3729.

[75] Plougastel B, Zucman J, Peter M, Thomas G, Delattre O. Genomic structure of the EWS gene and its relationship to EWSR1, a site of tumor-associated chromosome translocation. *Genomics*. 1993; 18:609-615.

[76] Rossow KL, Janknecht R. The Ewing's sarcoma gene product functions as a transcriptional activator. *Cancer Res*. 2001; 61:2690-2695.

[77] Sharrocks AD. The ETS-domain transcription factor family. *Nat Rev Mol Cell Biol.* 2001; 2:827-837.

[78] May WA, Lessnick SL, Braun BS, Klemsz M, Lewis BC, Lunsford LB, Hromas R, Denny CT, The Ewing's sarcoma EWS/FLI-1 fusion gene encodes a more transcriptional activator and is a more powerful transforming FLI-1. *Mol Cell Biol.* 1993; 13: 7393-7398.

[79] Bailly RA, Bosselut R, Zucman J, Cormier F, Delattre , Roussel M, Thomas G, Ghysdael J. DNA-binding and transcriptional activation properties of the EWS-FLI-1 fusion protein resulting from the t(11;22) translocation in Ewing sarcoma. *Mol Cell Biol.* 1994; 14:3230-3241.

[80] Ohno T, Rao VN, Reddy ES. EWS/Fli-1 chimeric protein is a transcriptional activator. *Cancer Res.* 1993; 53:5859-5863.

[81] Zucman J, Melot T, Desmaze C, Ghysdael J, Plougastel B, Peter M, Michel Zucker J, Triche TJ, Sheer D, Turc-Carel C, Ambros P, Combaret V, Lenoir G, Aurias A, Thomas G, Delattre O. Combinatorial generation of variable fusion proteins in the Ewing family of tumours. *EMBO J.* 1993; 12:4481-4487.

[82] de Alava E, Panizo A, Antonescu CR, Huvos AG, Javier pardo-Mindán F, Barr FG, Ladanyi M. Association of EWS-FLI1 type1 fusion with lower proliferative rate in Ewing's sarcoma. *Am J Pathol.* 2000; 156:849-855.

[83] Mao X, Miesfeldt S, Yang H, Leiden JM, Thomson CB. The FLI-1 and chimeric EWS-FLI-1 oncoproteins display similar specificities. *J Biol Chem.* 1994; 269:18216-18222.

[84] Lin PP, Brody RI, Hamelin AC, Bradner JE, Healey JH, Ladanyi M. Differential transactivation by alternative EWS-FLI1 fusion proteins correlates with clinical heterogeneity in Ewing's sarcoma. *Cancer Res.* 1999; 59:1428-1432.

[85] Zoubek, A, Dockhorn-Dworniczak B, Delattre O, Christiansen H, Niggli F, Gatterer-Menz, J, Smith TL, Jurgens H, Gadner H, Kovar H. Does expression of different EWS chimeric transcripts define clinically distinct risk groups of Ewing tumor patients? *J Clin Oncol.* 1996; 14:1245-1251.

[86] de Alava E, Kawai A, Healey JH, Fligman I, Meyers P, Huvos AG, Gerald WL, Jhanwar ER, Argani P, Antonescu CR, Pardo-Mindan FJ, Ginsberg J, Womer R, Lawlor ER, Wunder J, Andrulis I, Sorensen PHB, Barr FG, Ladanyi M. EWS-FLI1 fusion transcript structure is an independent determinant of prognosis in Ewing's sarcoma. *J Clin Oncol.* 1998; 16:1248-1255.

[87] Ohali A, Avigad S, Zaizov R, Ophir R, Horn-Saban S, Cohen IJ, Meller I, Kollender Y, Issakov J, Yaniv I. Prediction of high risk Ewing's sarcoma by gene expression profiling. *Oncogene.* 2004; 23:8997-9006.

[88] Siligan C, Ban J, Bachmaier R, Spahn L, Kreppel M, Shaefer KL, Poremba C, Aryee DNT, Kovar H. EWS-FLI1 target genes recovered from Ewing's sarcoma chromatin. *Oncogene.* 2005; 24:2512-2524.

[89] Hahm KB, Cho K, Lee C, Im YH, Chang J, Choi SG, Sorensen PH, Thiele CJ, Kim SJ. Repression of the gene encoding the TGF-beta type II receptor is a major target of the EWS-FLI1 oncoprotein. *Nat Genet.* 1999; 23:222-227.

[90] Dohjima T, Ohno T, Banno Y, Nozawa Y, Wen-yi Y, Shimizu K. Preferential down-regulation of phospholipase C-beta in Ewing's sarcoma cells transfected with antisense EWS-Fli-1. *Br J Cancer.* 2000; 82:16-19.

[91] Nishimori H, Sasaki Y, Yoshida K, Irifune H, Zembutsu H, Tanaka T, Aoyama T, Hosaka T, Kawaguchi S, Wada T, Hata J, Toguchida J, Nakamura Y, Tokino T. The Id2 gene is a novel target of transcriptional activation by EWS-ETS fusion proteins in Ewing family tumors. *Oncogene.* 2002; 21:8302-8309.

[92] Nakatani F, Tanaka K, Sakimura R, Matsumoto Y, Matsunobu T, Li X, Hanada M, Okada T, Iwamoto Y. Identification of p21WAF1/CIP1 as a direct target of EWS-Fli1 oncogenic fusion protein. *J Biol Chem.* 2003; 278:15105-15115.

[93] Fuchs B, Inwards CY, Janknecht R. Vascular endothelial growth factor expression is up-regulated by EWS-ETS oncoproteins and Sp1 and may represent an independent predictor of survival in Ewing's sarcoma. *Clin Cancer Res*. 2004; 10:1344-1353.

[94] Pieur A, Tirode F, Cohen P, Delattre O. EWS/FLI-1 silencing and gene profiling of Ewing cells reveal downstream oncogenic pathways and a crucial role for repression of insulin-like growth factor binding protein 3. *Mol Cell Biol*. 2004; 24:7275-7283.

[95] Nozawa S, Ohno T, Banno Y, Dohjima T, Wakahara K, Fan DG, Shimizu K. Inhibition of platelet-derived growth factor-induced cell growth signaling by a short interfering RNA for EWS-Fli1 via down-regulation of phospholipase D2 in Ewing sarcoma cells. *J Biol Chem*. 2005; 280:27544-27551.

[96] Matui Y, Chansky HA, Barahmand-Pour F, Zielinska-Kwiatkowska A, Tsumaki N, Myoui A, Yoshikawa H, Yang L, Eyre DR. COL11A2 collagen gene transcription is differentially regulated by EWS/ERG sarcoma fusion protein and wild-type ERG. *J Biol Chem*. 2003; 278: 11369-11375.

[97] Jeon IS, Davis JN, Braun BS, Sublett JE, Roussel MF, Denny CT, Shapiro DN. A variant Ewing's sarcoma translocation (7;22) fuses the EWS gene to the ETS gene ETV1. *Oncogene*. 1995; 10:1229-1234.

[98] Higashino F, Yoshida K, Fujinaga Y, Kamio K, Fujinaga K. Isolation of a cDNA encoding the adenovirus E1A enhancer binding protein: a new human member of the ets oncogene family. *Nucleic Acids Res*. 1993; 21:547-553.

[99] Urano F, Umezawa A, Yabe H, Hong W, Yoshida K, Fujiwara K, Hata J. Molecular analysis of Ewing's sarcoma: another fusion gene, EWS-E1AF, available for diagnosis. *Jpn J Cancer Res*. 1998; 89:703-711.

[100] Peter M, Couturier J, Pacquement H, Michon J, Thomas G, Magdelenat H, Delattre O. A new member of the ETS family fused to EWS in Ewing tumors. *Oncogene*. 1997; 14:1159-1164.

[101] Maurer P, T'Sas F, Coutte L, Callens N, Brenner C, van Lint C, de Launoit Y, Baert JL. FEV acts as a transcriptional repressor through its DNA-binding ETS domain and alanine-rich domain. *Oncogene*. 2003; 22:3319-3329.

[102] Tanaka K, Iwakuma T, Hariyama K, Sato H, Iwamoto Y. EWS-Fli1 antisense oligodeoxynucleotide inhibits proliferation of human Ewing's sarcoma and primitive neuroectodermal tumor cells. *J Clin Invest*. 1997; 99:239-247.

[103] Lambert G, Bertrand JR, Fattal E, Subra F, Pinto-Alphandary H, Malvy C, Auclair C, Couvreur P. EWS fli-1 antisense nanocapsules inhibits Ewing sarcoma-related tumor in mice. *Biochem Biophys Res Commun*. 2000; 279:401-406.

[104] Dohjima T, Lee NS, Li H, Ohno T, Rossi JJ. Small interfering RNAs expressed from a Pol III promoter suppress the EWS/Fli-1 transcript in an Ewing sarcoma cell line. *Mol Ther*. 2003; 7:811-816.

[105] Newton WA Jr, Gehan EA, Webber BL, Marsden HB, van Unnik AJM, Hamoudi AB, Tsokos MG, Shimada H, Harms D, Schmidt D, Ninfo V, Cazzavana AO, Gonzalez-Crussi F, Paham DM, Reiman HM, Asmar L, Beltangady MS, Sachs NE, Triche TJ, Maurer HM. Classification of rhabdomyosarcoma and related sarcomas: pathologic aspects and proposal for a new classification – an intergroup rhabdomyosarcoma study. *Cancer*. 1995; 6:1073-1085.

[106] Parham DM, Shapiro DN, Downing JR, Webber BL, Douglass EC. Solid alveolar rhabdomyosarcomas with the t(2;13). Report of two cases with diagnostic implications. *Am J Surg Pathol*. 1994; 18:474-478.

[107] Barr FG, Chatten J, D'Cruz CM, Wilson AE, Nauta LE, Nycum LM, Biegel JA, Womer RB. Molecular assays for chromosomal translocations in the diagnosis of pediatric soft tissue sarcomas. *JAMA*. 1995; 273:553-5517.

[108] Gordon A, McManus A, Anderson J, Fisher C, Abe S, Nojima T, Pritchard-Jones K, Shipley J. Chromosomal imbalances in pleomorphic rhabdomyosarcomas and identification of the alveolar rhabdomyosarcoma-associated PAX3-FOXO1A fusion gene in one case. *Cancer Genet Cytogenet*. 2003; 140:73-77.

[109] de Alava E, Lozano MD, Sola I, Panizo A, Idoate MA, Martinez-Isla C, Forteza J, Sierrasesumaga L, Pardo-Mindan FJ. Molecular features in a biphenotypic small cell sarcoma with neuroectodermal and muscle differentiation. *Human Pathol*. 1998; 29:181-184.

[110] Sorensen HB, Lynch JC, Qualman SJ, Tirabosco R, Lim JF, Maurer HM, Bridge JA, Crist WM, Triche TJ, Barr FG. PAX3-FKHR and PAX7-FKHR gene fusions are prognostic indicators in alveolar rhabdomyosarcoma: a report from the children's oncology group. *J Clin Oncol*. 2002; 20:2672-2679.

[111] Wachtel M, Dettling M, Koscielniak E, Stegmaier S, Treuner J, Simon-Klingenstein K, Buhlmann P, Niggli FK, Schafer BW. Gene expression signatures identify rhabdomyosarcoma subtypes and detect a novel t(2;2)(q35;p23) translocation fusing PAX3 to NCOA1. *Cancer Res*. 2004; 64:5539-5545.

[112] Tremblay P, Gruss P. Pax: genes for mice and men. *Pharmacol Ther*. 1994; 61: 205-226.

[113] Chi N, Epstain JA. Getting your Pax straight: Pax proteins in development and disease. *Trends Genet*. 2002; 18:41-47.

[114] Seale P, Sabourin LA, Girgis-Gabardo A, Mansouri A, Gruss P, Rudnicki MA. Pax7 is required for the specification of myogenic satellite cells. *Cell*. 2000; 102:777-786.

[115] RelaixF, Rocancourt D, Mansouri A, Buckingham M. Divergent functions of murine Pax3 and Pax7 in limb muscle development. *Genes Dev*. 2004; 18:1088-1105.

[116] Accili D, Arden KC. FoxOs at the crossroads of cellular metabolism, differentiation, and transformation. *Cell*. 2004; 117:421-426.

[117] Bois PR, Grosveld GC. FKHR (FOXO1a) is required for myotube fusion of primary mouse myoblasts. *EMBO J*. 2003; 22:1147-1157.

[118] Kaufmann E, Knochel W. Five years on the wings of fork head. *Mech Dev*. 1996; 57:3-20.

[119] Furuyama T, Nakazawa T, Nakano I, Mori N. Identification of the differential distribution patterns of mRNAs and consensus binding sequences for mouse DAF-16 homologues. *Biochem J*. 2000; 349:629-634.

[120] del Peso L, Gonzalez VM, Hernandez R, Barr FG, Nunez G. Regulation of the forkhead transcription factor FKHR, but not the PAX3-FKHR fusion protein, by the serine/threonine kinase Akt. *Oncogene*. 1999; 18:7328-7333.

[121] Bois PR, Izeradjene K, Houghton PJ, Cleveland JL, Houghton JA, Grosveld GC. FOXO1a acts as a selective tumor suppressor in alveolar rhabdomyosarcoma. *J Cell Biol*. 2005; 170:903-912.

[122] Bennicelli JL, Edwards RH, Barr FG. Mechanism for transcriptional gain of function resulting from chromosomal translocation in alveolar rhabdomyosarcoma. *Proc Natl Acad Sci U S A*. 1996; 93:5455-5459.

[123] Bennicelli JL, Advani S, Schafer BW, Barr FG. PAX3 and PAX7 exhibit conserved cis-acting transcription repression domains and utilize a common gain of function mechanism in alveolar rhabdomyosarcoma. *Oncogene*. 1999; 18:4348-4356.

[124] Ginsberg JP, Davis RJ, Bennicelli JL, Nauta LE, Barr FG. Up-regulation of MET but not neural cell adhesion molecule expression by the PAX3-FKHR fusion protein in alveolar rhabdomyosarcoma. *Cancer Res*. 1998; 58:3542-3546.

[125] Maroto M, Reshef R, Munsterberg AE, Koester S, Goulding M, Lassar AB. Ectopic Pax-3 activates MyoD and Myf-5 expression in embryonic mesoderm and neural tissue. *Cell*. 1997; 89:139-148.

[126] Scheidler S, Fredericks WJ, Rauscher FJ 3rd, Barr FG, Vogt PK. The hybrid PAX3-FKHR fusion protein of alveolar rhabdomyosarcoma transforms fibroblasts in culture. *Proc Natl Acad Sci U S A*. 1996; 93:9805-9809.

[127] Anderson J, Ramsay A, Gould S, Pritchard-Jones K. PAX3-FKHR induces morphological change and enhances cellular proliferation and invasion in rhabdomyosarcoma. *Am J Pathol*. 2001; 159:1089-1096.

[128] Keller C, Hansen MS, Coffin CM, Capecchi MR. Pax3:Fkhr interferes with embryonic Pax3 and Pax7 function: implications for alveolar rhabdomyosarcoma cell of origin. *Genes Dev*. 2004; 18:2608-2613.

[129] Keller C, Arenkiel BR, Coffin CM, El-Bardeesy N, DePinho RA, Capecchi MR. Alveolar rhabdomyosarcomas in conditional Pax3:Fkhr mice: cooperativity of Ink4a/ARF and Trp53 loss of function. *Genes Dev*. 2004; 18:2614-2626.

[130] Orvieto E, Furlanetto A, Laurino L, Dei Tos AP. Myxoid and round cell liposarcoma: a spectrum of myxoid adipocytic neoplasia. *Semin Diagn Pathol*. 2001; 18:267-273.

[131] Kilpatrick SE, Doyon J, Choong PFM, Sim FH, Nascimento AG. The clinicopathologic spectrum of myxoid and round cell liposarcoma. A study of 95 cases. *Cancer*. 1996; 77:1450-1458.

[132] Smith TA, Easley KA, Goldblum JR. Myxoid/round cell liposarcoma of the extremities. A clinicopathologic study of 29 cases with particular attention to extent of round cell liposarcoma. *Am J Surg Pathol*. 1996; 20:171-180.

[133] Rubin BP, Dal Cin P. The genetics of lipomatous tumors. *Semin Diagn Pathol*. 2001; 18:286-293.

[134] Meis-Kindblom JM, Sjögren H, Kindblom LG, Peydró-Mellquist A, Röijer E, Åman P, Stenman G. Cytogenetic and molecular genetic analyses of liposarcoma and its soft tissue simulators: recognition of new variants and differential diagnosis. *Virchows Arch*. 2001; 439:141-151.

[135] Hisaoka M, Morimitsu Y, Hashimoto H, Ishida T, Mukai H, Satoh H, Motoi T, Machinami R. Retroperitoneal liposarcoma with combined well-differentiated and myxoid malignant fibrous histiocytoma-like myxoid areas. *Am J Surg Pathol*. 1999; 23:1480-1492.

[136] Antonescu CR, Elahi A, Humphrey M, Lui MY, Healey JH, Brennan MF, Woodruff JM, Jhanwar SC, Ladanyi M. Specificity of TLS-CHOP rearrangement for classic myxoid/round cell liposarcoma. Absence in predominantly myxoid well-differentiated liposarcomas. *J Mol Diagn*. 2000; 2:132-138.

[137] Willeke F, Ridder R, Mechtersheimer G, Schwarzbach M, Duwe A, Weitz J, Lehnert T, Herfarth C, von Knebel Doeberitz M. Analysis of FUS-CHOP fusion transcripts in different types of soft tissue liposarcoma and their diagnostic implications. *Clin Cancer Res*. 1998; 4:1779-1784.

[138] Hosaka T, Nakashima Y, Kusuzaki K, Murata H, Nakayama T, Nakamata T, Aoyama T, Okamoto T, Nishijo K, Araki N, Tsuboyama T, Nakamura T, Toguchida J. A novel type of EWS-CHOP fusion gene in two cases of myxoid liposarcoma. *J Mol Diagn*. 2002; 4:164-171.

[139] Antonescu CR, Tschernyavsky SJ, Decuseara R, Leung DH, Woodruff JM, Brennan MF, Bridge JA, Neff JR, Goldblum JR, Ladanyi M. Prognostic impact of p53 status, TLS-CHOP fusion transcript structure, and histological grade in myxoid liposarcoma: a molecular and clinicopathologic study of 82 cases. *Clin Cancer Res*. 2001; 7:3977-3987.

[140] Crozat A, Åman P, Mandahl N, Ron D. Fusion of CHOP to a novel RNA-binding protein in human myxoid liposarcoma. *Nature*. 1993; 363:640-644.

[141] Kim J, Pelletier J. Molecular genetics of chromosome translocations involving EWS and related family members. *Physiol Genomics*. 1999; 1:127-138.

[142] Zinszner H, Sok J, Immanuel D, Yin Y, Ron D. TLS (FUS) binds RNA in vivo and engages in nucleo-cytoplasmic shuttling. *J Cell Sci*. 1997; 110:1741-1750.

[143] Prasad DD, Ouchida M, Lee L, Rao VN, Reddy ES. TLS/FUS fusion domain of TLS/FUS-erg chimeric protein resulting from the t(16;21) chromosomal translocation in human myeloid leukemia functions as a transcriptional activation domain. *Oncogene*. 1994; 9:3717-3729.

[144] Hallier M, Lerga A, Barnache S, Tavitian A, Moreau-Gachelin F. The transcription factor Spi-1/PU.1 interacts with the potential splicing factor TLS. *J Biol Chem*. 1998; 273:4838-4842.

[145] Ichikawa H, Shimizu K, Hayashi Y, Ohki M. An RNA-binding protein gene, TLS/FUS, is fused to ERG in human myeloid leukemia with t(16;21) chromosomal translocation. *Cancer Res*. 1994; 54:2865-2868.

[146] Batchvarova N, Wang XZ, Ron D. Inhibition of adipogenesis by the stress-induced protein CHOP (Gadd153). *EMBO J*. 1995; 14:4654-4661.

[147] Jousse C, Bruhat A, Carraro V, Urano F, Ferrara M, Ron D, Fafournoux P. Inhibition of CHOP translation by a peptide encoded by an open reading frame localized in the chop 5'UTR. *Nucleic Acids Res*. 2001; 29:4341-4351.

[148] Oyadomari S, Mori M. Roles of CHOP/GADD153 in endoplasmic reticulum stress. *Cell Death Diff*. 2004; 11:381-389.

[149] Zinszner H, Albalat R, Ron D. A novel effector domain from the RNA-binding protein TLS or EWS is required for oncogenic transformation by CHOP. *Genes Dev*. 1994; 8:2513-2526.

[150] Kuroda M, Ishida T, Takanashi M, Satoh M, Machinami R, Watanabe T. Oncogenic transformation and inhibition of adipocytic conversion of preadipocytes by TLS/FUS-CHOP type II chimeric protein. *Am J Pathol*. 1997; 151:735-744.

[151] Barone MV, Crozat A, Tabaee A, Philipson L, Ron D. CHOP (GADD153) and its oncogenic variant, TLS-CHOP, have opposing effects on the induction of G1/S arrest. *Genes Dev*. 1994; 8:453-464.

[152] Lae ME, Roche PC, Jin L, Lloyd RV, Nascimento AG. Desmoplastic small round cell tumor. A clinicopathologic, immunohistochemical, and molecular study of 32 tumors. *Am J Surg Pathol*. 2002; 26:823-835.

[153] Ladanyi M, Gerald W. Fusion of the EWS and WT1 genes in the desmoplastic small round cell tumor. *Cancer Res*. 1994; 54:2837-2840.

[154] Gerald WL, Rosai J, Ladanyi M. Characterization of the genomic breakpoint and chimeric transcripts in the EWS-WT1 gene fusion of desmoplastic small round cell tumor. *Proc Natl Acad Sci USA*. 1995; 92:1028-1032.

[155] Liu J, Nau MM, Yeh JC, Allegra CJ, Chu E, Wright JJ. Molecular heterogeneity and function of EWS-WT1 fusion transcripts in desmoplastic small round cell tumors. *Clin Cancer Res*. 2000; 6:3522-3529.

[156] Roberts SG. Transcriptional regulation by WT1 in development. *Curr Opin Genet Dev*. 2005; 15:542-547.

[157] Lee SB, Haber DA. Wilms tumor and the WT1 gene. *Exp Cell Res*. 2001; 264:74-99.

[158] Discenza MT, Pelletier J. Insights into the physiological role of WT1 from studies of genetically modified mice. *Physiol Genomics*. 2004; 16:287-300.

[159] Reynolds PA, Smolen GA, Palmer RE, Sgroi D, Yajnik V, Gerald WL, Haber DA. Identification of a DNA-binding site and transcriptional target for the EWS-WT1(+KTS) oncoprotein. *Genes Dev*. 2003; 17:2094-2107.

[160] Gerald WL, Haber DA. The EWS-WT1 gene fusion in desmoplastic small round cell tumor. *Semin Cancer Biol*. 2005; 15:197-205.

[161] Enzinger FM, Shiraki M. Extraskeletal myxoid chondrosarcoma: an analysis of 34 cases. *Hum Pathol*. 1972; 3:421-435.

[162] Chhieng DC, Erlandson RA, Antonescu C, Ladanyi M, Rosai J. Neuroendocrine differentiation in adult soft tissue sarcoma with features of extraskeletal myxoid chondrosarcoma: report of seven cases [abstract]. *Mod Pathol*. 1998; 11:8A.

[163] Meis-Kindblom JM, Bergh P, Gunterberg B, Kindblom LG. Extraskeletal myxoid chondrosarcoma: a reappraisal of its morphological spectrum and prognostic factors based on 117 cases. *Am J Surg Pathol*. 1999; 23:636-650.

[164] Panagopoulos I, Mertens F, Isaksson M, Domanski HA, Brosjö O, Heim S, Bjerkehagen B, Sciot R, Dal Cin P, Fletcher JA, Fletcher CD, Mandahl N. Molecular genetic characterization of the EWS/CHN and RBP56/CHN fusion genes in extraskeletal myxoid chondrosarcoma. *Genes Chromosomes Cancer*. 2002;35 :340-352.

[165] Sjögren H, Wedell B, Meis-Kindblom JM, Kindblom LG, Stenman G. Fusion of the NH2-terminal domain of the basic helix-loop-helix protein TCF12 to TEC in extraskeletal myxoid chondrosarcoma with translocation t(9;15)(q22;q21). *Cancer Res*. 2000; 60:6832-6835.

[166] Hisaoka M, Ishida T, Imamura T, Hashimoto H. TFG is a novel fusion partner of NOR1 in extraskeletal myxoid chondrosarcoma. *Genes Chromosomes Cancer*. 2004; 40:325-328.

[167] Maltais A, Labelle Y. Structure and expression of the mouse gene encoding the orphan nuclear receptor TEC. *DNA Cell Biol*. 2000; 19:121-130.

[168] Scearce LM, Laz TM, Hazel TG, Lau LF, Taub R. RNR-1, a nuclear receptor in the NGFI-N/Nur77 family that is rapidly induced in regenerating liver. *J Biol Chem*. 1993; 268:8855-8861.

[169] Hedvat CV, Irving SG. The isolation and characterization of MINOR, a novel mitogen-induced nuclear orphan receptor. *Mol Endocrinol.* 1995; 9:1692-1700.

[170] Cheng LEC, Chan FKM, Cado D, Winoto A. Functional redundancy of the Nur77 and Nor-1 orphan steroid receptors in T-cell apoptosis. *EMBO J.* 1997; 16:1865-1875.

[171] Ponnio T, Burton Q, Pereira FA, Wu DK, Conneely OM. The nuclear receptor Nor-1 is essential for proliferation of the semicircular canals of the mouse inner ear. *Mol Cell Biol.* 2002; 22:935-945.

[172] DeYoung RA, Baker JC, Cado D, Winoto A. The orphan steroid receptor Nur77 family member Nor-1 is essential for early mouse embryogenesis. *J Biol Chem.* 2003; 278:47104-47109.

[173] Wansa KDSA, Harris JM, Yan G, Ordentlich P, Muscat GEO. The AF-1 domain of NOR-1/NR4A3 mediates trans-activation, coactivator recruitment, and activation by the purine antimetabolite 6-mercaptopurine. *J Biol Chem.* 2003; 278:24776-24790.

[174] Mencinger M, Panagopoulos I, Andreasson P, Lassen C, Mittelman F, Åman P. Characterization and chromosomal mapping of the human TFG gene involved in thyroid carcinoma. *Genomics.* 1997; 41:327-331.

[175] Ohkura N, Yamaguchi H, Tsukada T, Yamaguchi K. The EWS/NR4A3 fusion gene product gains a novel activity affecting pre-mRNA splicing. *J Biol Chem.* 2002; 277:535-543.

[176] Mendenhall WM, Zlotecki RA, Scarborough MT. Dermatofibrosarcoma protuberans. *Cancer.* 2004; 101:2503-2508.

[177] Mentzel T, Beham A, Katenkamp D, Dei Tos AP, Fletcher CDM. Fibrosacomatous ("high-grade") dermatofibrosarcoma protuberans. Clinicopathologic and immunohistochemical study of a series of 41 cases with emphasis on prognostic significance. *Am J Surg Pathol.* 1998; 22:576-587.

[178] Bridge JA, Neff JR, Sandberg AA. Cytogenetic analysis of dermatofibrosarcoma protuberans. *Cancer Genet Cytogenet.* 1990; 49:199-202.

[179] Naeem R, Lux ML, Huang SF, Naber SP, Corson JM, Fletcher JA. Ring chromosomes in dermatofibrosarcoma protuberans are composed of interspersed sequences from chromosome 17 and 22. *Am J Pathol.* 1995; 147:1553-1558.

[180] Pedeutour F, Simon MP, Minolettic F, Barcelo G, Terrier-Lacombe MJ, Combemale P, Sozzi G, Ayraud N, Turc-Carel C. Translocation, t(17;22)(q22;q13), in dermatofibrosarcoma protuberans: a new tumor-associated chromosome rearrangement. *Cytogenet Cell Genet.* 1996; 72:171-174.

[181] Simon MP, Pedeutour F, Silvent N, Grosgeorge J, Minoletti F, Coindre JM, Terrier-Lacombe MJ, Mandahl N, Craber RD, Blin N, Sozzi G, Truc-Carel C, O'Brien KP, Kedra D, Fransson I, Guibaud C, Dumanski JP. Deregulation of the platelet-derived growth factor b-chain gene via fusion with collagen gene COL1A1 in dermatofibrosarcoma protuberans and giant-cell fibroblastoma. *Nat Genet.* 1997; 15: 95-98.

[182] Wang J, Hisaoka M, Shimajiri S, Morimitsu Y, Hashimoto H. Detection of COL1A1-PDGFB fusion transcripts in dermatofibrosarcoma protuberans by reverse transcription-polymerase chain reaction using archival formalin-fixed, paraffin-embedded tissues. *Diagn Mol Pathol.* 1999; 8:113-119.

[183] Wang J, Morimitsu Y, Okamoto S, Hisaoka M, Ishida T, Sheng W, Hashimoto H. COL1A1-PDGFB fusion transcripts in fibrosarcomatous areas of six dermatofibrosarcomas protuberans. *J Mol Diagn*. 2000; 2:47-52.

[184] Sheng WQ, Hashimoto H, Okamoto S, Ishida T, Meis-Kindblom JM, Kindblom LG, Hisaoka M. Expression of COL1A1-PDGFB fusion transcripts in superficial adult fibrosarcoma suggesting a close relationship to dermatofibrosarcoma protuberans. *J Pathol.* 2001; 194:88-94.

[185] Boedtker H, Fuller F, Tate V. The structure of collagen genes. *Int Rev Connect Tissue Res*. 1983; 10:1-63.

[186] Heldin CH, Östman A, Rönnstrand L. Signal transduction via platelet-derived growth factor receptors. *Biochim Biophys Acta*. 1998; 1378: F79-113.

[187] Heldin CH, Rönnstrand L. Growth factor receptors in cell transformation. In: Peters G, Vousden K (eds.), *Frontiers in Molecular Biology: Oncogenes and Tumor Suppressor Genes.* Oxford, Oxford University Press; 1997;55-85.

[188] Sandberg AA, Bridge JA. Updates on the cytogenetics and molecular genetics of bone and soft tissue tumors: dermatofibrosarcoma protuberans and giant cell fibroblastoma. *Cancer Genet Cytogenet.* 2003; 140:1-12.

[189] Shimizu A, O'Brien KP, Sjöblom T, Pietras K, Buchdunger E, Collins VP, Heldin CH, Dumanski JP, Östman A. The dermatofibrosarcoma protuberans-associated collagen type Iα1/platelet-derived growth factor (PDGF) B-chain fusion gene generates a transforming protein that is processed to functional PDGF-BB. *Cancer Res.* 1999; 59:3719-3723.

[190] Simon MP, Navarro M, Roux D, Pouysségur J. Structural and functional analysis of a chimeric protein COL1A1-PDGFB generated by the translocation t(17;22)(q22;q13.1) in dermatofibrosarcoma protuberans (DP). *Oncogene*. 2001; 20:2965-2975.

[191] Sjöblom T, Shimizu A, O'Brien KP, Pietras K, Dal Cin P, Buchdunger E, Dumanski JP, Östman A, Heldin CH. Growth inhibition of dermatofibrosarcoma protuberans tumors by the platelet-derived growth factor receptor antagonist STI571 through induction of apoptosis. *Cancer Res*. 2001; 61:5778-5783.

[192] Greco A, Roccato E, Miranda C, Cleris L, Formelli F, Pierotti MA. Growth-inhibitory effect of STI571 on cells transformed by the COL1A1/PDGFB rearrangement. *Int J Cancer.* 2001; 92:354-360.

[193] Rubin BP, Schuetze SM, Eary JF, Norwood TH, Mirza S, Conrad EU, Bruckner JD. Molecular targeting of platelet-derived growth factor B by imatinib mesylate in a patient with metastatic dermatofibrosarcoma protuberans. *J Clin Oncol*. 2002; 20:3586-3591.

[194] Maki RG, Awan RA, Dixon RH, Jhanwar S, Antonescu CR. Differential sensitivity to imatinib of 2 patients with metastatic sarcoma arising from dermatofibrosarcoma protuberans. *Int J Cancer*. 2002; 100:623-626.

[195] McArthur GA, Demetri GD, van Oosterom A, Heinrich MC, Debiec-Rychter M, Corless CL, Nikolova Z, Dimitrijevic S, Fletcher JA. Molecular and clinical analysis of locally advanced dermatofibrosarcoma protuberans treated with imatinib: Imatinib Target Exploration Consortium Study B2225. *J Clin Oncol*. 2005; 23:866-873.

[196] Chung EB, Enzinger FM. Infantile fibrosarcoma. *Cancer*. 1976; 38:729-739.

[197] Knezevich SR, McFadden DE, Tao W, Lim JF, Sorensen PH. A novel ETV6-NTRK3 gene fusion in congenital fibrosarcoma. *Nat Genet.* 1998;18:184-187.

[198] Rubin BP, Chen CJ, Morgan TW, Xiao S, Grier HE, Kozakewich HP, Perez-Atayde AR, Fletcher JA. Congenital mesoblastic nephroma t(12;15) is associated with ETV6-NTRK3 gene fusion: cytogenetic and molecular relationship to congenital (infantile) fibrosarcoma. *Am J Pathol.* 1998; 153:1451-1458.

[199] Knezevich SR, Garnett MJ, Pysher TJ, Beckwith JB, Grundy PE, Sorensen PH. ETV6-NTRK3 gene fusions and trisomy 11 establish a histogenetic link between mesoblastic nephroma and congenital fibrosarcoma. *Cancer Res.* 1998; 58:5046-5048.

[200] Eguchi M, Eguchi-Ishimae M, Tojo A, Morishita K, Suzuki K, Sato Y, Kudoh S, Tanaka K, Setoyama M, Nagamura F, Asano S, Kamada N. Fusion of ETV6 to neurotrophin-3 receptor TRKC in acute myeloid leukemia with t(12;15)(p13;q25). *Blood.* 1999; 93:1355-1363.

[201] Tognon C, Knezevich SR, Huntsman D, Roskelley CD, Melnyk N, Mathers JA, Becker L, Carneiro F, MacPherson N, Horsman D, Poremba C, Sorensen PH. Expression of the ETV6-NTRK3 gene fusion as a primary event in human secretory breast carcinoma. *Cancer Cell.* 2002;2:367-376.

[202] Lamballe F, Klein R, Barbacid M. trkC, a new member of the trk family of tyrosine protein kinases, is a receptor for neurotrophin-3. *Cell.* 1991; 66:967-979.

[203] Obermeier A, Bradshaw RA, Seedorf K, Choidas A, Schlessinger J, Ullrich A. Neuronal differentiation signals are controlled by nerve growth factor receptor/Trk binding sites for SHC and PLCγ. *EMBO J.* 1994; 13:1585-1590.

[204] Tessarollo L, Tsoulfas P, Martin-Zanca D, Gilbert DJ, Jenkins NA, Copeland NG, Parada LF. trkC, a receptor for neurotrophin-3, is widely expressed in the developing nervous system and in non-neuronal tissues. *Development.* 1993; 118:463-475.

[205] Shelton DL, Sutherland J, Gripp J, Camerato T, Armanini MP, Phillips HS, Carroll K, Spencer SD, Levinson AD. Human trks: molecular cloning, tissue distribution, and expression of extracellular domain immunoadhesins. *J Neurosci.* 1995; 15:477-491.

[206] Donovan MJ, Miranda RC, Kraemer R, McCaffrey TA, Tessarollo L, Mahadeo D, Sharif S, Kaplan DR, Tsoulfas P, Parada L, Toran-Allerand CD, Hajjar DP, Hempstead BL. Neurotrophin and neurotrophin receptors in vascular smooth muscle cells. Regulation of expression in response to injury. *Am J Pathol.* 1995; 147:309-324.

[207] Shibayama E, Koizumi H. Cellular localization of the Trk neurotrophin receptor family in human non-neuronal tissues. *Am J Pathol.* 1996; 148:1807-1818.

[208] Labouyrie E, Dubus P, Groppi A, Mahon FX, Ferrer J, Parrens M, Reiffers J, de Mascarel A, Merlio JP. Expression of neurotrophins and their receptors in human bone marrow. *Am J Pathol.* 1999; 154:405-415.

[209] Hisaoka M, Sheng WQ, Tanaka A, Hashimoto H. Gene expression of TrkC (NTRK3) in human soft tissue tumours. *J Pathol.* 2002; 197:661-667.

[210] Bohlander SK. ETV6: a versatile player in leukemogenesis. *Semin Cancer Biol.* 2005; 15:162-174.

[211] Baens M, Peeters P, Guo C, Aerssens J, Marynen P. Genomic organization of TEL: the human ETS-variant gene 6. *Genome Res.* 1996; 6:404-413.

[212] Wang LC, Swat W, Fujiwara Y, Davidson L, Visvader J, Kuo F, Alt FW, Gilliland DG, Golub TR, Orkin SH. The TEL/ETV6 gene is required specifically for hematopoiesis in the bone marrow. *Genes Dev.* 1998; 12:2392-2402.

[213] Wai DH, Knezevich SR, Lucas T, Jansen B, Kay RJ, Sorensen PHB. The ETV6-NTRK3 gene fusion encodes a chimeric protein tyrosine kinase that transforms NIH3T3 cells. *Oncogene.* 2000; 19:906-915.

[214] Tognon C, Garnett M, Kenward E, Kay R, Morrison K, Sorensen PH. The chimeric protein tyrosine kinase ETV6-NTRK3 requires both Ras-Erk1/2 and PI3-kinase-Akt signaling for fibroblast transformation. *Cancer Res.* 2001; 61:8909-8916.

[215] Morrison KB, Tognon CE, Garnett MJ, Deal C, Sorensen PHB. ETV6-NTRK3 transformation requires insulin-like growth factor 1 receptor signaling and is associated with constitutive IRS-1 tyrosine phosphorylation. *Oncogene.* 2002; 21:5684-5695.

[216] Tognon C, Knezevich SR, Huntsman D, Roskelley CD, Melnyk N, Mathers JA, Becker L, Carneiro F, MacPherson N, Horsman D, Poremba C, Sorensen PH. Expression of the ETV6-NTRK3 gene fusion as a primary event in human secretory breast carcinoma. *Cancer Cell.* 2002; 2:367-376.

[217] Jin W, Kim BC, Tognon C, Lee HJ, Patel S, Lannon CL, Maris JM, Triche TJ, Sorensen PH, Kim SJ. The ETV6-NTRK3 chimeric tyrosine kinase suppresses TGF-β signaling by inactivating the TGF-β type II receptor. *Proc Natl Acad Sci U S A.* 2005; 102:16239-16244.

[218] Coffin CM, Humphery PA, Dehner LP. Extrapulmonary inflammatory myofibroblastic tumor: a clinical and pathological survey. *Semin Diagn Pathol.* 1998; 15:85-101.

[219] Coffin CM, Watterson J, Priest JR, Dehner LP. Extrapulmonary inflammatory myofibroblastic tumor (inflammatory pseudotumor). A clinicopathologic and immunohistochemical study of 84 cases. *Am J Surg Pathol.* 1995; 19:859-872.

[220] Lawrence B, Perez-Atayde A, Hibbard MK, Rubin BP, Dal Cin P, Pinkus JL, Pinkus GS, Xiao S, Yi ES, Fletcher CD, Fletcher JA. TPM3-ALK and TPM4-ALK oncogenes in inflammatory myofibroblastic tumors. *Am J Pathol.* 2000; 157:377-384.

[221] Bridge JA, Kanamori M, Ma Z, Pickering D, Hill DA, Lydiatt W, Lui MY, Colleoni GW, Antonescu CR, Ladanyi M, Morris SW. Fusion of the ALK gene to the clathrin heavy chain gene, CLTC, in inflammatory myofibroblastic tumor. *Am J Pathol.* 2001; 159:411-415.

[222] Cools J, Wlodarska I, Somers R, Mentens N, Pedeutour F, Maes B, De Wolf-Peeters C, Pauwels P, Hagemeijer A, Marynen P. Identification of novel fusion partners of ALK, the anaplastic lymphoma kinase, in anaplastic large-cell lymphoma and inflammatory myofibroblastic tumor. *Genes Chromosomes Cancer.* 2002; 34:354-362.

[223] Ma Z, Hill DA, Collins MH, Morris SW, Sumegi J, Zhou M, Zuppan C, Bridge JA. Fusion of ALK to the Ran-binding protein 2 (RANBP2) gene in inflammatory myofibroblastic tumor. *Genes Chromosomes Cancer.* 2003; 37:98-105.

[224] Debiec-Rychter M, Marynen P, Hagemeijer A, Pauwels P. ALK-ATIC fusion in urinary bladder inflammatory myofibroblastic tumor. *Genes Chromosomes Cancer.* 2003; 38:187-190.

[225] Panagopoulos I, Nilsson T, Domanski HA, Isaksson M, Lindblom P, Mertens F, Mandahl N. Fusion of the SEC31L1 and ALK genes in an inflammatory myofibroblastic tumor. *Int J Cancer.* 2006; 118:1181-1186.

[226] Meech SJ, McGavran L, Odom LF, Liang X, Meltesen L, Gump J, Wei Q, Carlsen S, Hunger SP. Unusual childhood extramedullary hematologic malignancy with natural

killer cell properties that contains tropomyosin 4--anaplastic lymphoma kinase gene fusion. *Blood.* 2001; 98:1209-1216.

[227] Wasik MA. Expression of anaplastic lymphoma kinase in non-Hodgkin's lymphomas and other malignant neoplasms. Biological, diagnostic, and clinical implications. *Am J Clin Pathol.* 2002; 118 Suppl:S81-92.

[228] Gascoyne RD, Lamant L, Martin-Subero JI, Lestou VS, Harris NL, Muller-Hermelink HK, Seymour JF, Campbell LJ, Horsman DE, Auvigne I, Espinos E, Siebert R, Delsol G. ALK-positive diffuse large B-cell lymphoma is associated with Clathrin-ALK rearrangements: report of 6 cases. *Blood.* 2003; 102:2568-2573.

[229] Maes B, Vanhentenrijk V, Wlodarska I, Cools J, Peeters B, Marynen P, de Wolf-Peeters C. The NPM-ALK and the ATIC-ALK fusion genes can be detected in non-neoplastic cells. *Am J Pathol.* 2001;158:2185-2193.

[230] Duyster J, Bai RY, Morris SW. Translocation involving anaplastic lymphoma kinase (ALK). Oncogene. 2001; 20:5623-5637.

[231] Iwahara T, Fujimoto J, Wen D, Cupples R, Bucay N, Arakawa T, Mori S, Ratzkin B, Yamamoto T. Molecular characterization of ALK, a receptor tyrosine kinase expressed specifically in the nervous system. *Oncogene.* 1997; 14:439-449.

[232] Stoica GE, Kuo A, Aigner A, Sunitha I, Souttou B, Malerczyk C, Caughey DJ, Wen D, Karavanov A, Riegel AT, Wellstein A. Identification of anaplastic lymphoma kinase as a receptor for the growth factor pleiotrophin. *J Biol Chem.* 2001; 276:16772-16779.

[233] Stoica GE, Kuo A, Powers C, Bowden ET, Sale EB, Riegel AT, Wellstein A: Midkine binds to anaplastic lymphoma kinase (ALK) and acts as a growth factor for different cell types. *J Biol Chem.* 2002; 277:35990-35998.

[234] Lees-Miller JP, Helfman DM. The molecular basis for tropomyosin isoform diversity. *Bioessays.* 1991; 13:429-437.

[235] Sugita T, Aya H, Ueno M, Ishizuka T, Kawashima K. Characterization of molecularly cloned human 5-aminoimidazole-4-carboxamide ribonucleotide transformylase. *J Biochem (Tokyo).* 1997; 122:309-313.

[236] Cruzen ME, Arfin SM. Nucleotide and deduced amino acid sequence of human cysteinyl-tRNA synthetase. *DNA Seq.* 1994; 4:243-248.

[237] Goldstein JL, Anderson RG, Brown MS. Coated pits, coated vesicles, and receptor-mediated endocytosis. *Nature.* 1979; 279:679-685.

[238] Tang BL, Zhang T, Low DY, Wong ET, Horstmann H, Hong W. Mammalian homologues of yeast sec31p. An ubiquitously expressed form is localized to endoplasmic reticulum (ER) exit sites and is essential for ER-Golgi transport. *J Biol Chem.* 2000; 275:13597-13604.

[239] Graadt van Roggen JF, Mooi WJ, Hogendoorn PCW. Clear cell sarcoma of tendons and aponeuroses (malignant melanoma of soft parts) and cutaneous melanoma: exploring the histogenetic relationship between these two clinicopathological entities. *J Pathol.* 1998; 186:3-7.

[240] Jacobs IA, Chang CK, Guzman G, Salti GI. Clear cell sarcoma: an institutional review. *Am Surg.* 2004; 70:300-303.

[241] Patel RM, Downs-Kelly E, Weiss SW, Folpe AL, Tubbs RR, Tuthill RJ, Goldblum JR, Skacel M. Dual-color, break-apart fluorescence in situ hybridization for EWS gene rearrangement distinguishes clear cell sarcoma of soft tissue from malignant melanoma. *Mod Pathol.* 2005; 18:1585-1590.

[242] Hallor KH, Mertens F, Jin Y, Meis-Kindblom JM, Kindblom LG, Behrendtz M, Kalén A, Mandahl N, Panagopoulos I. Fusion of the EWSR1 and ATF1 genes without expression of the MITF-M transcript in angiomatoid fibrous histiocytoma. *Genes Chromosomes Cancer.* 2005; 44:97-102.

[243] Somers GR, Viero S, Nathan PC, Teshima I, Pereira C, Zielenska M. Association of the t(12;22)(q13;q12) EWS/ATF1 rearrangement with polyphenotypic round cell sarcoma of bone: a case report. *Am J Surg Pathol.* 2005; 29:1673-1679.

[244] Lee KA, Masson N. Transcriptional regulation by CREB and its relatives. *Biochim Biophys Acta.* 1993; 1174:221-233.

[245] Antonescu CR, Nafa K, Dal Cin P, Ladanyi M. EWS-CREB1: a recurrent variant fusion in clear cell sarcoma associated with gastrointestinal location and absence of melanocytic differentiation (abstract). *Mod Pathol.* 2006; 19 (suppl 1):8A

[246] Panagopoulos I, Mertens F, Debiec-Rychter M, Isaksson M, Limon J, Kardas I, Domanski HA, Sciot R, Perek D, Crnalic S, Larsson O, Mandahl N. Molecular genetic characterization of the EWS/ATF1 fusion gene in clear cell sarcoma of tendons and aponeuroses. *Int J Cancer.* 2002; 99:560-567.

[247] Brown AD, Lopez-Terrada D, Denny C, Lee KA. Promoters containing ATF-binding sites are de-regulated in cells that express the EWS/ATF1 oncogene. *Oncogene.* 1995; 10:1749-1756.

[248] Fujimura Y, Ohno T, Siddique H, Lee L, Rao VN, Reddy ES. The EWS-ATF-1 gene involved in malignant melanoma of soft parts with t(12;22) chromosome translocation, encodes a constitutive transcriptional activator. *Oncogene.* 1996;12:159-167.

[249] Li KK, Lee KA. MMSP tumor cells expressing the EWS/ATF1 oncogene do not support cAMP-inducible transcription. *Oncogene.* 1998; 16:1325-1331.

[250] Fujimura Y, Siddique H, Lee L, Rao VN, Reddy ES. EWS-ATF-1 chimeric protein in soft tissue clear cell sarcoma associates with CREB-binding protein and interferes with p53-mediated trans-activation function. *Oncogene.* 2001; 20:6653-6659.

[251] Jishage M, Fujino T, Yamazaki Y, Kuroda H, Nakamura T. Identification of target genes for EWS/ATF-1 chimeric transcription factor. *Oncogene.* 2003; 22:41-49.

[252] Schaefer KL, Brachwitz K, Wai DH, Braun Y, Diallo R, Korsching E, Eisenacher M, Voss R, Van Valen F, Baer C, Selle B, Spahn L, Liao SK, Lee KA, Hogendoorn PC, Reifenberger G, Gabbert HE, Poremba C. Expression profiling of t(12;22) positive clear cell sarcoma of soft tissue cell lines reveals characteristic up-regulation of potential new marker genes including ERBB3. *Cancer Res.* 2004; 64:3395-3405.

[253] Mathew T. Evidence supporting neural crest origin of an alveolar soft part sarcoma. An ultrastructural study. *Cancer.* 1982; 50:507-514.

[254] Sciot R, Dal Cin P, De Vos R, Van Damme B, De Wever I, Van den Berghe H, Desmet VJ. Alveolar soft-part sarcoma: evidence for its myogenic origin and for the involvement of 17q25. *Histopathology.* 1993; 23:439-444.

[255] Lieberman PH, Brennan MF, Kimmel M, Erlandson RA, Garin-Chesa P, Flehinger BY. Alveolar soft-part sarcoma. A clinico-pathologic study of half a century. *Cancer.* 1989; 63:1-13.

[256] Sandberg A, Bridge J. Updates on the cytogenetics and molecular genetics of bone and soft tissue tumors: alveolar soft part sarcoma. *Cancer Genet Cytogenet.* 2002; 136:1-9.

[257] Argani P, Antonescu CR, Illei PB, Lui MY, Timmons CF, Newbury R, Reuter VE, Garvin AJ, Perez-Atayde AR, Fletcher JA, Beckwith JB, Bridge JA, Ladanyi M.

Primary renal neoplasms with the ASPL-TFE3 gene fusion of alveolar soft part sarcoma: a distinctive tumor entity previously included among renal cell carcinomas of children and adolescents. *Am J Pathol.* 2001; 159:179-192.

[258] Huang HY, Lui MY, Ladanyi M. Nonrandom cell-cycle timing of a somatic chromosomal translocation: The t(X;17) of alveolar soft-part sarcoma occurs in G2. *Genes Chromosomes Cancer.* 2005; 44:170-176.

[259] Ladanyi M, Lui MY, Antonescu CR, Krause-Boehm A, Meindl A, Argani P, Healey JH, Ueda T, Yoshikawa H, Meloni-Ehrig A, Sorensen PH, Mertens F, Mandahl N, van den Berghe H, Sciot R, Cin PD, Bridge J. The der(17)t(X;17)(p11;q25) of human alveolar soft part sarcoma fuses the TFE3 transcription factor gene to ASPL, a novel gene at 17q25. *Oncogene.* 2001; 20:48-57.

[260] Buchberger A. From UBA to UBX: new words in the ubiquitin vocabulary. *Trends Cell Biol.* 2002; 12:216-221.

[261] Bogan JS, Hendon N, McKee AE, Tsao TS, Lodish HF. Functional cloning of TUG as a regulator of GLUT4 glucose transporter trafficking. *Nature.* 2003; 425:727-733.

[262] Wong SC, Moffat MA, Coker GT, Merlie JP, O'Malley KL. The 3' flanking region of the human tyrosine hydroxylase gene directs reporter gene expression in peripheral neuroendocrine tissues. *J Neurochem.* 1995; 65:23-31.

[263] Verastegui C, Bertolotto C, Bille K, Abbe P, Ortonne JP, Ballotti R. TFE3, a transcription factor homologous to microphthalmia, is a potential transcriptional activator of tyrosinase and TyrpI genes. *Mol Endocrinol.* 2000; 14:449-456.

[264] Hershey CL, Fisher DE. Mitf and Tfe3: members of a b-HLH-ZIP transcription factor family essential for osteoclast development and function. *Bone.* 2004; 34:689-696.

[265] Hua X, Miller ZA, Benchabane H, Wrana JL, Lodish HF. Synergism between transcription factors TFE3 and Smad3 in transforming growth factor-beta-induced transcription of the Smad7 gene. *J Biol Chem.* 2000; 275:33205-33208.

[266] Argani P, Ladanyi M. Recent advances in pediatric renal neoplasia. *Adv Anat Pathol.* 2003; 10:243-260.

[267] Argani P, Lal P, Hutchinson B, Lui MY, Reuter VE, Ladanyi M. Aberrant nuclear immunoreactivity for TFE3 in neoplasms with TFE3 gene fusions: a sensitive and specific immunohistochemical assay. *Am J Surg Pathol.* 2003; 27:750-761.

[268] Evans HL. Low-grade fibromyxoid sarcoma: a report of two metastasizing neoplasms having a deceptively benign appearance. *Am J Clin Pathol.* 1987; 88:615-619.

[269] Folpe AL, Lane KL, Paull G, Weiss SW. Low-grade fibromyxoid sarcoma and hyalinizing spindle cell tumor with giant rosettes. A clinicopathologic study of 73 cases supporting their identity and assessing the impact of high-grade areas. *Am J Surg Pathol.* 2000; 24:1353-1360.

[270] Storlazzi CT, Mertens F, Nascimento A, Isaksson M, Wejde J, Brosjö O, Mandahl N, Panagopoulos I. Fusion of the FUS and BBF2H7 genes in low grade fibromyxoid sarcoma. *Hum Mol Genet.* 2003; 12:2349-2358.

[271] Reid R, de Silva MV, Paterson L, Ryan E, Fisher C. Low-grade fibromyxoid sarcoma and hyalinizing spindle cell tumor with giant rosettes share a common t(7;16)(q34;p11) translocation. *Am J Surg Pathol.* 2003; 27:1229-1236.

[272] Panagopoulos I, Storlazzi CT, Fletcher CD, Fletcher JA, Nascimento A, Domanski HA, Wejde J, Brosjö O, Rydholm A, Isaksson M, Mandahl N, Mertens F. The chimeric

FUS/CREB3L2 gene is specific for low-grade fibromyxoid sarcoma. *Genes Chromosomes Cancer.* 2004; 40:218-228.

[273] Mertens F, Fletcher CD, Antonescu CR, Coindre JM, Colecchia M, Domanski HA, Downs-Kelly E, Fisher C, Goldblum JR, Guillou L, Reid R, Rosai J, Sciot R, Mandahl N, Panagopoulos I. Clinicopathologic and molecular genetic characterization of low-grade fibromyxoid sarcoma, and cloning of a novel FUS/CREB3L1 fusion gene. *Lab Invest.* 2005; 85(3):408-415.

[274] Matsuyama A, Hisaoka M, Shimajiri S, Hayashi T, Imamura T, Ishida T, Fukunaga M, Fukuhara T, Minato H, Nakajima T, Yonezawa S, Kuroda M, Yamasaki F, Toyoshima S, Hashimoto H. Molecular detection of FUS-CREB3L2 fusion transcripts in low grade fibromyxoid sarcoma using formalin-fixed, paraffin-embedded tissue specimens. *Am J Surg Pathol.* 2006 (in press).

[275] Vinson C, Myakishev M, Acharya A, Mir AA, Moll JR, Bonovich M. Classification of human B-ZIP proteins based on dimerization properties. *Mol Cell Biol.* 2002; 22:6321-6335.

[276] Kondo S, Murakami T, Tatsumi K, Ogata M, Kanemoto S, Otori K, Iseki K, Wanaka A, Imaizumi K. OASIS, a CREB/ATF-family member, modulates UPR signalling in astrocytes. *Nat Cell Biol.* 2005; 7:186-194.

[277] Enzinger FM. Angiomatoid malignant fibrous histiocytoma. A distinct fibrohistiocytic tumor of children and young adults simulating a vascular neoplasm. *Cancer.* 1979; 44:2147-2157.

[278] Costa MJ, Weiss SW. Angiomatoid malignant fibrous histiocytoma. A follow-up study of 108 cases with evaluation of possible histologic predictors of outcome. *Am J Surg Pathol.* 1990; 14:1126-1132.

[279] Fanburg-Smith JC, Miettinen M. Angiomatoid "malignant" fibrous histiocytoma: a clinicopathologic study of 158 cases and further exploration of the myoid phenotype. *Hum Pathol.* 1999; 30:1336-1343.

[280] Waters BL, Panagopoulos I, Allen EF. Genetic characterization of angiomatoid fibrous histiocytoma identifies fusion of the FUS and ATF-1 genes induced by a chromosomal translocation involving bands 12q13 and 16p11. *Cancer Genet Cytogenet.* 2000; 121:109-116.

[281] Raddaouri E, Donner LR, Panagopoulos I. Fusion of the FUS and ATF1 genes in a large, deep-seated angiomatoid fibrous histiocytoma. *Diagn Mol Pathol.* 2002; 11: 157-162.

[282] Antonescu CR. The role of genetic testing in soft tissue sarcoma. *Histopathology.* 2006; 48:13-21.

[283] Panagopoulos I, Lassen C, Isaksson M, Mitelman F, Mandahl N, Åman P. Characteristic sequence motifs at the breakpoints of the hybrid genes FUS/CHOP, EWS/CHOP and FUS/ERG in myxoid liposarcoma and acute myeloid leukemia. *Oncogene.* 1997; 15:1357-62.

[284] Parada LA, Misteli T. Chromosome positioning in the interphase nucleus. *Trends Cell Biol* 2002; 12:425-432.

[285] Kuroda M, Tanabe H, Yoshida K, Oikawa K, Saito A, Kiyuna T, Mizusawa H, Mukai K. Alteration of chromosome positioning during adipocyte differentiation. *J Cell Sci.* 2004; 117:5897-5903.

[286] Trinei M, Lanfrancone L, Campo E, Pulford K, Mason DY, Pelicci PG, Falini B. A new variant anaplastic lymphoma kinase (ALK)-fusion protein (ATIC-ALK) in a case of ALK-positive anaplastic large cell lymphoma. *Cancer Res.* 2000; 60:793-798.

[287] Lamant L, Dastugue N, Pulford K, Delsol G, Mariame B. A new fusion gene TPM3-ALK in anaplastic large cell lymphoma created by a (1;2) (q25;p23) translocation. *Blood.* 1999; 93:3088-3095.

[288] Touriol C, Greenland C, Lamant L, Pulford K, Bernard F, Rousset T, Mason DY, Delsol G: Further demonstration of the diversity of chromosomal changes involving 2p23 in ALK-positive lymphoma: 2 cases expressing ALK kinase fused to CLTCL (clathrin chain polypeptide-like). *Blood* 2000; 95:3204-3207.

[289] Hu-Lieskovan S, Zhang J, Wu L, Shimada H, Schofield DE, Triche TJ. EWS-FLI1 fusion protein up-regulates critical genes in neural crest development and is responsible for the observed phenotype of Ewing's family of tumors. *Cancer Res.* 2005; 65:4633-4644.

[290] Keller C, Capecchi MR. New genetic tactics to model alveolar rhabdomyosarcoma in the mouse. *Cancer Res.* 2005; 65:7530-7532.

[291] Åman P. Fusion genes in solid tumors. *Semin Cancer Biol.* 1999; 9:303-318.

[292] Thompson AD, Teitell MA, Arvand A, Denny CT. Divergent Ewing's sarcoma EWS/ETS fusions confer a common tumorigenic phenotype on NIH3T3 cells. *Oncogene.* 1999; 18:5506-5513.

[293] Davis RJ, Barr FG. Fusion genes resulting from alternative chromosomal translocations are overexpressed by gene-specific mechanisms in alveolar rhabdomyosarcoma. *Proc Natl Acad Sci USA.* 1997; 94:8047-8051.

[294] Armstrong F, Duplantier MM, Trempat P, Hieblot C, Lamant L, Espinos E, Racaud-Sultan C, Allouche M, Campo E, Delsol G, Touriol C. Differential effects of X-ALK fusion proteins on proliferation, transformation, and invasion properties of NIH3T3 cells. *Oncogene.* 2004; 23:6071-6082

In: Genetic Translocations and Other Chromosome Aberrations
Editor: Guy T. Leyden, pp. 117-144
ISBN: 978-1-60456-344-3
© 2008 Nova Science Publishers, Inc.

Chapter 3

Translocations as Tools for Understanding Control of Sex Determination in the Silkworm, *Bombyx mori*

Hiroaki Abe[1,a] and Tsuguru Fujii[2,b]

[1]Department of Biological Production, Faculty of Agriculture, Tokyo University of Agriculture and Technology, Saiwai-cho 3-5-8, 183-8509, Fuchu, Tokyo, Japan;
[2]Department of Agricultural and Environmental Biology, Graduate School of Agricultural and Life Sciences, The University of Tokyo, Yayoi 1-1-1, 113-8657, Bunkyo-ku, Tokyo, Japan;

Abstract

The sex chromosomes of the silkworm, *Bombyx mori*, are designated ZW (XY) for females and ZZ (XX) for males. The female mode of development is determined by the presence of a single W chromosome. Therefore, it is presumed that the female-determining gene (*Fem*) is present on the W chromosome. This chromosome does not recombine with either the Z chromosome or autosomes and as yet, no genes for morphological traits have been mapped to the W chromosome. The silkworm has been reared in Japan for more than 1,700 years. As heterosis helps to ensure stable cocoon production, almost all silkworm strains reared at sericultural farms are F_1 hybrids, a practice that has been in place since about 1910. Therefore, discriminating the sexes of would-be silkworm parents for the purpose of controlled hybridization is very important. This was traditionally accomplished by inspecting the post-ventral surface of final instar larvae. However, the method was problematic and inconvenient. For this reason, identification of a translocation to the W chromosome that carries one or more dominant markers would be helpful. Several researchers undertook the challenge of inducing translocation of an autosomal fragment containing dominant alleles of various genes (including genes that control egg color, larval marking, and cocoon color) to the W

[a] E-mail address: wfem@cc.tuat.ac.jp. Phone and Fax: +81-42-367-5848, (Author for correspondence)
[b] E-mail address: fujii@ss.ab.a.u-tokyo.ac.jp

chromosome using X-ray or γ-ray irradiation, and were successful. One of the W-translocation chromosomes, the sex-limited Zebra-W strain or T(W;3)*Ze*, was thought to be the product of reciprocal translocation between W and chromosome 3. Thus, the W chromosome in this strain should contain deletions; however this cannot be confirmed using conventional genetic methods. To further test for the presence of a deletion, we compared the T(W;3)*Ze* chromosome to the parental W chromosome using 12 W-specific random amplified polymorphic DNA (RAPD) markers. The T(W;3)*Ze* chromosome lacked two W-specific RAPD markers. These results strongly indicate that the region containing these two W-specific RAPD markers was deleted. Additionally, we investigated T(W;10)+$^{w-2}$ (the sex-limited Black-egg-W strain) and T(W;2)*Y* (the sex-limited Yellow-cocoon-W strain). The T(W;10)+$^{w-2}$ chromosome lacks one W-specific RAPD marker and the T(W;2)*Y* chromosome lacks 11 W-specific RAPD markers. These results indicate that the W chromosomes of sex-limited strains are not the results of simple fusion of an autosomal fragment to the end of an unaltered W chromosome but rather, they are the products of reciprocal translocation accompanied by deletion. Furthermore, we investigated a female-killing chromosome derived from the translocation-carrying W chromosome and found that this chromosome was generated by translocation between the W and Z chromosomes. Our analysis of W chromosome variants helped to elucidate the relative positions of molecular markers and the putative *Fem* gene(s).

Introduction

The silkworm, *Bombyx mori,* belongs to the family Bombycidae in the order Insecta. Sericulture of *B. mori* dates back about 4,500 years. It is generally thought that sericulture began in ancient times on the Asian continent. The silkworm has been completely domesticated and is no longer found in the wild. However, a presumed ancestral species, *Bombyx mandarina,* can still be found in nature.

Silkworms have been reared in Japan for more than 1,700 years. Since the latter half of the 19th century, raw silk has been one of the most important export goods of Japan. Although sericulture in Japan has at present decreased to its lowest level in recent history, the silkworm is a convenient experimental model system for the following reasons:

(1) Rearing silkworms is very easy. Silkworms do not escape from the rearing bed, even in the absence of bait. The silkworm has lost the ability to look for bait by itself. The adult moths cannot fly. Controlling pairings between females and males is easy.
(2) Useful marker strains exist. That is, strains with mutations that result in heritable variations, many of which are observable at larval stages, have been isolated. Variant traits include egg color, larval markings, body color, body shape, blood color and cocoon color [1, 2].
(3) A single female moth can lay a large number of eggs (from 200 to 500 eggs in her lifetime).

The silkworm has been used in genetic studies since the birth of genetics in the early 1900s, primarily in Japan. The chromosomes of the silkworm are small, holocentric, and numerous (2n=56). Thus, cytogenetic analysis of the silkworm is difficult in comparison with *Drosophila*. Moreover, for the silkworm, cytogenetic detection of chromosomal translocations under the microscope is very difficult. However, stable inheritance of translocations that bring along or cause observable phenotypes can be used to monitor changes in chromosome structure via conventional linkage studies.

History of Isolation and Analysis of Translocations between W and Autosomes

In this section, we discuss the history of study of the W chromosome in *B. mori*, which provides a foundation for understanding W chromosome translocations, which is described later.

In general, Japanese sericulture was stable in the spring season in Japan but became unstable in summer and autumn, both because the quality of mulberry leaves deteriorates after the spring and because of unfavorable weather conditions (namely, high temperature and humidity). Therefore, isolation of more vigorous silkworm strains for sericulture during the summer and autumn seasons became a goal to be reached. In 1906, Toyama, one of the founders of silkworm genetics, rediscovered Mendel's laws in a study of inheritance of cocoon color. Then, he recognized heterosis in hybrids between Japanese and Thai strains (Toyama, 1906 cited in Tazima, 1964 [3]). Following his advice, the Japanese government decided to encourage the rearing of hybrid silkworms and began distributing hybrid eggs in 1914. Thereafter, to take advantage of heterosis, almost all silkworm strains reared at Japanese sericultural farms have been F_1 hybrids. At the same time, determining the sex of silkworms for hybridization became very important to silkworm egg producers because for the production of F_1 eggs, males and females must be separated at or before the cocoon stage. In the absence of separation, sibling emerged moths copulate mutually and the females cannot be used for production of F_1 hybrid eggs. Traditionally, separation was accomplished by inspecting Ishiwata's germal discs in the female and Herald's gland in the male on the post-ventral surface on the 5th (final) instar larvae. However, use of this method was problematic and inconvenient, and was usually carried out by a professional 'discriminator'. Therefore, development of a convenient method of discriminating males from females that even an amateur could do was desired.

Tanaka (1916) determined the chromosomal makeup of the silkworm in terms of sex linked (Z-linked) inheritance; namely, that females are ZW (equivalent to XY) and males are ZZ (XX) [4]. However, at that time, it was not clear which of the two chromosomes plays the principal role in sex determination. Subsequently, Hasimoto (1933) obtained a very interesting result related to the W chromosome. He found that the W chromosome harbors a strong positive female-determining gene that directs sex determination irrespective of the number of autosomes or Z chromosomes present, as polyploid individuals, such as ZZW and ZZWW individuals, are phenotypically normal females. Therefore, Hasimoto (1933) postulated that in *B. mori*, the female mode of development is determined by the presence of a single W chromosome [5]. Based on these studies, the existence and function of the W chromosome was almost certain. However, as mentioned above, there were no methods at that time to identify the W chromosome cytologically. Moreover, it was very difficult to prove the existence of W by genetic experimentation, as despite the fact that many morphological gene loci had been found on the Z chromosome, no genes for morphological traits have been found on the W chromosome. Furthermore, recombination or 'crossing over' is restricted to males in *B. mori*.

Tazima proposed that "If we can translocate any one of the autosomes which carry dominant genes for noticeable traits to the W chromosome, we are able to discriminate the female and male by these traits easily. Moreover, we can verify the Hasimoto's view easily and conclude that the femaleness is determined by the presence or absence of the W chromosome in *B. mori*." In this way, the effort to isolate translocations between the W and autosomes—that is, isolation of the "sex-limited strains" —had begun.

p^{Sa} $+^{p}$ W Chromosome (T(W;2) p^{Sa} $+^{p}$ Chromosome)

In 1941, in the course of investigating chromosomal aberrations induced by X-rays, Tazima discovered a translocation between W and chromosome 2 that marked the W chromosome with two different alleles of a gene that are recognizable in the larva ($+^{p}$ and p^{Sa}= normal larval marking and sable marking). In only one of the batches isolated in the study, the sex ratio was distorted, with all of the females displaying $+^{p}$ p^{Sa} traits and all of the males, *p* (plain). There were no $+^{p}$ p^{Sa} males. Tazima had already obtained a strain in which the two copies of chromosome 2 were attached (marked by $+^{p}$ and p^{Sa}, respectively). He concluded that the attached chromosome in the sex-seggregating strain (marked by $+^{p}$ and p^{Sa}) had attached to the W chromosome. Moreover, he further hypothesized that this W chromosome could be used as a marker for sex separation of silkworm larvae. Tazima tested his assertion by crossing this female with a *p* male and observing the pattern of segregation in the next generation. Tazima named the strain with this chromosome, which we now refer to technically as (T(W;2)p^{Sa} $+^{p}$), the W-PSa strain [6].

Tazima next tested if it was appropriate to use the T(W;2)p^{Sa} $+^{p}$ for commercial breeding by determining the growth curves for T(W;2)p^{Sa} $+^{p}$ larvae in the final instar. In normal strains, the growth curve for females is 20% higher than that for males. However, in the W-PSa strain, the growth curve for females was lower than that for males. Tazima reasoned that the cause of this handicap is the extra chromosome 2 translocated to the W chromosome

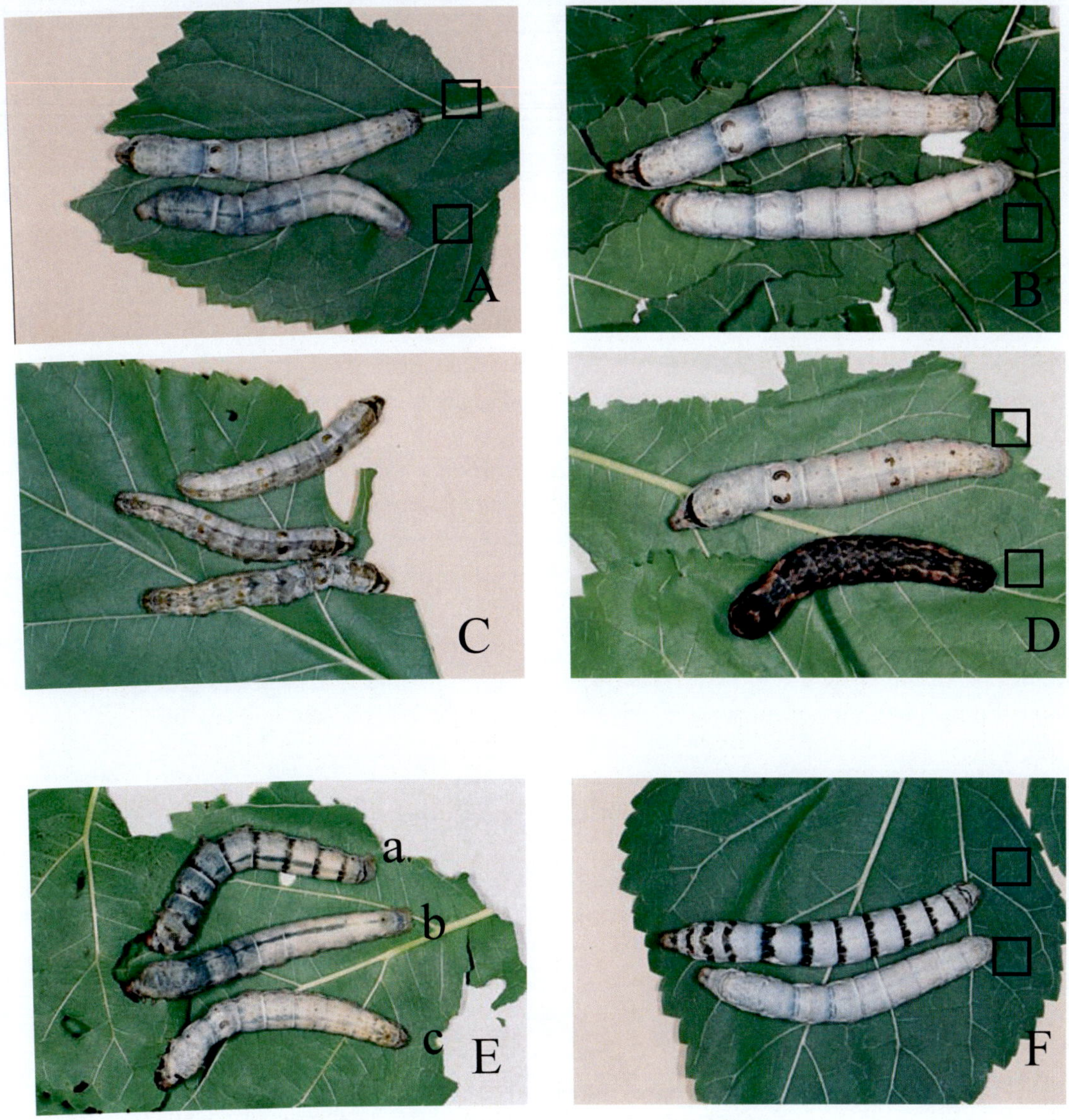

Figure 1. Visible markers translocated from autosomes make it possible to distinguish *B. mori* males from females in larval stages. (A) Fifth instar larvae of the sex-limited strain carrying the p^{Sa} $+^{p}$ W $+^{od}$ chromosome. The female has normal and sable skin due to the genotype $od/$ p^{Sa} $+^{p}$ W $+^{od}$, p/p, whereas the male has translucent and plain skin due to the genotype od/od, p/p. (B) Sex-limited W-P strain (Chinese137). The female has normal larval marking (Z/T(W;2)$+^{p}$, p/p). whereas the male has white, plain skin (Z/Z, p/p). (C) Mosaic larvae. The female region has brown skin (Z/T(W;2)p^{Sa}, $+^{p}/+^{p}$) and the male region has white skin (Z/Z, $+^{p}/+^{p}$). (D) Sex-limited W-B strain. The female has black skin (Z/T(W;2)p^{B}, $+^{p}/+^{p}$), whereas the male has white normal larval marking skin (Z/Z, $+^{p}/+^{p}$). (E) Fifth-instar larvae of the Z101 strain. (a): Translucent and zebra marking female (Z(*od*)/ T(W;3)*Ze*, p/p). (b): Translucent and plain skin male (Z(*od*)/Z(*od*), p/p). (c): Normal skin and sable male (Z (*od*) /Df(p^{Sa} $+^{p}$ W $+^{od}$), p/p). (F) Sex-limited Zebra-W strain. The female has zebra marking (Z/T(W;3)*Ze*, p/p) whereas the male lacks markings (Z/Z, p/p).

(partial triploid status). Therefore, he planned to isolate a strain in which the excess part of the translocated chromosome had been removed by irradiation with X-rays. In the course of the experiment, several new larval marking females (W-P (T(W;2)$+^{p}$), W-Sa (T(W;2)p^{Sa} and W-B (T(W;2)p^{B}) were isolated (Fig. 1B, C and D). These processes are described in detail in Tazima (2001) [7]. Tazima worked toward the practical use of the W-translocation strain for many years. In 1976, selection of a strain appropriate for commercial breeding was finally completed [7].

p^{Sa} $+^{p}$ W $+^{od}$ Chromosome

After exposing female pupae to 40℃ for 24 hours, Tazima (1944, 1948) obtained a complex translocation, p^{Sa} $+^{p}$ W $+^{od}$, which was presumably the product of a connection between the short fragment of the Z chromosome that includes the $+^{od}$ gene (*od*; distinct translucent, 1-49.6) and the p^{Sa} $+^{p}$ W chromosome (Fig. 1A). From the results of a dissociation experiment with the p^{Sa} $+^{p}$ W $+^{od}$ chromosome, Tazima (1944, 1948, 1964) reached the conclusion that the chromosome 2 fragment containing p^{Sa} $+^{p}$ was translocated to the left end of the W chromosome, whereas the Z chromosome fragment containing the $+^{od}$ gene was translocated to the right end of W via an exceptional crossover between Z and the p^{Sa} $+^{p}$ W chromosome (right-end model) [3, 8, 9]. However, Tazima did not exclude the possibility that the Z chromosome fragment containing the $+^{od}$ gene had translocated to the end of the chromosome 2 fragment and attached to the left end of W (left-end model). As will be mentioned later, our recent results appear to support the left-end model [10].

Df(p^{Sa} $+^{p}$ W $+^{od}$)*Fem* Chromosome (Female-Killing Chromosome)

In a series of dissociation experiments using the p^{Sa} $+^{p}$ W $+^{od}$ chromosome, Tazima (1948) found an exceptional animal that was phenotypically male and marked with $+^{p}$, p^{Sa}, and $+^{od}$ among the progeny of an X-ray irradiated Z/p^{Sa} $+^{p}$ W $+^{od}$ female [9]. Tazima (1952) presumed that this exceptional male had a mutant W chromosome that was generated by deletion of the female-determining region of the p^{Sa} $+^{p}$ W $+^{od}$ chromosome [11]. For convenience, the mutant W chromosome was designated Df(p^{Sa} $+^{p}$ W $+^{od}$)*Fem* [10]. Tazima (1952) showed that a female with Df(p^{Sa} $+^{p}$ W $+^{od}$)*Fem* dies during embryogenesis whereas a male with Df(p^{Sa} $+^{p}$ W $+^{od}$)*Fem* exhibits normal viability. Based on genetic analyses, it was presumed that Df(p^{Sa} $+^{p}$ W $+^{od}$)*Fem* is a female-killing chromosome and that female killing is independent of Z [11].

Recently, Fujii et al. (2006) analyzed the genetic behavior of the Df(p^{Sa} $+^{p}$ W $+^{od}$)*Fem* chromosome during male meiosis using phenotypic and molecular markers (Z101 strain)(Fig. 1E). In this analysis, it was revealed that Df(p^{Sa} $+^{p}$ W $+^{od}$)*Fem* is connected to the deleted Z chromosome and that this fused chromosome behaves as a Z chromosome during male meiosis (Fig. 3). Furthermore, a ZZW-type triploid female with Df(p^{Sa} $+^{p}$ W $+^{od}$)*Fem* is viable whereas a ZW-type diploid female with Df(p^{Sa} $+^{p}$ W $+^{od}$)*Fem* dies during embryogenesis [10]. Therefore, we conclude that the Df(p^{Sa} $+^{p}$ W $+^{od}$)*Fem* chromosome does not have a female-killing factor but rather, a partial deletion of the Z chromosome causes the death of ZW-type diploid females with Df(p^{Sa} $+^{p}$ W $+^{od}$)*Fem* [10].

Based on these results, it was assumed that the fused chromosome composed of the deleted Z and the Df(p^{Sa} $+^{p}$ W $+^{od}$)*Fem* chromosome was generated by a single translocation event between p^{Sa} $+^{p}$ W $+^{od}$ and Z. This complicated chromosome is designated DfZ-DfW [10] (Fig. 3). This assumption strongly supports the left end model for the structure of the p^{Sa} $+^{p}$ W $+^{od}$ chromosome. We will describe the W chromosomal region of the Df(p^{Sa} $+^{p}$ W $+^{od}$)*Fem* chromosome in the following section.

T(W;3)*Ze* Chromosome

Encouraged by Tazima's successful experimentation, several researchers undertook the challenge of inducing new translocations between the W chromosome and an autosome carrying dominant alleles of genes with noticeable traits (egg color, larval marking, and cocoon or blood color). Hasimoto (1948) was able to isolate a translocation of a chromosome 3 fragment bearing the *Ze* (zebra) marker to the W chromosome (T(W;3)*Ze*) after X-ray irradiation [12]. In the resulting strain with the T(W;3)*Ze* chromosome, female larvae have zebra markings, whereas male larvae have whitish skin (Fig. 1F).

Male-Killing Factor Derived from the T(W;3)*Ze* Chromosome

One intriguing observation for *B. mori* is that part of the W chromosome appears to be able to convey a male-killing activity [13]. As described above, Hasimoto (1948) isolated a translocation between the W chromosome and chromosome 3 bearing the *Ze* gene (Fig. 1F). As a result, T(W;3)*Ze* and Df(3)*Ze* chromosomes were generated. Hasimoto (1953) showed that Z/T(W;3)*Ze*, 3/Df(3)*Ze* females are viable whereas Z/Z, 3/Df(3)*Ze* males die during embryogenesis. Based on these observations, Hasimoto (1953) concluded that T(W;3)*Ze* is a product of reciprocal translocation between W and chromosome 3 [13]. Furthermore, he suggested that T(W;3)*Ze* lacks the part of W that normally contains the putative male-killing factor, which he proposed was instead connected to the Df(3)*Ze* chromosome. Unfortunately, the Df(3)*Ze* chromosome was lost before detailed analyses could be conducted.

Figure 2. Visible markers translocated from autosomes make it possible to distinguish males from females and identify mosaic animals in embryonic stages. (A) Mosaic eggs laid by a Z/T(W;10)$+^{w\text{-}2}$, *mo*/*mo*, *w-2*/*w-2* female crossed with a *w-2*/*w-2* male. Black eggs and black regions in the egg are female, whereas white eggs and white regions are male. (B) Mosaic moths with the egg (eye) color marker gene ($+^{w\text{-}2}$/*w-2* and *w-2*/*w-2*). Black eyes are due to $+^{w\text{-}2}$/*w-2*, whereas white eye are due to *w-2*/*w-2*. (C) a: Bilateral mosaic larva with "multilunar" larval markings. b: Bilateral mosaic larva with "striped" larval markings. (D) Mottled larvae due to detachment of the chromosome 2 fragment from T(W;2)p^{B}. (E) Several cocoon colors. (F) Cocoons spun by the sex-limited yellow cocoon strain. The yellow cocoon is female (Z/T(W;2)*Y*), whereas the white cocoon is male (Z/Z).

T(W;10)$+^{w\text{-}2}$ Chromosome

The abundance of egg color (serosa color) is a unique feature of the silkworm. Egg color results from pigments produced in cells of the serosa, a single-layer membrane that covers the embryo. The cells of the serosa are derived from cleavage nuclei such that the genotype of serosa cells is identical to that of the embryo. Normal eggs have dark brown (black) pigmentation, whereas one of the mutants, *w-2* (white 2 gene), has yellowish-white pigmentation. Tazima et al (1951) was able to isolate a translocation of a chromosome 10 fragment bearing the $+^{w-2}$ gene to the W chromosome (T(W;10)$+^{w-2}$) after irradiation with X-rays. In this strain (T(W;10)$+^{w-2}$), white eggs are males of the genotype Z/Z, *w-2/w-2* and black eggs are females of the genotype Z/T(W;10)$+^{w-2}$, *w-2/w-2* [14]. In addition to being helpful for sex determination, the T(W;10)$+^{w-2}$ chromosome is also useful to detect mosaic eggs (Fig. 2A).

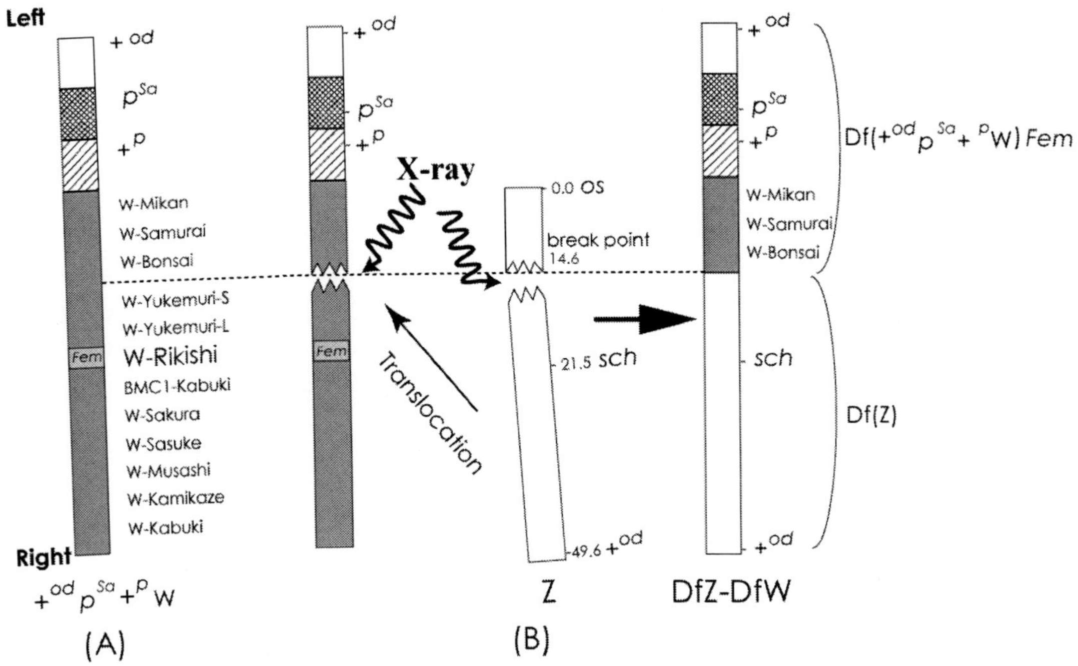

Figure 3. Low-resolution mapping of molecular markers and the putative *Fem* gene on the W chromosome. (A) The $+^{od}$ p^{Sa} $+^{p}$ W chromosome. The positions of W-specific RAPD markers are indicated to the right of the W chromosome. *Fem*; putative female-determining gene. (B) Schematic representation of chromosomal translocation between the $+^{od}$ p^{Sa} $+^{p}$ W chromosome and the Z chromosome, and generation of the DfZ-DfW chromosome.

T(W;2)*Y* Chromosome

Many cocoon color mutants have been maintained in Japan. A white (colorless) cocoon is conventionally considered to be normal, as almost of silkworms reared in Japan for industrial purposes have white cocoons. However, during the long history of sericulture, animals with other cocoon colors have been observed and maintained, including animals with yellow, golden yellow, pinkish or green cocoons (Fig. 2E). Yellow and pinkish colors are attributable to the presence of carotenoids, carotenes and xanthophylls derived from mulberry leaves. Green color is due to flavonoids [15]. However, expression of cocoon color is complicated, as there appear to be many genes related to permeability of pigments in the alimentary canal and in the silk gland. In general, yellow-blooded larvae are also yellow cocoon spinners [15]. *Y* is an allele of the yellow blood gene (*Y*, 2-25.6). In a larva with the *Y* allele, the hemolymph is deep yellow because carotenoids from mulberry pass through the digestive organs.

To establish a sex-limited yellow cocoon strain, Kimura et al (1971) used γ-rays or X-rays to irradiate a total of 707 female pupae of a strain with a normal W chromosome and with the *Y* allele over the course of nine years (from 1961 to 1969). After irradiation of the female pupae, the researchers crossed the emerged female moths to white cocoon males (white blood, $+^Y/+^Y$), reared female individuals, crossed them in the same manner and investigated the subsequent generations. A total of 4,502 batches of eggs were analyzed over a period of nine years and yet, only one batch was observed in which all the females made yellow cocoons and all the males made white cocoons. Thus, Kimura et al (1971) were able to isolate a translocation of a chromosome 2 fragment bearing the *Y* allele to the W chromosome, and obtained a sex-limited yellow cocoon strain (T(W;2)*Y*) [16]. In this strain, females have yellow cocoons (Z/T(W;2)*Y*, $+^Y/+^Y$,) and males have white cocoons (Z/Z, $+^Y/+^Y$) (Fig. 2F). Close to the time it was isolated, this sex-limited yellow cocoon strain did not have observable physiological defects [16]. During a subsequent breeding process aimed at commercial use of this strain, however, several physiological defects (less healthy, a lighter weight cocoon shell and proctocele-larvae) were recognized in females containing the T(W;2)*Y* chromosome [17]. To alleviate these defects, Niino et al (1988) γ-ray irradiated the strain and isolated a derivative in which non-*Y* portions of the translocated chromosome 2 fragment were deleted [18].

Mottling Due to a Chromosome Fragment

In *B. mori,* there are several "mottling" strains (those with small patches of each of two allelic characters in the same individual). Mottling of the silkworm is thought to be the result of retention and elimination of a chromosome fragment in different cells during the course of somatic cell division. In *Drosophila*, chromosome pieces without centromeres are usually eliminated during a first round of cell division. In Lepidoptera, however, the chromosomes lack localized centromeres; rather, the whole chromosome appears to have centromeric properties (that is, the chromosomes are holocentric) [19]. When the chromosomes of *B. mori* become fragmented, the fragments can attach to the spindle and are often transmitted to the next generation, which then exhibits mottling [20, 21].

In rare instances during maintenance of the W-B strain, we discover mottled female individuals, which presumably arise via detachment of the chromosome 2 fragment from the T(W;2)p^B chromosome (Fig. 2D). Among the progeny of a mottled female are white males, white females, mottled males and mottled females. These results indicate that the translocated chromosome 2 fragment can be transmitted to the next generation separately from the W chromosome in this strain. Thus, even though this chromosome fragment can sometimes be lost during cell division, transmission to the next generation is relatively stable.

Why Is the W Chromosome Analyzed?

The W chromosome is interesting from the point of view of sex determination. One example of fruitful analysis of sex determination in *B. mori* comes from study of a heritable mosaic strain with an interesting allele of the *mo* gene [22]. In the mosaic strain, the eggs laid by a female (*mo*/*mo*) contain two nuclei, due to a failure to eliminate one of the polar bodies, and both nuclei are fertilized. This mosaic strain has been used to study developmental biology, genetics pathology, and other topics [23,24]. Using a version of mosaic strain with the T(W;2)p^{Sa}, T(W;2)p^{B} or T(W;10)$+^{w-2}$ chromosomes, sexual mosaics such as gynandromorphs that contain distinct boundaries between female and male tissues (cells) can easily be obtained at the embryo, larva and moth stages (Fig. 1C, 2A, B, and C). This strongly suggests that no diffusible substances, such as sex hormones found in mammals, influence sex determination or sexual development in *B. mori*. Moreover, the data suggest that a female or male is genetically determined only by the presence or absence of the W chromosome. Based on these observations, it seems that the sex determination mechanism of *B. mori* is straightforward and rests solely on the presence of a master control gene or genes on the W chromosome.

Additionally, an interesting feature of the W chromosome is that it does not recombine with the Z chromosome or autosomes. Without recombination, the W chromosome remains static, undergoing no changes during transmission from mother to daughter. Even assuming there is a gradual accumulation of nucleotide changes over time, it seems unlikely that more extensive restructuring would occur, such that the structure of W should be unusually stable. Another interesting feature of the sex chromosomes in *B. mori* is that although there is evidence for the presence of many genes on Z [1,25], the W chromosome is, by contrast, devoid of functional genes except for the putative *Fem* (female-determining) gene. Therefore, we were interested to learn the nucleotide sequence of the W chromosome, with the idea that it might reveal aspects of W chromosome function and sex determination.

As mentioned above, the main focus of study and modification of the *B. mori* W chromosome was for practical use of the resulting strains. As a consequence, molecular biological study of the W chromosome was not performed for many years even after it became practical to do so. More recently, however, molecular studies have been initiated. Promboon et al (1995) constructed a linkage map of random amplified polymorphic DNAs (RAPDs), but no W-specific RAPD markers were obtained [26]. Subsequently, Yasukochi (1998) created a dense genetic map constructed based on 1018 molecular markers but again, no W-specific markers were found [27]. Still later, Tan et al (2001) constructed an amplified fragment length polymorphism (AFLP) map of *B. mori* and once again, no W-specific markers were reported [28]. Recently, a draft sequence of the genome of the p50 strain of *B. mori* was constructed by three-fold whole-genome shotgun sequencing by Mita et al (2004) in Japan [29] and in addition, a draft sequence of p50 strain was constructed by 5.9-fold whole-genome shotgun sequencing Xia et al. (2004) in China [30]. However, these sequencing efforts were undertaken using only the male genome. Thus, systematic molecular analysis of the W chromosome of *B. mori* has not yet been reported.

Complete molecular biological analysis of the W chromosome rests on first obtaining the W-specific nucleotide sequence. However, with the exception of translocated fragments, the nucleotide sequence of the W chromosome has not been determined. As a first step, Abe et al (1995), who felt that the acquisition of a W-specific nucleotide sequence was very important, attempted to obtain a RAPD marker on the W chromosome using a large set of arbitrary 10-mer primers [31]. Perhaps unexpectedly, the W-translocation strains have also proved very useful for molecular biological analysis of the W chromosome (see below).

W-Specific RAPD Markers

We initially used the T(W;2)$+^{p}$ chromosome rather than the normal W chromosome as template DNA for PCR aimed at detection of W-specific markers for two reasons. First, if there are polymorphisms in the normal W chromosome at the molecular level, we can not discriminate the differences based on the larval appearance. Therefore, it is necessary to determine the standard type of W chromosome. Second, even if a given W-specific RAPD marker is derived from the chromosome 2 fragment and not the W chromosome region, it should be possible to access the W chromosome region via a chromosome walk. For these reasons, we used genomic DNA from males and females of the Chinese137 strain (Fig. 1B) with the T(W;2)$+^{p}$ chromosome as a PCR template and screened about 400 arbitrary 10-mer primers for their ability to serve as RAPD markers. We were successful in finding one female-specific RAPD marker, Female-218 [31, 32]. The Female-218 RAPD marker resulted in amplification of a product from the T(W;2)p^{Sa} and T(W;2)p^{B} chromosomes but not from the "normal" W chromosome [33]. At that time, whether the Female-218 RAPD marker was derived from the W region or the chromosome 2 fragment region was not known. Recently, it became clear that the Female-218 RAPD marker was localized to the chromosome 2 fragment region of the translocated W chromosomes (T(W;2)$+^{p}$, T(W;2)p^{Sa} and T(W;2)p^{B}) from the results of detachment experiments [34]. More recently, we found another chromosome 2 fragment-specific RAPD marker, W-Maji, during another screen of arbitrary 10-mer primers [10].

Thus, it is possible to find female-specific RAPD markers using genomic DNA from the Chinese137 strain (with T(W;2)$+^{p}$) as a PCR template. However, the markers we isolated did not recognize W chromosome regions and obtaining a RAPD marker for W sequences remained a goal for our group, as these would allow study of the normal W chromosome. In subsequent analyses, then, we used genomic DNA from several strains with a normal W chromosome as templates and to date, we have been able to identify 12 W-specific RAPD markers from the normal W chromosome [35, 36].

DNA Sequence of the W-Specific RAPD Markers

The deduced amino acid sequences of 11 of the 12 W-specific RAPD markers show similarity to previously reported amino acid sequences of retrotransposon from various organisms (the exception is the marker W-Musashi; [35, 36]). Moreover, almost all amino acid sequences of W-specific RAPD markers contain boundaries of two or three retrotransposable elements [35, 36].

W-Specific Lambda Phage and BAC Clones

As a next step, we constructed genomic DNA lambda phage libraries of *B. mori*. From these libraries, we obtained two lambda phage clones containing the W-Kabuki and W-Samurai RAPD sequences, as assayed by PCR and sequencing. The DNA sequences of the two phage clones comprise the nested structure of many retrotransposable elements [36, 37, 38].

A *B. mori* BAC (bacterial artificial chromosome) library, the RPCI-96 library, has been constructed in Japan using genomic DNA extracted from female and male mixed population of the p50 silkworm strain (Mita, unpublished data). In addition, two other *B. mori* BAC libraries, the p50 library and the C108 library, were also constructed using genomic DNA of the p50 and C108 strains, respectively [39]. We used the W-specific RAPD marker sequence to identify W-specific BAC clones and subjected these clones to shotgun sequencing (unpublished data). However, due to the presence of many repetitive DNA elements, in particular the non-LTR retrotransposons BMC1 [40] and *Kendo* [41], assembly of the shotgun clone sequence did not result in a single contig. However, although assembly of the shotgun sequence proved difficult, sub-sections of the clone sequences could be assembled into contigs and have been analyzed in detail [42]. As shown in Fig. 7, the structural features of the W chromosome of *B. mori* are quite different from structures typical of other chromosomes. Many LTR and non-LTR retrotransposons, retroposns (Bm1), DNA transposons and their derivatives have accumulated on the W chromosome.

Presence of W-Specific RAPD Markers in Silkworm Strains Maintained in Japan

Abe et al.(2005b) determined the presence or absence of W-specific RAPD markers in silkworm strains maintained in Japan. All strains with normal W chromosomes used in this study were positive for all of the W-specific RAPD markers, with the exception of the W-Kamikaze marker. These results strongly indicate that the level of diversity in the W chromosome is low in silkworm strains maintained in Japan [36].

Presence of the W-Specific RAPD Markers in W Chromosome Variants [T(W;3)*Ze* Chromosome]

As mentioned above, it was Hasimoto (1948) who isolated the T(W;3)*Ze* chromosome and after much study, Hasimoto (1953) hypothesized that T(W;3)*Ze* is a product of reciprocal translocation between W and chromosome 3 [13]. If reciprocal translocation occurred, then the W region in the T(W;3)*Ze* chromosome should be shorter than that in the original W chromosome. Moreover, if the W region was shortened, the fact that it still functions as a W (i.e. female-determining) chromosome would suggest that the putative *Fem* is still present. Thus, we decided that it was important to determine whether or not the W region of T(W;3)*Ze* is indeed shorter than that in the original W chromosome. However, deletions of the W chromosome cannot be detected using conventional genetic methods. To confirm this deletion, then, a molecular biological comparison between the T(W;3)*Ze* chromosome and the original W chromosome used in the translocation experiment would be useful. However, the original strain used in the translocation experiment is not available for comparison because there is no precise rearing record. Fortunately, however, the normal W chromosomes of strains maintained in Japan (the Japanese-W-Eve type) are almost identical, suggesting that they provide a suitable point of comparison to the putative W deletion strain T(W;3)*Ze* [36]. Partial deletion of the W in T(W;3)*Ze* can thus be detected by the disappearance of W-specific RAPD markers that were isolated on the basis of recognition of the normal W. The W chromosome region of the T(W;3)*Ze* chromosome lacks two of 12 W-specific RAPD markers (W-Mikan and W-Samurai) [36]. These results strongly indicate that the W chromosome region containing the W-Samurai and W-Mikan markers was deleted in the T(W;3)*Ze* chromosome due to reciprocal translocation (Fig. 4).

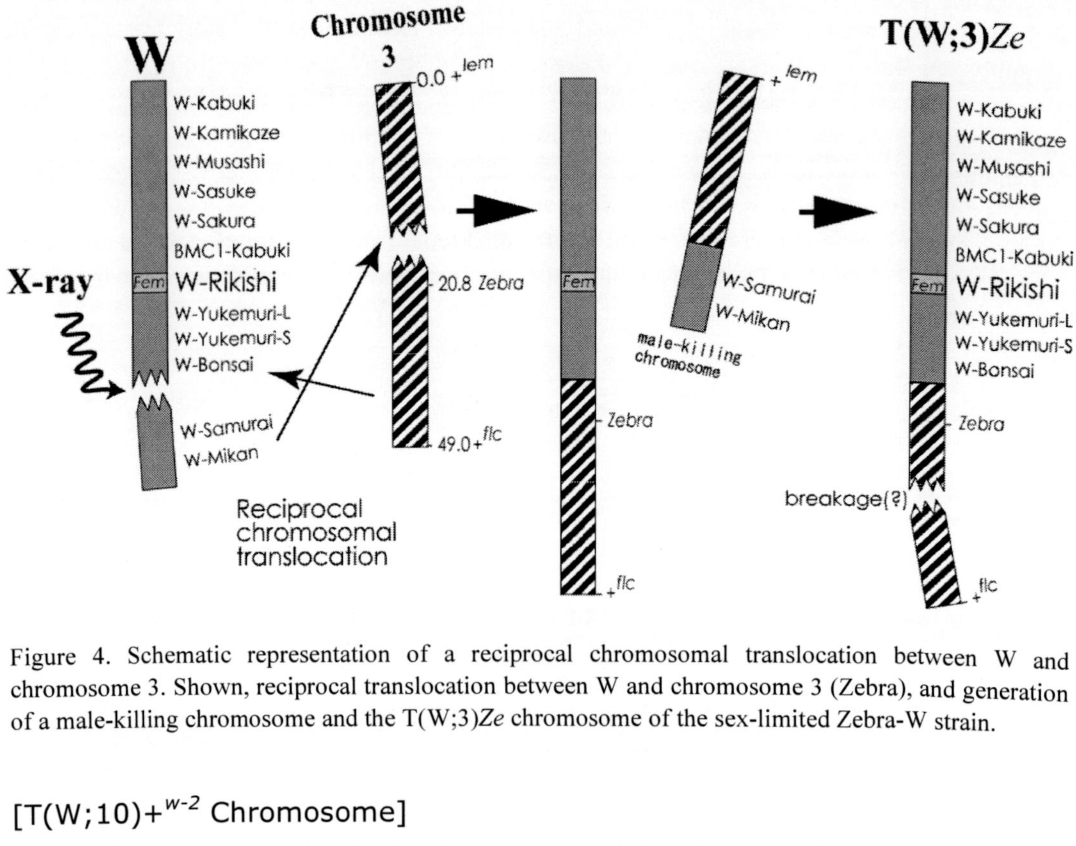

Figure 4. Schematic representation of a reciprocal chromosomal translocation between W and chromosome 3. Shown, reciprocal translocation between W and chromosome 3 (Zebra), and generation of a male-killing chromosome and the T(W;3)*Ze* chromosome of the sex-limited Zebra-W strain.

[T(W;10)+$^{w-2}$ Chromosome]

Based on the results discussed above, we reasoned that the W chromosomes of other sex-limited strains are not the result of simple fusion of an autosomal fragment to the end of an unaltered W chromosome but rather, they are the products of reciprocal translocation accompanied by the breakage (deletion) of the W chromosome. Therefore, we next looked at the presence or absence of the W-specific RAPD markers on the T(W;10)+$^{w-2}$, T(W;2)+^{p}, and T(W;2)*Y* chromosomes.

The W chromosome region of the T(W;10)+$^{w-2}$ chromosome lacked one of 12 W-specific RAPD markers (W-Mikan) [36]. This result strongly indicates that the region containing the W-Mikan RAPD marker was deleted in the T(W;10)+$^{w-2}$ chromosome (Fig. 5).

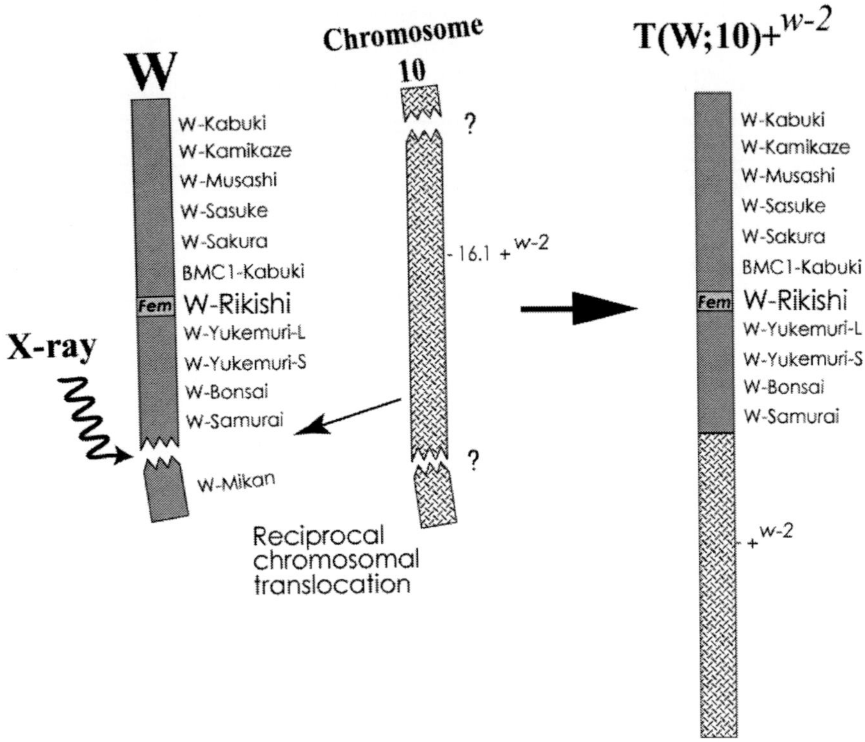

Figure 5. Schematic representation of reciprocal chromosomal translocation between W and chromosome 10. Shown, reciprocal translocation between W and chromosome 10 ($+^{w-2}$), and the generation of the T(W;10)$+^{w-2}$ chromosome of the sex-limited Black-egg-W strain.

[T(W;2)$+^{p}$ Chromosome]

We did not detect deletion of the W chromosome region in the T(W;2)$+^{p}$ chromosome using W-specific RAPD markers [36]. We think the breakpoint of this T(W;2)$+^{p}$ chromosome is close to the telomere of the W chromosome.

[T(W;2)*Y* Chromosomes]

We next obtained strains with each of three T(W;2)*Y* chromosomes (T(W;2)*Y*-Abe, T(W;2)*Y*-Chu and T(W;2)*Y*-Ban) maintained by three independent research groups. We then determined the presence or absence of the W-specific RAPD markers on the three T(W;2)*Y* chromosomes. T(W;2)*Y*-Chu contained six of the 12 W-specific RAPD markers (W-Rikishi, W-Yukemuri-L, W-Yukemuri-S, W-Bonsai, W-Samurai, and W-Mikan). However, both T(W;2)*Y*-Abe and T(W;2)*Y*-Ban contained only one of the 12 W-specific RAPD markers (W-Rikishi) [43]. Based on these results, we suspected that the original W chromosome of T(W;2)*Y*-Abe and -Ban differed from the Japanese-W-Eve type. However, the original W chromosome of the T(W;2)*Y* was the W chromosome of the C125 strain [16], which does have the normal Japanese-W-Eve type [43]. Moreover, the original T(W;2)*Y* chromosome was clearly produced by only one event (translocation) in a single female isolated from the C125 strain [16]. Therefore, we examined another possibility; namely, that all of the marker-positive regions were deleted from the T(W;2)*Y*-Abe and -Ban chromosomes with the exception of a region that contains the W-Rikishi RAPD marker and the putative *Fem* gene.

To determine the presence or absence of regions around the W-Rikishi RAPD marker sequence in the T(W;2)*Y*-Abe and -Ban chromosomes, we obtained a BAC clone (1C7C) positive for the W-Rikishi RAPD marker. Following shotgun sequencing of the 1C7C BAC clone, no single contig could be constructed due to the presence of repetitive DNA elements. However, we identified many regions containing boundaries between retrotransposable elements, which were used to design seven new W-specific PCR markers. The T(W;2)*Y*-Abe and -Ban types were positive for all seven W-specific PCR markers [43]. These results indicate that almost all regions of the W chromosome were deleted from the T(W;2)*Y*-Abe and -Ban chromosomes but that the region containing the sequence from the 1C7C BAC clone and the putative *Fem* gene, remained.

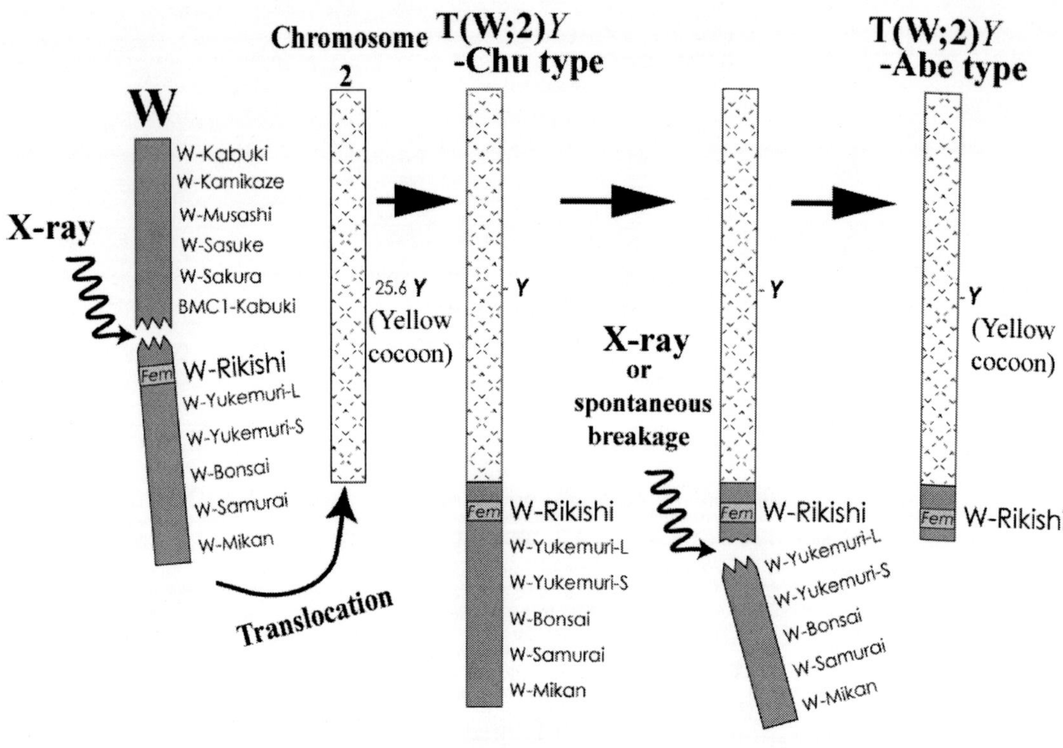

Figure 6. Schematic representation of generation of sex-limited Yellow-cocoon-W strains. Shown, generation of the T(W;2)*Y*-Chu and T(W;2)*Y*-Abe. *Y,* the yellow blood gene on chromosome 2.

The results of marker analysis help us to learn about the origins of these translocations. The T(W;2)*Y*-Chu, -Abe and -Ban chromosomes were thought to be produced in the following manner. First, the region of the W chromosome containing the six W-specific RAPD markers (W-Kabuki, W-Kamikaze, W-Musashi, W-Sasuke and W-BMC1-Kabuki) was deleted by X-ray irradiation. Then, the chromosome 2 fragment containing the *Y* gene translocated to a region that contains the putative *Fem* gene and is positive for a different set of six W-specific RAPD markers (W-Mikan, W-Samurai, W-Bonsai, W-Yukemuri-L, W-Yukemuri-S, and W-Rikishi). The resulting chromosome is thought to be the T(W;2)*Y*-Chu chromosome. It is also thought that the region of the W chromosome containing the five W-specific RAPD markers (W-Mikan, W-Samurai, W-Bonsai, W-Yukemuri-L, and W-Yukemuri-S) was further deleted following irradiation during the breeding process (Fig. 6). The resulting changes are thought to be present in the T(W;2)*Y*-Abe and -Ban chromosomes.

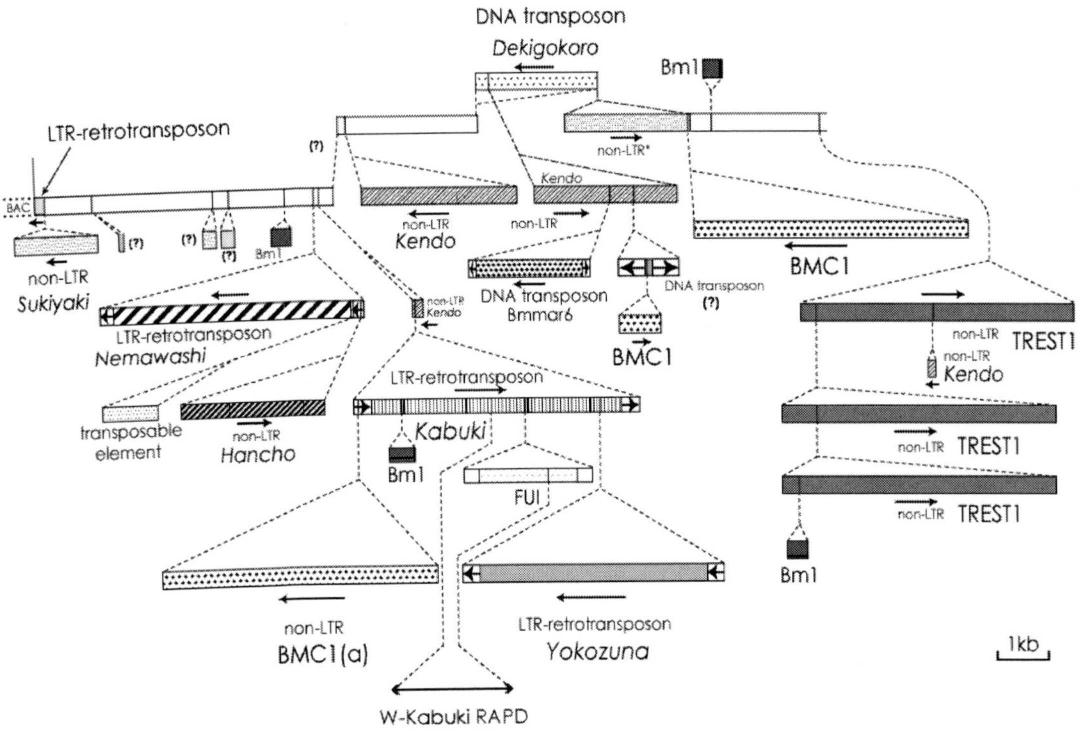

Figure 7. The presence of transposable elements in W chromosome sequence. Nested transposable elements in the BAC clone that includes the W-Kabuki RAPD marker sequence. This figure is based on DNA sequence information. See Abe et al (2005a) [42] for details.

[Df(*pSa* +^{p} W +^{od})*Fem* Chromosome (DfZ-DfW Chromosome)]

As described above, Abe et al (2005b) revealed that the T(W;3)*Ze* chromosome lacks the left side of W that is positive for two of 12 W-specific RAPD markers (W-Mikan and W-Samurai) [36]. Moreover, Fujii et al (2006) revealed that Df(*pSa* +^{p} W +^{od})*Fem* (DfZ-DfW) does contain the left side of W that includes three of 12 W-specific RAPD markers (W-Mikan, W-Samurai, and W-Bonsai) [10](Fig. 3). These results indicate (1) that the W-specific RAPD markers are arranged in the order W-Mikan, W-Samurai, and W-Bonsai (starting from the left end of the W chromosome), and (2) that the male-killing factor [13] is located between the left end and the W-Bonsai marker region. However, we think that there is not a true 'male-killing factor' between the left end and the W-Bonsai RAPD marker because males with the DfZ-DfW chromosome have normal viability. The DfZ-DfW chromosome includes a fragment of the left side of the W chromosome that includes the W-Mikan, W-

Samurai, and W-Bonsai RAPD markers. If we assume that a male-killing factor does not exist on the W chromosome of *B. mori*, then why would the Z/Z, 3/Df(3)*Ze* males die before completing embryogenesis? It is well established that some genes are haploinsufficient, such that the presence of only one normal copy of the gene is not sufficient for normal function. Therefore, one possible explanation of the embryonic death observed in the Z/Z, 3/Df(3)*Ze* males is that there is a haploinsufficient gene required during embryogenesis on the chromosome 3 around the *Ze* locus.

[W(B-YL-YS)*Ze* Chromosome]

We reasoned that if more W chromosome variants could be isolated, such as following X-ray irradiation-induced chromosome breakage, then we could map W-specific RAPD markers and the putative *Fem* gene on the W chromosome to a smaller region. Towards this goal, we X-ray irradiated female pupae of the sex-limited Zebra-W strain, which have the T(W;3)*Ze* chromosome, and we isolated an exceptional zebra male larva. After the male was allowed to copulate with females, genomic DNA was extracted from the legs of the male and used as template for PCR. Three W-specific RAPD markers (W-Bonsai, W-Yukemuri-L and W-Yukemuri-S) were detectable in this male. These results indicate that breakage occurred on the T(W;3)*Ze* chromosome and that the resulting derivative W;3 translocated fragment contains the W-Bonsai, W-Yukemuri-L and W-Yukemuri-S RAPD markers as shown in Fig 8. We designated this chromosome the W(B-YL-YS)*Ze* chromosome [43].

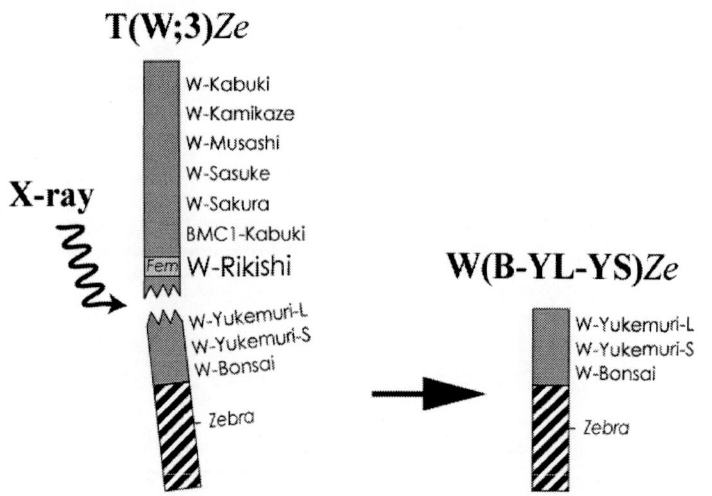

Figure 8. Schematic representation of X-ray induced generation of the W(B-YL-YS)*Ze* chromosome from the T(W;3)*Ze* chromosome.

[Z_1 and $Ze^W Z_2$ Chromosomes]

When we used X-ray irradiation to break the T(W;3)*Ze* chromosome, we also obtained a fragment of the W chromosome we designate Ze^W. We were able to determine that the Ze^W

fragment is positive for three of 12 W-specific RAPD markers (W-Bonsai, W-Yukemuri-L and W-Yukemuri-S). Unexpectedly, it was revealed that the Z chromosome was also broken into a large fragment (Z_1) with $+^{sch}$ (1-21.5) and a small fragment (Z_2) with $+^{od}$ (1-46.9). Moreover, we found that the Ze^W fragment was connected to the Z_2 fragment. We designate this joined chromosomal fragment Ze^WZ_2 (Fig 9) [44].

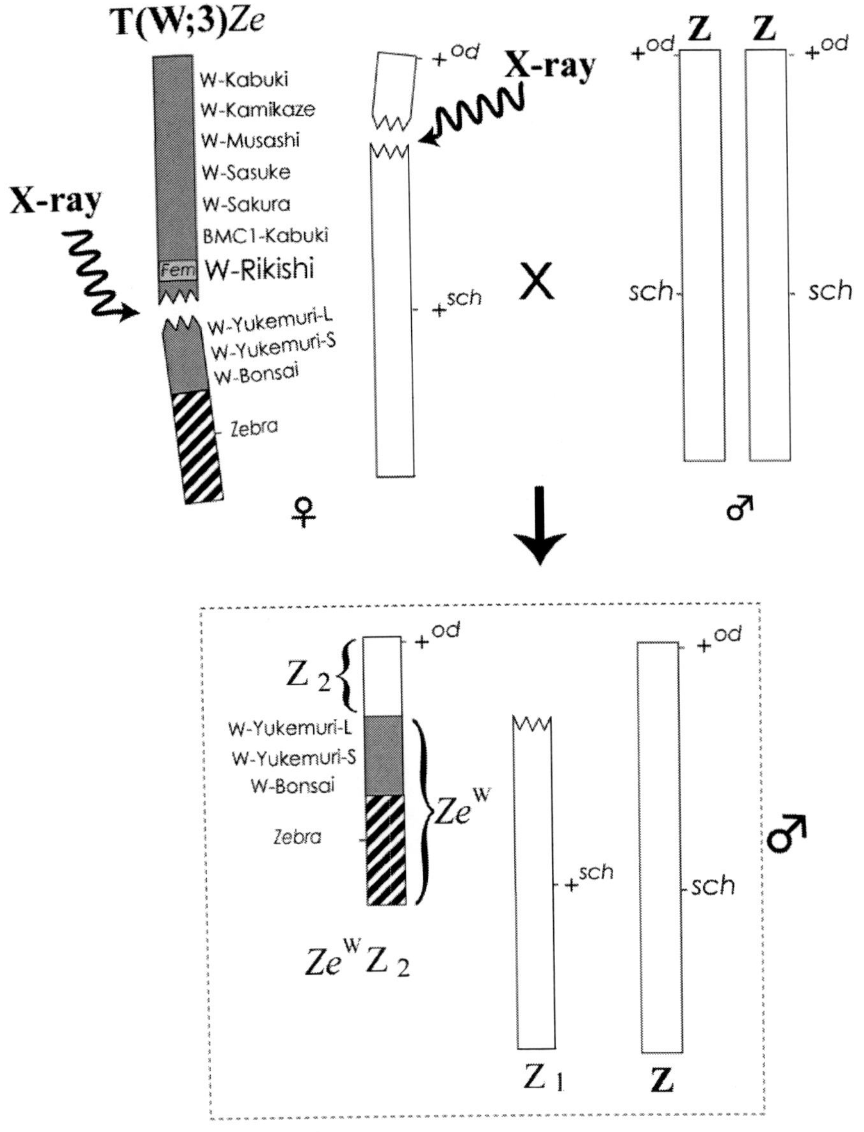

Figure 9. Isolation of changes to the T(W;3)*Ze* chromosome induced by X-ray irradiation. Males of the genotype *sch* $+^{od}$/*sch* $+^{od}$ were crossed to X-ray-irradiated females. A chromosomal fragment dissociated from the T(W;3)*Ze* chromosome is designated as the Ze^W fragment. The Ze^W fragment is positive for three of 12 W-specific RAPD markers (W-Yukemuri-L, W-Yukemuri-S, and W-Bonsai). The long fragment of the Z chromosome bearing the $+^{sch}$ locus (1-21.5) is designated Z_1, and the short fragment of Z bearing the $+^{od}$ locus (1-49.6) is designated Z_2. The chromosome composed of the Ze^W fragment and the Z_2 fragment is designated Ze^WZ_2. See Fujii et al (2007) [44] for details.

We next analyzed genetic behavior of the Z_1 and $Ze^W Z_2$ fragments during meiosis in males (Z_1 $Ze^W Z_2$/Z) and females (Z_1 $Ze^W Z_2$/W). The Z_1 fragment and the Z or W chromosomes segregated properly but non-disjunction was observed between the $Ze^W Z_2$ fragment and the Z chromosome and between the $Ze^W Z_2$ fragment and the W chromosome. Furthermore, the 2A:Z/Z_1males, which were the products of non-disjunction between $Ze^W Z_2$ and W, had observable phenotypic defects. Namely, Z/Z_1 male moths can walk and can copulate with females but they cannot flap their wings. To analyze why the Z/Z_1 males cannot flap their wings, we looked at the indirect flight muscles in these males. In the "flapless" males, the indirect flight muscles are broken or decayed [44]. The flapless phenotype was not caused by Z_1 as Z/Z_1 $Ze^W Z_2$ males can flap their wings vigorously.

In humans, balanced translocation carriers have no clinical symptoms (reviewed in Emanuel and Shaikh, 2001 [45]). Similarly, Z/Z_1 $Ze^W Z_2$ is thought to be a balanced translocation in *B. mori* [44]. Moreover, in *Drosophila melanogaster*, the genes that encode major myofibrillar proteins such as actin, myosin, and tropomyosin are haploinsufficient for flight. For example, haploidy for the myosin heavy chain (MHC) gene results in a reduction of MHC in adult thoracic, leg and larval muscles of the fly. This affects the function of the flight muscles in haploid animals, although the leg and larval muscles function normally [46, 47]. In *B. mori*, the absence of dosage compensation has been suggested based on gene expression patterns. Twelve of 15 genes on the Z chromosome are expressed more abundantly in males (ZZ) than in females (ZW) [25, 48, 49]. However, the mRNA levels of three of 15 genes are equivalent in males and females, or are higher in females [25]. Thus, it is likely that expression is regulated gene-by-gene. Therefore, we presume that there is a haploinsufficient gene(s) involved in indirect flight muscle development that is present on the Z_2 region of the Z chromosome in *B. mori*. However, if there are haploinsufficient gene(s) on the Z chromosome, then expression of the gene(s) must necessarily be up-regulated during female development in order to compensate for the presence of only one copy of the gene(s) in females (Z/W).

Order of W-specific Markers and Position of the Putative *Fem* Gene on the W Chromosome

The presence or absence of molecular markers in W chromosome variants makes it possible to map the positions of the markers relative to the putative *Fem* gene (Fig. 10). The T(W;10)$+^{w-2}$ chromosome does not contain the W-Mikan RAPD marker [36]. The T(W;3)*Ze* chromosome does not contain the W-Samurai and W-Mikan RAPD markers [36]. The female-killing DfZ-DfW chromosome contains the W-Bonsai, W-Samurai, and W-Mikan RAPD markers [10]. Moreover, the W(B-YL-YS)*Ze* chromosome contains the W-Yukemuri-L, W-Yukemuri-S, and W-Bonsai RAPD markers [43]. From these results, the order of W-Yukemuri-L and W-Yukemuri-S could not be determined but the order of five W-specific RAPD markers could be determined: from one end (left end) of W the order is W-Mikan, W-Samurai, W-Bonsai, and W-Yukemuri-S or W-Yukemuri-L (Fig. 10). Moreover, the evidence suggests that the regions containing these five W-specific RAPD markers do not contain the putative *Fem* gene, as their loss has no effect on sex determination.

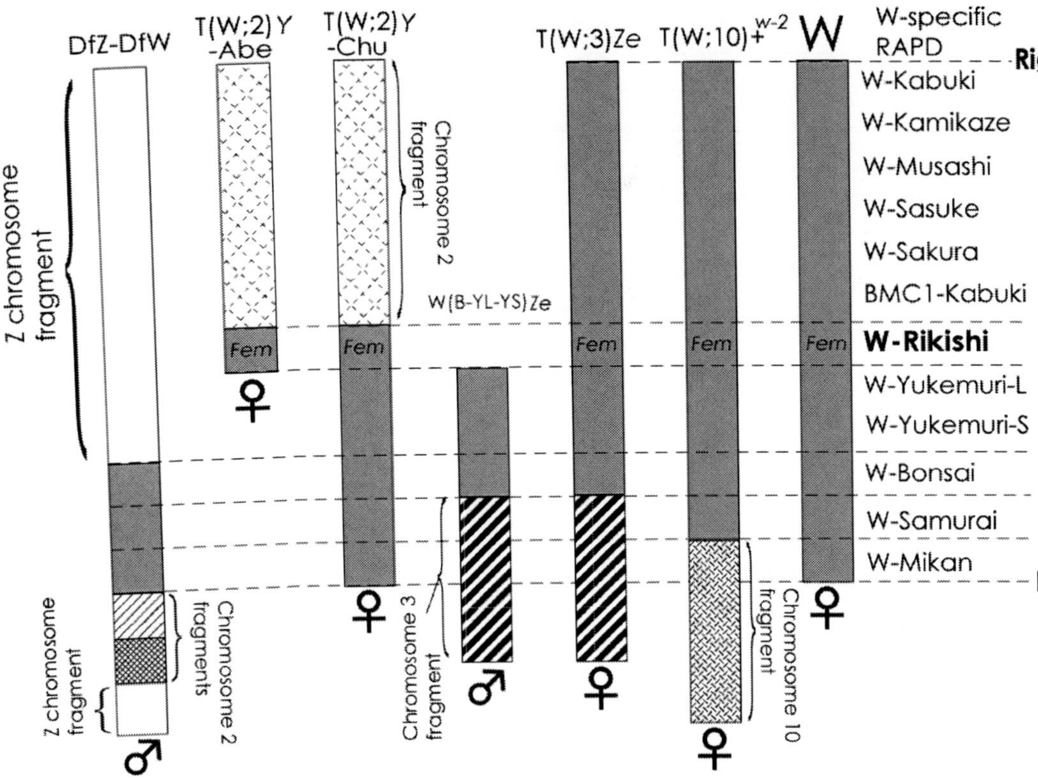

Figure 10. Higher-resolution mapping of W-specific RAPD markers and the putative *Fem* gene on the W chromosome using W chromosome variants. The order of six W-specific RAPD markers (W-Kabuki, W-Kamikaze, W-Musashi, W-Sasuke, W-Sakura and BMC1-Kabuki) and two W-specific RAPD markers (W-Yukemuri-L and W-Yukemuri-S) could not be determined. Mapping of the other markers is shown. For the T(W;3)*Ze*, T(W;10)+$^{w-2}$ and DfZ-DfW chromosomes, see Abe et al (2005b) [36] and Fujii et al (2006) [10].

The T(W;2)*Y*-Abe and -Chu chromosomes do not contain the W-Kabuki, W-Kamikaze, W-Musashi, W-Sasuke, W-Sakura, and BMC1-Kabuki RAPD markers. However, the T(W;2)*Y*-Abe chromosome contains the W-Rikishi RAPD marker. From these results, the order of W-Kabuki, W-Kamikaze, W-Musashi, W-Sasuke, W-Sakura, and BMC1-Kabuki RAPD markers could not be determined but we can conclude that the W-Rikishi RAPD marker is distal to one end (right end) of the W chromosome (Fig. 10). This positional information should be of significant help towards the goal of cloning the putative *Fem* gene.

Translocated Chromosomes in Other Lepidopteran Insects

In the other lepidopteran insects, isolation of chromosomal translocations seems to be difficult because there are few genetic markers useful for identifying translocations when they occur.

Similar to *B. mori*, the sex chromosome constitution of *Ephestia kuehniella* is ZW for females and ZZ for males. The role of the Z and W chromosomes in sex determination in *E. kuehniella* is not clear. Chromosomes in Lepidoptera are generally small and numerous and lack differentiating features. However, using a light microscope, Z-W bivalents can be recognized in the pachytene stage by the presence of a heterochromatic thread that is a feature of the *E. kuehniella* W chromosome [50]. A sex chromatin body (SB), deduced to be a condensed W chromosome, has also been observed in the highly polyploidy interphase nuclei of Lepidopteran females [51, 52]. In *E. kuehniella*, Traut and Rathjens (1973) found that fusion of the W chromosome with an autosome is accompanied by fragmentation of the otherwise cohesive heterochromatin body (SB). Therefore, SB morphology was successfully used as cytogenetic marker to isolate T(W;A) chromosomes induced by γ-irradiation [50, 53]. The structure of the each T(W;A) chromosome was analyzed at the pachytene stage of oocytes [54]. Moreover, some T(W;A) chromosomes have been characterized using the autosomal marker genes located on the translocated autosome [54].

To isolated changes in the T(W;Z) chromosome in *E. kuehniella*, Marec and Mirchi (1990) crossed γ-ray irradiated Z^{dz+}/W females to *dz*/*dz* males and screened for Z^{dz+} females, which are exceptions to the normal pattern of sex-linked inheritance in this species [55]. From the Z^{dz+} females, Marec and Mirchi (1990) isolated 12 sex chromosome mutant strains with Z chromosomes bearing additional sequences that include $+^{dz}$ gene (Z^{dz+}segment). It was revealed that Z^{dz+} segments were translocated onto the W chromosome in four of the 12 lines [55]. Additionally, the Z^{dz+} segment was connected with a terminal segment of the W chromosome in two of the 12 lines [56]. Curiously, males with the Z^{dz+} segment do not survive embryogenesis in these two lines whereas males with any of the other Z^{dz+} chromosomes are viable. Based on these results, Marec et al. (2001) suggested that a part of the *E. kuehniella* W chromosome includes a male-killing factor [56].

Conclusions

The W chromosome of *B. mori* was artificially modified for practical use [3, 6, 8, 9, 12, 14, 16]. In the past, the intension was to isolate translocations of autosomal fragments to the W chromosome, rather than to delete the W chromosome. However, in the process of generating these strains, deletions of the W chromosome were unconsciously selected. From these strains, we can conclude that even large deletions of the W chromosome or large translocations to W do not affect female sex determination. Therefore, we think that a full-length, "normal" W chromosome is not essential for femaleness. Instead, we concluded from test of the translocation and deletion strains that (1) putative *Fem* genes are not distributed evenly over the W chromosome, (2) no gene governing femaleness is located in the regions deleted from W in T(W;2)*Y*-Abe, and (3) although the exact length of the region can not be determined, only what appears to be an extremely limited region around the W-Rikishi RAPD marker is required to determine the female mode of development.

We have proposed that the W chromosome evolves more slowly than other chromosomes, as crossing-over is restricted to males in *B. mori* and W is recombinationally isolated from the Z chromosome and autosomes. Furthermore, a higher mutation rate is predicted for Z or autosomes relative to W because of the greater number of cell divisions that occur in spermatogenesis than in oogenesis. Indeed, many transposable elements, which have undergone little to no changes, are found on the W chromosome [42]. However, it is likely that large-scale changes such as large deletions or translocations occur more frequently on the

W chromosome than on Z or on the autosomes as these rearrangements do not affect female sex determination.

Chromosome fragments that have translocated to W lose the opportunity to undergo chromosomal recombination and behave like part of the W chromosome. Conversely, chromosome fragments translocated from W to the Z chromosome gain the opportunity to undergo chromosomal recombination. Interestingly, lack of recombination and genetic degeneration of the W chromosome are closely connected in the evolution of heteromorphic sex chromosomes (reviewed in Charlesworth et al., 2005 [57]). In Diptera, the relationship between transposable elements and generation of a non-recombining Y chromosome has been studied extensively [58, 59]. In *Drosophila miranda*, for example, chromosome degeneration is driven by accumulation of transposable elements [60, 61, 62, 63, 64]. Apart from these well-studied examples from Drosophila, the fate of chromosome fragments translocated onto sex chromosomes is not well understood (reviewed in Charlesworth et al., 2005 [57]). Therefore, *B. mori* strains with autosomal or Z chromosome fragments translocated to the W chromosome (such as p^{Sa} $+^{p}$ W $+^{od}$, T(W;2)$p^{Sa}$$+^{p}$, T(W;10)$+^{w\text{-}2}$, T(W;2)*Y* and T(W;3)*Ze*) and with W chromosome fragments translocated to the Z chromosome (such as DfZ-DfW and $Ze^{W}Z_2$) will be useful for molecular biological analysis aimed at understanding how the W chromosome has degenerated and/or evolved over time.

In the course of Drosophila genome projects, researchers were not able to assemble Y chromosome sequence data into a single contig due to the presence of repetitive sequences [65]. To identify new genes on the Y chromosome of *D. melanogaster*, another strategy (TBLASTN search) was developed [66, 67]. Recently, numerous Y-linked repetitive sequences in the Y chromosome of the African malaria mosquito *Anopheles gambiae* were obtained [68, 69]. However, no Y chromosome genes were identified based on a TBLASTN search [70]. Thus, both in Drosophila and Anopheles, molecular biological analysis of the Y chromosome is made very difficult by the presence of unusual chromosomal features.

Whole genome shotgun sequencing of *B. mori* and the continuing effort to annotate the sequence provide valuable biological information about the autosomes and the Z chromosome [29, 30]. However, no information about the W chromosome was obtained in these sequencing efforts. In addition, it has been very difficult to address complex issues such as dosage compensation or haploinsufficiency with only the existing shotgun sequence data. Thus, it seems likely that isolation of additional chromosomal translocations followed by genetic and molecular analyses, including tests with molecular markers, will be important for study of sex determination, dosage compensation, and evolution of the W chromosome in the future.

References

[1] Fujii H, Banno Y, Doira H, Kihara H, Kawaguchi Y (1998) Genetical stocks and mutations of *Bombyx mori*: Important genetic resources, 2nd edn. Institute of Genetic Resources, Faculty of Agriculture, Kyusyu University, Fukuoka.

[2] Goldsmith MR, Shimada T, Abe H (2005) The genetics and genomics of the silkworm, *Bombyx mori*. *Annu Rev Entomol* **50**: 71-100.

[3] Tazima Y (1964) The Genetics of the Silkworm (Logos Press and Prentice Hall, London and Englewood Cliffs).

[4] Tanaka Y (1916) Genetic studies in the silkworm. J Coll Agric Sapporo 6: 1-33.

[5] Hasimoto H (1933) The role of the W-chromosome in the sex determination of *Bombyx mori* (in Japanese). *Jpn J Genet* **8:** 245-247.

[6] Tazima Y (1941) A simple method of sex discrimination by means of larval markings in *Bombyx mori* (in Japanese). *J Seric sci Jpn* **12**: 184-188.

[7] Tazima Y (2001) Improvement of Biological Functions in the Silkworm (Science Publishers, Inc. Enfield (NH), USA) (Translated from Japanese).

[8] Tazima Y (1944) Studies on chromosome aberrations in the silkworm. II. Translocation involving second and W-chromosomes (in Japanese with English summary). *Bull Seric Exp Stn* **12**: 109-181.

[9] Tazima Y (1948) Translocation of the Z chromosome to the W chromosome of the silkworm, Bombyx mori. In *Ogura Commemoration Volume on Cytology and Genetics*. 88-91 (in Japanese).

[10]Fujii T, Tanaka N, Yokoyama T, Ninaki O, Oshiki T, Ohnuma A, Tazima Y, Banno Y, Ajimura M, Mita K, Seki M, Ohbayashi F, Shimada T, Abe H (2006) The female-killing chromosome of the silkworm, *Bombyx mori*, was generated by translocation between the Z and W chromosomes. *Genetica* **127**: 253-265.

[11]Tazima Y (1952) Inheritance of sex. In *Silkworm Genetics*, ed. Y. Tanaka. 351-372 Shokabo, Tokyo (in Japanese).

[12] Hasimoto H (1948) Sex-limited zebra, an X-ray mutation in the silkworm (in Japanese with English summary). *J Seric Sci Jpn* **16**: 62-64.

[13]Hasimoto H (1953) Genetical studies of *Bombyx mori* L. on the lethal gene which affects the male (in Japanese with English summary). *J Seric Sci Jpn* **22**: 200-204.

[14]Tazima Y, Harada C, Ohta N (1951) On the sex discriminating method by colouring genes of silkworm eggs. I. Induction of translocation between the W and the tenth chromosomes (in Japanese with English summary). *Jpn J breed* **1**: 47-50.

[15] Doira H (1978) Genetic stocks of the silkworm. In The Silkworm, an Important Laboratory Tool (Tazima, Y., ed), pp. 53-81. Kodansha Ltd, Tokyo, Japan.

[16]Kimura K, Harada C, Aoki H (1971) Studies on the W-translocation of yellow blood gene in the silkworm (*Bombyx mori*) (in Japanese with English summary). *Jpn J Breed* **21**: 199-203.

[17]Niino T, Kanda T, Eguchi R, Shimazaki A, Shibukawa A (1987) *Defects and structure of translocated chromosome in the sex-limited yellow cocoon strain of the silkworm, Bombyx mori* (inJapanese with English summary). 56: 240-246.

[18]Niino T, Eguchi R, Shimazaki A, Shibukawa A (1988) Breakage by g-rays of the +i-lem locus on the translocated 2nd chromosome in the sex-limited yellow cocoon silkworm. *J Seric Sci Jpn* **57**: 75-76.

[19]Murakami A, Imai HT (1974) Cytological evidence for holocentric chromosomes of the silkworm, *Bombyx mori* and *B. mandarina* (Bombycidae, Lepidoptera). *Chromosoma* **47**: 167-178.

[20]Fujiwara H, Ninaki O, Kobayashi M, Kusuda J, Maekawa H (1991) Chromosomal fragment responsible for genetic mosaicism in larval body marking of the silkworm, *Bombyx mori*. *Genet Res* **57**: 11-16.

[21]Fujiwara H, Maekawa H (1994) RFLP analysis of chromosomal fragments in genetic mosaic strain of *Bombyx mori*. *Chromosoma* **103**: 468-474.

[22]Goldschmidt R and Katsuki K (1931) Vierte Mitteilung über erblichen Gynandromorphism von *Bombyx mori L. Biol. Zbl.,* **48**: 685-699.

[23]Abe H, Kobayashi M, Watanabe H (1990) Mosaic infection with a densonucleosis virus in the midgut epithelium of the silkworm, *Bombyx mori. J Invertebr Pathol* **55**: 112-117.

[24]Shimada T, Kobayashi M (1992) Homeotic transformation in E^{nk} mutant cells is not affected by neighboring normal cells in genetic mosaics of the silkworm, *Bombyx mori. J Seric Sci Jpn* **61**: 444-450.

[25]Koike Y, Mita K, Suzuki MG, Maeda S, Abe H, Osoegawa K, deJong PJ, Shimada T (2003) Genomic sequence of a 320-kb segment of the Z chromosome of *Bombyx mori* containing a *kettin* ortholog. *Mol Genet Genomics* **269**: 137-149.

[26]Promboon A, Shimada T, Fujiwara H, Kobayashi M (1995) Linkage map of random amplified polymorphic DNAs (RAPDs) in the silkworm, *Bombyx mori. Genet Res Camb* **66**: 1-7.

[27]Yasukochi Y (1998) A dense genetic map of the silkworm, *Bombyx mori*, covering all chromosomes based on 1,018 molecular markers. *Genetics* **150**: 1513-1525.

[28]Tan YD, Wan C, Zhu Y, Lu C, Xiang Z, Deng HW (2001) An amplified fragment length polymorphism map of the silkworm. *Genetics* **157**: 1277-1284.

[29]Mita K, Kasahara M, Sasaki S, Nagayasu Y, Yamada T, Kanamori H, Namiki N, Kitagawa M, Yamashita H, Yasukochi Y, Kadono-Okuda K, Yamamoto K, Ajimura M, Ravikumar G, Shimomura M, Nagamura Y, Shin-I T, Abe H, Shimada T, Morishita S, Sasaki T (2004) The genome sequence of silkworm, *Bombyx mori. DNA Res* **11**: 27-35.

[30]Xia Q, Zhou Z, Lu C, Cheng D, Dai F, Li B, et al. (2004) A draft sequence for the genome of the domesticated silkworm (*Bombyx mori*). *Science* **306**: 1937-1940.

[31]Abe H, Shimada T, Yokoyama T, Oshiki T, Kobayashi M (1995) Identification of random amplified polymorphic DNA on the W chromosome of the Chinese 137 strain of the silkworm, *Bombyx mori* (in Japanese with English summary). *J Seric Sci Jpn* **64**: 19-22.

[32]Abe H, Shimada T, Kawai S, Ohbayashi F, Harada T, Yokoyama T, Oshiki T, Kobayashi M (1996) Nucleotide sequence of the random amplified polymorphic DNA (RAPD) on the W chromosome of the silkworm, *Bombyx mori* (Lepidoptera: Bombycidae). *Appl Entomol Zool* **31**: 633-637.

[33]Ohbayashi F, Abe H, Shimada T, Kawai S, Yokoyama T, Oshiki T, Kobayashi M (1996) A common random amplified polymorphic DNA in the silkworm, *Bombyx mori* is shared by W chromosomes onto which the normal marking, Sable and Black genes are translocated respectively (in Japanese). *J Seric Sci Jpn* **65**: 395-398.

[34]Yokoyama T, Abe H, Irobe Y, Saito K, Tanaka N, Kawai S, Ohbayashi F, Shimada T, Oshiki T (2003) Detachment analysis of the translocated W chromosome shows that the female-specific randomly amplified polymorphic DNA (RAPD) marker, Female-218, is derived from the second chromosome fragment region of the translocated W chromosome of the sex-limited p^B silkworm (*Bombyx mori*) strain. *Hereditas* **138**: 148-153.

[35]Abe H, Kanehara M, Terada T, Ohbayashi F, Shimada T, Kawai S, Suzuki M, Sugasaki T, Oshiki T (1998a) Identification of novel random amplified polymorphic DNAs (RAPDs) on the W chromosome of the domesticated silkworm, *Bombyx mori*, and the

wild silkworm, *B. mandarina*, and their retrotransposable element-related nucleotide sequences. *Genes Genet Syst* **73**: 243-254.

[36]Abe H, Seki M, Ohbayashi F, Tanaka N, Yamashita J, Fujii T, Yokoyama T, Takahashi M, Banno Y, Sahara K, Yoshido A, Ihara J, Yasukochi Y, Mita K, Ajimura M, Suzuki MG, Oshiki T, Shimada T (2005b) Partial deletions of the W chromosome due to reciprocal translocation in the silkworm *Bombyx mori*. *Insect Mol Biol* **14**: 339-352.

[37]Ohbayashi F, Shimada T, Sugasaki T, Kawai S, Mita K, Oshiki T, Abe H (1998) Molecular structure of the *copia*-like retrotransposable element *Yokozuna* on the W chromosome of the silkworm, *Bombyx mori*. *Genes Genet Syst* **73**: 345-352.

[38] Abe H, OhbayashiF, Shimada T, Sugasaki T, Kawai S, Mita K, Oshiki T (2000) Molecular structure of a novel *gypsy*-Ty3-like retrotransposon (*Kabuki*) and nested retrotransposable elements on the W chromosome of the silkworm *Bombyx mori*. *Mol Gen Genet* **263**: 916-924.

[39]Wu C, Asakawa S, Shimizu N, Kawasaki S Yasukochi Y (1999) Construction and characterization of bacterial artificial chromosome libraries from the silkworm, *Bombyx mori*. *Mol Gen Genet* **261**: 698-706.

[40] Abe H, Ohbayashi F, Shimada T, Sugasaki T, Kawai S, Oshiki T (1998b) A complete full-length non-LTR retrotransposon, BMC1, on the W chromosome of the silkworm, *Bombyx mori*. *Genes Genet Syst* **73**: 353-358.

[41]Abe H, Sugasaki T, Terada T, Kanehara M, Ohbayashi F, Shimada T, Kawai S, Mita K, Oshiki T(2002) Nested retrotransposons on the W chromosome of the wild silkworm *Bombyx mandarina*. *Insect Mol Biol* **11**:307-314.

[42]Abe H, Mita K, Yasukochi Y, Oshiki T, Shimada T (2005a) Retrotransposable elements on the W chromosome of the silkworm *Bombyx mori*. *Cytogenet Genome Res* **110**: 144-151.

[43]Abe H, Fujii T, Tanaka N, Yokoyama T, Kakehashi H, Ajimura, M, Mita K, Banno Y, Yasukochi Y, Oshiki T, Nenoi M, Ishikawa T, Shimada T (2008) Identification of the female-determining region of the W chromosome in *Bombyx mori*. *Genetica* (in press).

[44]Fujii T, Yokoyama T, Ninagi O, Kakehashi K, Obara Y, Nenoi M, Ishikawa T, Mita K, Shimada T, Abe H (2007) Isolation and characterization of sex chromosome rearrangements generating male muscle dystrophy and female abnormal oogenesis in the silkworm, *Bombyx mori*. *Genetica* **130**: 267-280.

[45]Emanuel BS, Shaikh TH (2001) Segmental duplications: an 'expanding' role in genomic instability and disease. *Nature Review Genetics* **2:** 791-800.

[46]Bernstein AI, Mogami K, Donady JJ, Emerson CP (1983) *Drosophila* muscle myosin heavy chain encoded by a single gene in a cluster of muscle mutations. *Nature* **302**: 393-397.

[47]Mogami K, O'Donnell PT, Bernstein SI, Wright TRF, Emerson CP (1986) Mutations of the *Drosophila* myosin heavy-chain gene: Effects on transcription, myosin accumulation, and muscle function. *Proc Natl Acad Sci USA* **83**:1393-1397.

[48]Suzuki MG, Shimada T, Kobayashi M (1998) Absence of dosage compensation at the transcription level of a sex-linked gene in a female heterogametic insect, *Bombyx mori*. *Heredity* **81**: 275-283.

[49]Suzuki MG, Shimada T, Kobayashi M (1999) *Bm kettin*, homologue of the *Drosophila kettin* gene, is located on the Z chromosome in *Bombyx mori* and is not dosage compensated. *Heredity* **82**: 170-179.

[50]Traut W, Rathjens B (1973) Das W-Chromosom von *Ephestia kuehniella* (Lepidoptera) und die Arbeitung des Geschlechtschromatins. *Chromosoma* **41**: 437-446.
[51]Ennis TJ (1976) Sex chromatin and chromosome numbers in Lepidoptera. *Can J Cytol* **18**: 119-130.
[52]Traut W, Marec F (1996) Sex chromatin in Lepidoptera. *Quart Rev Biol* **71**: 239-256.
[53]Rathjens B (1974) Zur Funktion des W-Chromatins bei *Ephestia kuehniella* (Lepidoptera). Isolierung und Charaktesisierung von W-Chromatin-Mutanten. *Chromosoma* **47**: 21-44.
[54]Traut W, Weith A, Traut G (1986) Structural mutants of the W chromosome in *Ephestia* (Insecta, Lepidoptera). *Genetica* **70**: 69-79.
[55]Marec F, Mirchi R (1990) Genetic control of the pest Lepidoptera: gamma-ray induction of translocations between sex chromosomes of *Ephestia kuehniella* Zeller (Lepidoptera: Pyralidae). *J Stored Prod Res* **26**: 109-116.
[56]Marec F, Tothova A, Sahara K, Traut W (2001) Meiotic pairing of sex chromosome fragments and its relation to atypical transmission of a sex-linked marker in *Ephestia kuehniella* (Insecta: Lepidoptera). *Heredity* **87**: 659-671.
[57]Charlesworth D, Charlesworth B, Marais G (2005) Steps in the evolution of heteromorphic sex chromosomes. *Heredity* **95**: 118-128.
[58]Charlesworth B, Sniegowski P, Strphan W (1994) The evolutionary dynamics of repetitive DNA in eukaryotes. *Nature* **371**: 215-220.
[59]Charlesworth B (2001) Genome analysis: More *Drosophila* Y chromosome genes. *Current Biol* **11**: R182-R184.
[60]Steinemann M, Steinemann S (1992) Degenerating Y chromosome of *Drosophila miranda*: a trap for retrotransposons. *Proc Natl Acad Sci USA* **89**: 7591-7595.
[61]Steinemann M, Steinemann S Lottspeich S (1993) How Y chromosomes become genetically inert. Proc Natl Acad Sci USA 90: 5737-5741.
[62]Steinemann M, Steinemann S (1997) The enigma of Y chromosome degeneration: *TRAM*, a novel retrotransposon is preferentially located on the *neo-Y* chromosome of *Drosophila miranda. Genetics* **145**: 261-266.
[63]Steinemann M, Steinemann S (1998) Enigma of Y chromosome degeneration: *Neo-Y* and *Neo-X* chromosomes of *Drosophila miranda* a model for sex chromosome evolution. *Genetica* **102** (103): 409-420.
[64]Steinemann S, Steinemann M (2005) Y chromosome: born to be destroyed. *BioEssays* **27**: 1076-1083.
[65]Adams MD, Celniker SE, Holt RA, et al (2000) The genome sequence of *Drosophila melanogaster. Science* **287**: 2185-2195.
[66]Carvalho AB, Dobo BA, Vibranovski MD, Clark AG (2001) Identification of five new genes on the Y chromosome of *Drosophila melanogaster. Proc Natl Acad Sci USA* **98**: 132225-13230.
[67]Carvalho AB, Clark AG (2005) Y chromosome of *D. pseudoobscura* is not homologous to the ancestral *Drosophila Y. Science* **307**: 108-110.
[68] Krzywinski J, Nusskern DR, Kern MK, Besansky NJ (2004) Isolation and characterization of Y chromosome sequences from the African malaria mosquito *Anopheles bambiae. Genetics* **166**: 1291-1302.
[69]Krzywinski J, Sangare D, Besansky NJ (2005) Satellite DNA from the Y chromosome of the malaria vector *Anopheles gambiae. Genetics* **169**: 185-196.

[70]Krzywinski J, Chrystal MA, Besansky NJ (2006) Gene finding on the Y: fruitful strategy in *Drosophila* does not deliver in *Anopheles*. *Genetica* **126**: 369-375.

In: Genetic Translocations and Other Chromosome Aberrations ISBN: 978-1-60456-344-3
Editor: Guy T. Leyden, pp. 145-159 © 2008 Nova Science Publishers, Inc.

Chapter 4

Augmented Translocation of Nuclear Factor-κB into Nuclei and Down-Regulation of Apoptosis Related Gene Expression by Nipradilol, a Nonselective β- and Selective α_1-Blocker, is Related to Cytoprotection

Akira Ando[1,*], Shiho Kaneko[1], Naoko Kiriyama[1], Sawako Unezaki[2], Emiko Okuda-Ashitaka[2], Tadayoshi Okumura[2], Tetsuya Nishimura[1], Miyo Matsumura[1] and Seiji Ito[2]

Departments of [1]Ophthalmology and [2]Medical Chemistry, Kansai Medical University
10-15 Fumizono-cho, Moriguchi, Osaka 570-8507, Japan

Abstract

Apoptosis is a programmed cell death that plays an important role in the physiological process of development in animals and plants, and has also been reported to be involved in the pathogenesis and pathophysiology of various diseases. It is expected that control of apoptosis will be an important goal of new therapies for apoptosis related diseases. Glaucoma is a progressive optic neuropathy related to apoptosis that finally leads to blindness. Clinical features of the disease include visual field loss and a large cupping of the optic nerve head due to a glaucoma-specific deterioration of the retinal nerve fibers, which consist of ganglion cell axons, in association with elevated intraocular pressure (IOP). Although abnormally elevated IOP is a risk factor, apoptosis of retinal ganglion cells is a cause of the disease. The relationships between neuro-retinal tissue damage and apoptosis have been investigated and clarified in several experimental models, such as N-methyl-D-aspartate (NMDA)-induced retinal damage and secondary cell death caused by transection of the optic nerve, as well as in

* Correspondence to:Akira Ando, M.D., Ph. D.; Department of Ophthalmology, Kansai Medical University, 10-15 Fumizono-cho, Moriguchi, Osaka 570-8507, Japan. Tel: +81-6-6992-1001 (ext. 3324). Fax: +81-6-6993-2222

ocular hypertensive rats. Thus, the prevention of apoptosis of retinal ganglion cells is considered to be a novel strategy for treatment of glaucoma. Nipradilol [3,4-dihydro-8-(2-hydroxy-3-isopropyl-amino)-propoxy-3-nitroxy-2H-1-beozopyran]is a nonselective β- and selective α_1-adrenoceptor antagonist that has been used as an anti-glaucomatous agent due to its ability to lower IOP. Recently, it was reported that nipradilol also had a cytoprotective effect toward neuronal cells by nitric oxide (NO) donative action. NO has been shown to regulate apoptosis in both positive and negative directions in several experimental models, and many reports have shown that NO inhibits apoptosis by increasing caspase-3-like activity and cGMP-mediated mechanisms, as well as by inhibiting Bcl-2 cleavage and cytochrome *c* release *in vitro*. The gene expressions of some apoptosis associated factors are also regulated by the activity of the transcription factor nuclear factor-κB (NF-κB). However, whether nipradilol regulates the expression of apoptosis associated genes and/or NF-κB activity remains unclear. In the present study, we found that nipradilol down-regulated Bax, Caspase-9, and Smac/DIABLO gene expression using a real-time polymerase chain reaction method, and increased levels of translocation of NF-κB into nuclei were shown using an electrophoresis mobility shift assay of PC12 cells during apoptosis induced by serum deprivation. Activation of NF-κB was seen prior to the alterations of gene expression, therefore, the down-regulation of some apoptosis related genes in PC12 cells by nipradilol may be the result of NF-κB activation. Our results suggest that nipradilol regulates apoptosis associated gene expression and the translocation of NF-κB into nuclei as a means of neuro-protection against apoptosis via NO donative activity.

Key words: nuclear factor-κB, nitric oxide, cytoprotection, nipradilol, apoptosis.

Clinical Background

Apoptosis is a programmed cell death that plays important roles in the physiological process of development in animals and plants, and has also been reported to be involved in the pathogenesis and pathophysiology of various diseases. Various signals and factors contribute to apoptosis, and activation or inactivation of FasL, Fas, Bax, Bcl, a series of molecules known as caspases that form a complicated pathway, promotes or inhibits apoptosis. Control of apoptosis is expected to be an important goal of new therapies for apoptosis related diseases.

Glaucoma is a progressive optic neuropathy that is a leading cause of blindness, whose characteristic symptom is a progressive visual field defect due to damaged retinal nerve fibers accompanied by a decrease in glial tissue, which is observed as enlargement of cupping of the optic nerve head. It has been reported that apoptosis is associated with the mechanism of deterioration of the retinal nerve fiber and retinal ganglion cells in an experimental glaucoma model [1] as well as glaucoma patients [2, 3]. Various factors are considered to be causes of the progression of neuro-retinal deterioration in glaucoma, with intraocular pressure (IOP) known to be closely related to disease progression. Abnormally elevated IOP causes deprivation of nutritional factors, mechanical compression, ischemic cellular damage to retinal ganglion cells, and dysfunction of the glutamic acid transporter, after which the axon filaments of the retinal ganglion cells become damaged. Presently, the only effective treatment for glaucoma is IOP reduction, even in cases with normal tension glaucoma that exhibit IOP levels within the normal range. Therefore, anti-glaucomatous drugs are used to reduce IOP in order to protect the retinal nerve fibers and retinal ganglion cells. Relationships between neural retinal tissue disorders and apoptosis have been demonstrated in some experimental models, such as a retinal degenerative disease model using N-methyl-D-aspartate (NMDA) [4], a delayed retinal cell death model created by incision of the optic nerve [5], and an ocular hypertensive rat model [6]. However, no evidence-based therapy beside IOP reduction has been presented, therefore, a pharmaceutical glaucoma treatment that displays anti-apoptosis and neuroprotective actions is an important goal of research.

Neuro-protection and Nitric Oxide

It has been reported that nitric oxide (NO) is involved in apoptotic signal transduction, as NO regulates the process of apoptosis under certain conditions. Nitric oxide synthase (NOS) and its activities have been observed in the optic nerve head of glaucoma patients [7], as well as in the retina and front part of the uvea in rabbits [8]. NO inhibits apoptosis by preventing increases in caspase activity [9, 10] and via cGMP signaling [11-13], and by inhibition of Bcl-2 cleavage and cytochrome *c* release from mitochondria [14], suggesting that it has a cytoprotective effect. When inducible NOS (iNOS) activity was inhibited in a glaucomatous rat model, retinal ganglion cell damage was inhibited. Further, that study suggested that excessive NO produced by iNOS promoted retinal ganglion cell damage [15]. On the other hand, it has also been suggested that an increase in NO synthase by α-lipoic acid is involved in retinal protection [16]. It has been shown that NO plays an important role in neuro-retinal degeneration, including conditions other than glaucoma, and also clarified that NO contributes to apoptosis by both inhibition and acceleration in some experimental models [17-19].

Nipradilol (3,4-dihydro-8-(2-hydroxy-3-isopropyl-amino) propoxy-3-nitroxy-2H-1-beozopyran) (HYPADIL Kowa®) is a nonselective β- and selective α_1-adrenoceptor antagonist [20, 21], which has been used as an anti-glaucomatous drug in Japan since 1999, due to its IOP reduction effect to inhibit aqueous production and an accelerate aqueous drainage via uveoscleral outflow [22, 23]. Recently, a neuro-protective effect of nipradilol via NO donative action was reported [4, 5, 24, 25] owing to its nitroxy moiety [26]. In addition, NO increases the expression of Bcl-2 and has been shown to activate the transcription factor nuclear factor-kappaB (NF-κB) and regulate the gene expression of apoptosis related molecules. Tomita et al. [25] reported that a decrease in caspase-3 activity by nipradilol via *S*-nitrosylation and a cGMP-dependent pathway was implied in the mechanisms of cytoprotection in PC12 cells. Apoptosis is easily induced in PC12 cells by serum deprivation, which has been used to study glaucomatous neuro-retinal degeneration, even though the PC12 cell line was originally established from a rat pheochromocytoma. Kim et al. [13] reported that NO inhibited apoptosis by decreasing caspase-3-like activity via cGMP-mediated mechanisms in PC12 cells. However, it has also been shown that NO alters apoptosis associated gene expression during the inhibition of apoptosis, e.g. by elevation of Bcl-2 expression in splenic B lymphocytes [27]. In addition, the gene expressions of some apoptosis associated factors are also regulated by the activities of NF-κB [28-31].

Transcription Factor NF-κB

NF-κB regulates the expression of genes coding various kinds of protein, including apoptosis-related molecules. NF-κB activity has been shown to regulate apoptosis related gene expression in an anti-apoptotic direction in several cancer cell lines [28-31]. In contrast, when NF-κB is inactivated, Bax gene expression is up-regulated and Bcl-2, A1, and cIAP-2 gene expression is down-regulated, resulting in the induction of apoptosis [31]. From the results of experiments using a cancer cell line, suppression of NF-κB activity was demonstrated during anoxia induced apoptosis, while NF-κB was reactivated and the gene expression of Bcl-2 and Bcl-x was up-regulated and Bax gene expression was down-regulated by reoxygenation [29]. Thus, NF-κB contributes to apoptosis by controlling the gene expression of various molecules, though the mechanism remains to be elucidated.

Experimental Background

It was reported that nipradilol and NO decreased the activity of caspase-3 with *S*-nitrosylation via a pathway by which the cGMP cascade inhibits apoptosis [25], however, it remaines unclear whether nipradilol regulates apoptosis related genes or NF-κB activity. Therefore, we attempted to clarify the effects of nipradilol on the expression of apoptosis associated genes and activation of NF-κB in PC12 cells during apoptosis induced by serum deprivation, a well-established *in vitro* model of apoptosis, in order to investigate the mechanism of the neuronal cytoprotective effect of nipradilol. We found that nipradilol inhibited the expression of apoptosis related genes via NO donative action and activation of NF-κB. Next, we added different concentrations of nipradilol and denitronipradilol, which cancels the influence of NO donative action due to lack of nitroxy moiety, and incubated the cells for 12, 24, and 48 hours to examine the effects on cell viability as well as ladder formation of genomic DNA using agarose gel electrophoresis. In addition, we performed quantitative real-time PCR and an electrophoresis mobility shift assay (EMSA) to investigate the gene expression of the apoptosis related molecules Bax, Bcl-2, Fas, FasL, Caspase-1, 2, 3 and 9, p53, and Smac/DIABLO, and activity of transcription factor NF-κB.

Cell Culture

The PC12 cell line, established from a rat pheochromocytoma, can be induced into apoptosis by culturing in serum-free culture media. Cells were cultured in RPMI1640 medium supplemented with 10% horse serum, 5% fetal bovine serum, 2 mM of L-glutamine, 100 U/ml of penicillin, and 100 mg/ml of streptomycin in an atmosphere of 5% CO_2/95% air at 37 C in a 75-cm^2 flask and supplied with fresh medium every 3 days. When the cells reached sub-confluence, they were treated with 0.01% trypsin in a 0.002% EDTA solution, after which 1 x 10^5 cells were plated onto 6-well culture plates.

Cellular Viability

To examine the effects of nipradilol on the viability of PC12 cells during serum-deprivation, the cells were incubated in serum-free medium for 12, 24, or 48 hours with or without 0.01, 0.1, 1, or 10 μM of nipradilol or denitronipradilol, after which the number of cells was counted using a hemocytometer.

Electrophoresis Mobility Shift Assay (EMSA)

Translocation into nuclei of NF-κB was examined using an EMSA according to the method reported by Schreiber et al. [33]. This EMSA method can visualize the translocation of NF-κB in order to evaluate NF-κB activity by electrophoresis using polyacrylamide gel with a ^{32}P labeled NF-κB consensus sequenced DNA probe. Briefly, PC12 cells were incubated for 2 hours with or without 0.1 or 1 μM of nipradilol or denitronipradilol, followed by a brief washing 3 times with Tris-buffered saline (TBS, pH 7.4) on ice, after which they were collected with 0.5 ml of TBS. The cells were then lysed with 400 μl of lysis buffer [10 mM of HEPES-KOH (pH 7.9), 10 mM of KCl, 0.1 mM of ethylene diamine tetraric acid (EDTA), 0.1 mM of ethylene glycine tetraric acid (EGTA), 500 U/ml of Trasyrol, 1 mM of dithiothreitol (DTT), 0.5 mM of phenylmethylsulfonyl fluoride (PMSF)], after which 25 μl of 10% Nonidet P-40 was added. The tubes were then shaken vigorously for 1 minute at room temperature, followed by centrifugation at 15 000 rpm for 1 minute. The nuclear pellet was dissolved with 75 μl of lysis buffer (20 mM HEPES-KOH, 400 mM NaCl, 1 mM EDTA, 1 mM EGTA, 500 U/ml Trasyrol, 1 mM DTT, 0.5 mM PMSF) and shaken vigorously for 20 minutes at 4°C. After centrifugation at 15 000 rpm for 5 minutes at 4°C, the nuclear extract was collected into a new tube. Each nuclear extract containing 4 μg of protein was analyzed by electrophoresis on a 4.8% Tris-glycine-EDTA (TGE) polyacrylamide gel. For competition experiments, cold NF-κB (without ^{32}P labeling, 250-fold of ^{32}P-labeled probe) was added. The gels were dried using a gel dryer for 1 hour at 80°C and then exposed to X-ray film.

Genomic DNA Ladder Formation

Cells in the other tube were lysed with 200 μl of lysis buffer (150 mM NaCl, 10 mM Tris-HCl, 10 mM EDTA, 0.1% SDS) and 20 μg of proteinase K for 2 hours at 55°C. DNA was extracted using 200 μl of neutralized phenol (pH 7.0) and treated with 10 μg of RNase A for 30 minutes at 37°C. DNA was extracted using neutralized phenol-chloroform-isoamyl alcohol (50:49:1), precipitated using isopropyl alcohol, and dissolved with 20 μl of Tris-EDTA buffer (pH 8.0). Genomic DNA was separated by electrophoresis on a 2% agarose gel to examine ladder formation.

Principles of Real-Time Polymerase Chain Reaction

The semi-quantitative real time polymerase chain reaction (PCR) method used was developed to quantitatively analyze and evaluate the gene expressions of various molecules

[34]. With this method, the intensity of SYBR green dye fluorescence is increased in proportion to the quantity of the product during a PCR reaction. The number of PCR reaction cycles during which the emission of fluorescence was strong was defined as the threshold cycle. We calculated the ratio to the threshold cycle using an internal standard, such as S16 or GAPD, that was ubiquitously and stably expressed without varying by condition, and then evaluated the differences of threshold cycles between the samples. A difference of 1 cycle in the PCR reaction equaled a doubled gene expression between two samples. For example, the Bax gene expression in PC12 cells cultured in serum-containing medium was 1 and was doubled in cells cultured in serum-free medium and decreased by 1.08 times by culturing with 1 mM of nipradilol (Figure 5).

RNA Extraction and Real-Time Polymerase Chain Reaction for Apoptosis Related Genes

Cells were incubated in serum-free medium for 12, 24, or 48 hours with or without 0.1 or 1 μM of nipradilol or denitronipradilol. After a brief washing 3 times with PBS, each sample was separated into 2 tubes. One tube was used for RNA extraction using TRIzol® according to the manufacturer's instructions. Briefly, cells were homogenized with 200 μl of TRIzol®, after which 40 μl of chloroform was added and the mixture was shaken vigorously, followed

Table 1. Primer sequences used for real-time PCR

Bax	forward 5'- TGCAGAGGATGATTGCTGAC -3' reverse 5'- GGAGGAAGTCCAGTGTCCAG -3'
Bcl-2	forward 5'- CGACTTTGCAGAGATGTCCA -3' reverse 5'- CATCCACAGAGCGATGTTGT -3'
Caspase-1	forward 5'- GTGGTTCCCTCAAGTTTTGC -3' reverse 5'- TGCAGCAGCAACTTCATTTC -3'
Caspase-2	forward 5'- CACCCTCTTCAAGCTTTTGG -3' reverse 5'- AACCTCTTGGAGCTGAAGCA -3'
Caspase-3	forward 5'- TACTCTACCGCACCCGGTTA -3' reverse 5'- CGGGATCTGTTTCTTTGCAT -3'
Caspase-9	forward 5'- TTTGAGGTGGCCTTCACTTC -3' reverse 5'- CAGGAACCGCTCTTCTTGTC -3'
Fas	forward 5'- CTCTTAGGGGGAACCTTTGC -3' reverse 5'- GACGGTCATCCCTCAAAGAA -3'
Fas-L	forward 5'- ATCATGAGCCAGATGGGAAG -3' reverse 5'- CTAAAAGACGGCCTCCTGTG -3'
p53	forward 5'- GTCTACGTCCCGCCATAAAA -3' reverse 5'- AGGCAGTGAAGGGACTAGCA -3'
Smac/DIABLO	forward 5'- CTCGGAGCGTAACCTTTCTG -3' reverse 5'- TCCTCATCAGTGCTTCGTTG -3'
S16	forward 5'- CACTGCAAACGGGGAAATGG-3' reverse 5'- TGAGATGGACTGTCGGATGG -3'

by centrifugation at 12 000 rpm for 15 minutes. Total RNA was precipitated by isopropyl alcohol and then dissolved with 10 μl of Rnase-free water. The gene expressions of Bax, Bcl-2, Fas, FasL, Caspase-1, 2, 3, and 9, p53, and Smac/DIABLO were examined using a semi-quantitative real time PCR method. Total RNA (1 μg) was reverse transcribed into cDNA. The first strand cDNA was amplified using Taq polymerase and anti-Taq antibody, with oligonucleotide primers specific for each molecule associated with apoptosis. The oligonucleotide primer sequences used for the PCR reactions are indicated in Table 1. The PCR conditions were as follows: hold at 94°C for 30 seconds, 40 cycles of amplification (94°C for 1 minute, 60°C for 1 minute, 72°C for 1 minute), and a final extension at 72°C for 5 minutes. The presence of corresponding PCR products was determined by electrophoresis on a 2% agarose gel in 1x Tris-acetate-EDTA buffer.

Results of Experiments

Suppression of Serum Deprivation Induced Apoptosis in PC12 Cells by Nipradilol

After 48 hours, the number of cells in the serum-free samples (Figure 1B) was significantly lower than that in the serum-added controls (Figure 1A). Nipradilol (0.1 μM) (Figure 1C), but not denitronipradilol (Figure 1D), increased the number of cells after 48 hours of incubation without serum ($p = 0.0285$ at 0.1 μM and $p = 0.0235$ at 1 μM) (Figure 2).

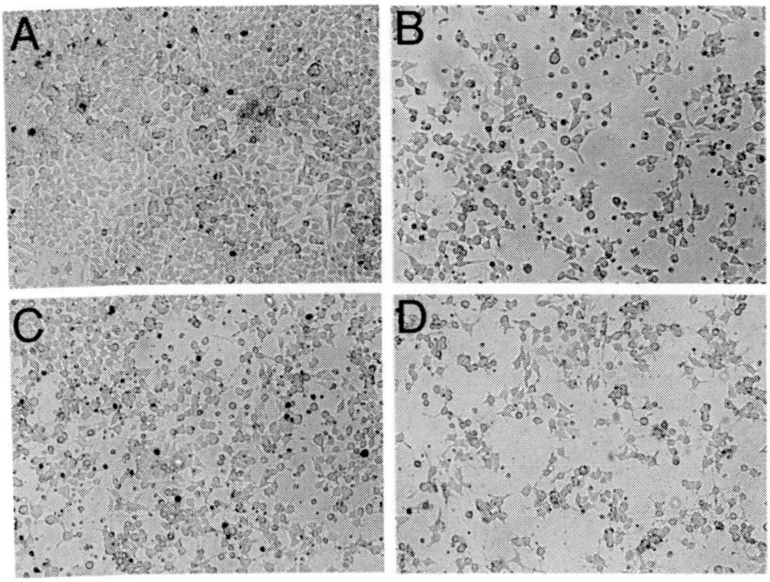

Figure 1. As compared to the serum-added control (A), PC12 cells were damaged and detached from the culture substrate of the dish 48 hours after serum deprivation (B). The damage was limited by the addition of 0.1 μM of nipradilol (C), while denitronipradilol (0.1 μM) showed no effects (D). Phase-contrast photomicrographs, original magnification x 100.

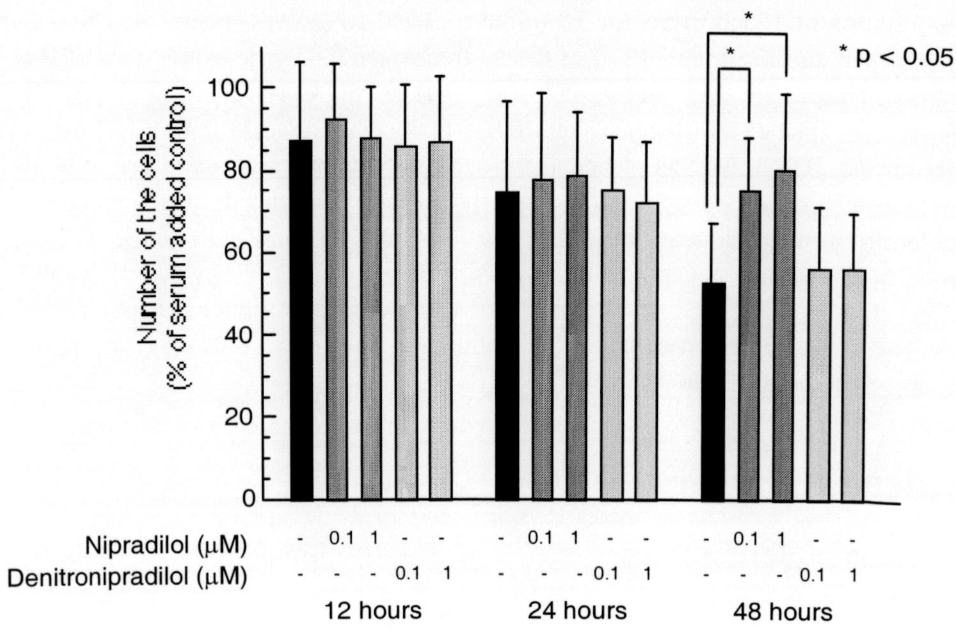

Figure 2. Following serum deprivation for 48 hours, the number of cells was half (closed column) that of the serum-added control. Nipradilol prevented the decrease in number of cells (hatched column) (p = 0.0285 at 0.1 μM and p = 0.0235 at 1 μM), while denitronipradilol showed no effect (dotted column).

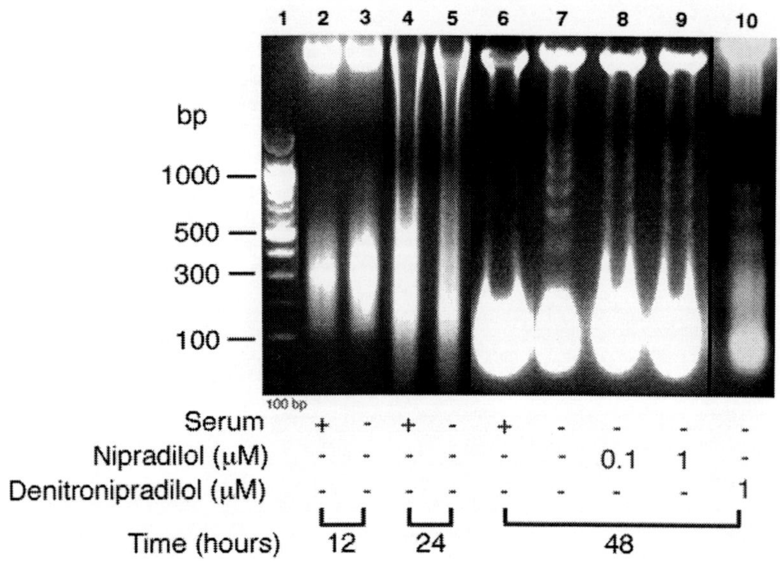

Figure 3. Genomic DNA ladder formation was clearly observed in PC12 cells at 48 hours after serum deprivation (lane 7), though not at 12 or 24 hours (lanes 3 and 5), or in the serum-added control (lanes 2, 4, and 6), as shown by 2% agarose gel electrophoresis. Nipradilol (0.1 and 1 μM, lanes 8 and 9) decreased ladder formation at 48 hours, whereas denitronipradilol (1 μM, lane 10) did not. Lane 1 shows the 100 bp DNA size marker. Each image is representative of 3 independent experiments.

Genomic DNA ladder formation was clearly observed between approximately 300 and 1200 bp at 48 hours (Figure 3, lane 7), though not at 12 or 24 hours (Figure 3, lanes 3 and 5), using 2% agarose gel electrophoresis, and was suppressed at 48 hours by the addition of 0.1 or 1 μM of nipradilol (Figure 3, lanes 8 and 9). Nipradilol (0.01 and 10 μM) did not exhibit a significant effect on PC12 cell viability or genomic DNA ladder formation (data not shown). Further, denitronipradilol showed no inhibitory effect toward DNA ladder formation (Figure 3, lane 10).

Translocation into Nuclei of NF-κB Increased by Nipradilol

Figure 4 shows the EMSA results, which indicate that constitutive NF-κB activity in PC12 cells (lane 1) was slightly suppressed by serum deprivation after 2 hours (lane 2). Further, nipradilol (0.1 and 1 μM) increased the translocation into nuclei of NF-κB (lane 3, 4) as compared to the serum-free controls at 2 hours (lane 2), whereas denitronipradilol had no effect (lanes 5, 6) and an excessive cold probe diminished the band (lane 7).

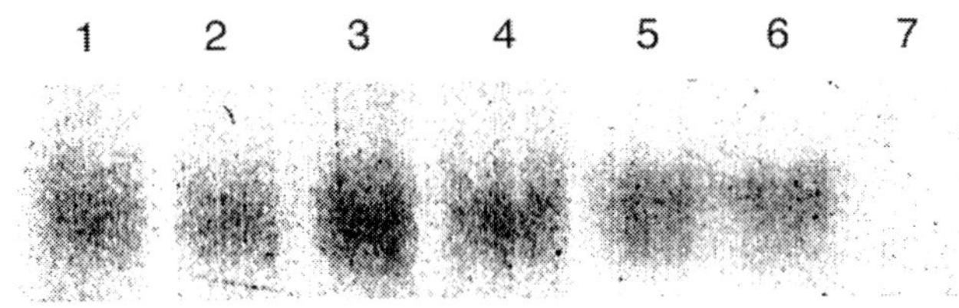

	1	2	3	4	5	6	7
Serum	+	-	-	-	-	-	+
Nipradilol (μM)	-	-	0.1	1	-	-	-
Denitronipradilol (μM)	-	-	-	-	0.1	1	-
Cold probe	-	-	-	-	-	-	+

Figure 4. Results of electrophoresis mobility shift assays showing that the activity of NF-κB was slightly suppressed by serum deprivation after 2 hours (lane 2), as compared to the serum-added sample (lane 1). Nipradilol (0.1 μM in lane 3 and 1 μM in lane 4) increased NF-κB activity, whereas denitronipradilol (0.1 μM in lane 5 and 1 μM in lane 6) had no effect. Further, the band was diminished by use of an excessive cold probe (lane 7). Each image is representative of 3 independent experiments.

Down-regulation of Bax, Caspase-9, and Smac/DIABLO Gene Expression by Nipradilol

Figure 5 shows a comparison of threshold cycle values. These results indicate that Bax gene expression was significantly increased ($p < 0.05$) in samples incubated with 0.1 and 1 μM of nipradilol in serum-free medium at 12 hours, and Caspase-9 and Smac/DIABLO gene expressions were also significantly increased ($p < 0.05$) in samples incubated with 0.1 μM of nipradilol in serum-free medium for 24 hours. Table 2 indicates the effects of nipradilol on apoptosis associated gene expression levels during serum deprivation in PC12 cells using quantitative real time PCR. Each value shows the ratio of gene expression as compared to the

Table 2. Effects of nipradilol on apoptosis associated gene expression (ratio to serum-free control, mean ± SE)

	12 hours			24 hours			48 hours		
Nipradilol (μM)	0	0.1	1	0	0.1	1	0	0.1	1
Bax	1.88 ± 0.88	0.96* ± 0.34	1.07* ± 0.21	1.40 ± 0.99	1.06 ± 0.63	1.31 ± 0.58	1.16 ± 0.43	0.89 ± 0.13	1.19 ± 0.52
Bcl-2	1.25 ± 0.53	0.65 ± 0.38	0.78 ± 0.53	0.75 ± 0.43	0.71 ± 0.53	1.09 ± 0.72	0.68 ± 0.23	0.81 ± 0.51	0.62 ± 0.47
Caspase-1	1.06 ± 0.71	1.72 ± 1.23	1.20 ± 1.14	1.17 ± 1.52	0.89 ± 1.05	1.46 ± 1.68	1.94 ± 1.64	2.71 ± 3.56	2.98 ± 2.81
Caspase-2	1.15 ± 0.75	0.43 ± 0.41	0.86 ± 0.56	0.98 ± 0.63	0.72 ± 0.63	1.60 ± 1.23	1.33 ± 0.66	1.05 ± 0.58	0.59 ± 0.32
Caspase-3	0.76 ± 0.16	0.39 ± 0.25	0.72 ± 0.24	1.35 ± 0.62	0.71 ± 0.38	1.07 ± 0.71	1.52 ± 1.32	1.13 ± 0.42	1.66 ± 1.62
Caspase-9	0.82 ± 0.10	0.42 ± 0.25	0.86 ± 0.26	1.34 ± 0.42	0.77* ± 0.45	1.18 ± 0.54	1.08 ± 0.38	0.80 ± 0.28	0.87 ± 0.49
Fas	1.01 ± 0.68	1.38 ± 1.66	2.22 ± 2.12	1.65 ± 1.31	1.74 ± 1.37	2.47 ± 2.40	2.06 ± 1.44	0.84 ± 0.47	1.31 ± 1.16
FasL	0.72 ± 0.49	1.40 ± 1.87	0.97 ± 0.85	1.13 ± 1.15	1.05 ± 0.86	1.75 ± 1.05	1.48 ± 1.10	1.07 ± 0.73	1.32 ± 0.73
p53	1.36 ± 0.31	1.25 ± 0.58	1.32 ± 0.44	1.00 ± 0.20	0.79 ± 0.17	1.23 ± 0.46	1.11 ± 0.33	1.17 ± 0.23	1.58 ± 0.98
Smac/DIABLO	1.20 ± 0.51	0.71 ± 0.51	1.23 ± 0.59	1.25 ± 0.34	0.52* ± 0.25	1.06 ± 0.31	1.28 ± 0.51	1.14 ± 0.29	1.31 ± 0.79

(n=4, * $p < 0.05$, ANOVA)

Table 3. Effects of denitronipradilol on apoptosis associated gene expression (ratio to serum-added control, mean ± SE)

	12 hours			24 hours		
Denitronipradilol (μM)	0	0.1	1	0	0.1	1
Bax	1.47 ± 0.42	1.24 ± 0.34	1.59 ± 0.38	1.33 ± 0.45	1.72 ± 0.68	1.46 ± 0.65
Caspase-9	0.79 ± 0.12	0.79 ± 0.20	0.85 ± 0.27	0.90 ± 0.18	0.84 ± 0.40	1.09 ± 0.57
Smac/DIABLO	0.89 ± 0.27	0.90 ± 0.22	0.71 ± 0.35	0.86 ± 0.22	0.61 ± 0.35	0.80 ± 0.27

(n=3)

serum-added controls. Nipradilol at 0.1 and 1 μM suppressed Bax gene expression to 0.96 and 1.07, respectively, compared to the serum-free (1.88) and serum-added samples (1.00) after 12 hours (Table 2). Further, 0.1 μM of the agent suppressed Capase-9 and Smac/DIABLO gene expressions to 0.77 and 0.52, respectively, compared to the serum-free (1.34 and 1.25, respectively) and serum-added samples (1.00) after 24 hours (Table 2), whereas it did not exhibit any effects at the other concentrations and periods tested. The expressions of the other genes examined in this experiment were not significantly affected by nipradilol (data not shown). In contrast, denitronipradilol did not have a statistically significant effect on the expressions of genes examined in these experiments, including with those molecules that exhibited a significant change by nipradilol (Table 3).

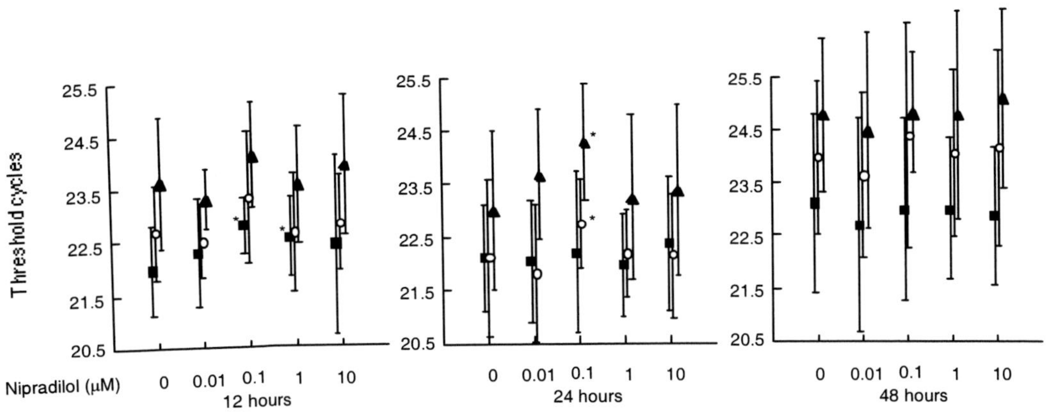

Figure 5. A comparison of threshold cycles using a real-time PCR method. Values were corrected by the S16 gene (internal control) in each experimental series. The number of threshold cycles that indicated Bax (■) gene expression in the cells was significantly increased with the addition of 0.1 and 1 μM of nipradilol in serum-free medium at 12 hours, and 0.1 μM of nipradilol in serum-free medium significantly increased that of Caspase-9 (○) and Smac/DIABLO (▲) at 24 hours (* $p < 0.05$).

Conclusion

NF-κB exists ubiquitously in the cytoplasm of cells combined with IκB in an inactivated form. When an apoptosis signal comes to the cell surface, apoptotic signal transduction begins, and phosphorylation of IkB dissembles the NF-κB/IκB combined construct and activated NF-κB shifts to the nucleus. Activated NF-κB regulates the transcription of messenger RNA that finally leads to protein production of various genes, which involves the apoptosis-related genes that exist downstream of the signal transduction cascade. NF-κB activity has been shown to regulate apoptosis related gene expression in an anti-apoptotic direction in several cancer cell lines [28-31]. Further, Poulaki et al. [31] reported that inhibition of constitutive NF-κB activity up-regulated Bax, whereas it down-regulated Bcl-2, A1, and cIAP-2, resulting in apoptosis induction. It was also reported that activation and inhibition of NF-κB down- and up-regulated, respectively, Bax gene expression in several cancer cell lines [30, 31], while activation of NF-κB up-regulated both Bcl-2 and Bcl-x during apoptosis induced by hypoxia followed by reoxygenation [29]. In the present study, activation of NF-κB was seen prior to the alteration of gene expression, therefore the down-regulation of some apoptosis related genes in PC12 cells by nipradilol might have been because of that activation. Also, NO may be located upstream of the process of activation of NF-κB.

NO plays a crucial role in neuro-retinal degeneration, including that involved with glaucoma [7, 15, 35-37]. It has been reported that NO regulated apoptosis in both positive and negative directions in several experimental models [17-19], while many authors also found that NO inhibited apoptosis by decreasing caspase-3-like activity and cGMP-mediated mechanisms, as well as inhibition of Bcl-2 cleavage and cytochrome c release *in vitro* [9-14]. Nipradilol down-regulates Bax, Caspase-9, and Smac/DIABLO gene expressions, and increases the translocation of NF-κB into the nuclei of PC12 cells during apoptosis induced by serum deprivation [32]. Bax induces apoptosis by contributing to the breakdown of the mitochondrial potential, through formation of an integral part of the anionic pores in the mitochondrial membrane [38]. Caspase-9, which is localized in the intermembrane space of mitochondria, appears to play a pivotal role in the downstream activation of other caspases [39]. Further, Smac/DIABLO is a mitochondrial protein that is released together with cytochrome c during apoptosis and promotes cytochrome c-dependent caspase activation by neutralizing the inhibitor of apoptosis proteins (IAPs) [40, 41]. Together with the results of those previous studies, our findings suggest that nipradilol plays roles at various points of the apoptosis pathway, such as inhibition of Caspase-3 and gene expression, especially around the mitochondria.

Glaucoma patients exhibit visual field loss and a large cupping of the optic nerve head due to disease-specific deterioration of the retinal nerve fibers, which consist of axons of ganglion cells. Elevated IOP may be involved in the pathogenesis of glaucoma and might induce trophic factor deprivation, mechanical and/or ischemic lesions, and glutamate transporter dysfunction, finally resulting in apoptosis of ganglion cells. Recently, apoptosis was reported to be involved in the pathogenesis of glaucoma, which has been shown to cause damage to the retinal nerve fibers in an experimental glaucoma model and clinical investigations of glaucoma patients [1-3]. In addition, the relationship between neuro-retinal deterioration and apoptosis has been examined and clarified in several experimental models, such as with N-methyl-D-aspartate (NMDA)-induced retinal damage [4], secondary cell death caused by transection of the optic nerve [5], and ocular hypertensive rat models [6].

Our results suggest that the NO donative action of nipradilol is critical for the prevention of serum deprivation induced apoptosis, down-regulation of apoptosis related gene expression, and increase of translocation of NF-κB into nuclei, because denitronipradilol did not exhibit any of those effects. Although further experiments are required, the present findings imply that nipradilol has an effect on apoptosis associated gene expression and translocation of NF-κB into nuclei during the prevention of apoptosis via NO donative action.

Acknowledgements

This study was supported in part by a grant from Kowa Company, Ltd. (Nagoya, Japan).

References

[1] Quigley H.A., Nickells R.W., Kerrigan L.A., Pease M.E., Thibault D.J., Zack D.J., 1995. Retinal ganglion cell death in experimental glaucoma and after axotomy occurs by apoptosis. *Invest. Ophthalmol. Vis. Sci.* **36**, 774-786.

[2] Kerrigan L.A., Zack D.J., Quigley H.A., Smith S.D., Pease M.E., 1997. TUNEL-positive ganglion cells in human primary open-angle glaucoma. *Arch. Ophthalmol.* **115**, 1031-1035.

[3] Nickells R.W., 1999. Apoptosis of retinal ganglion cells in glaucoma: an update of the molecular pathways involved in cell death. *Surv. Ophthalmol.* **43** Suppl 1, S151-61.

[4] Mizuno K., Koide T., Yoshimura M., Araie M., 2001. Neuroprotective effect and intraocular penetration of nipradilol, a beta-blocker with nitric oxide donative action. *Invest. Ophthalmol. Vis. Sci.* **42**, 688-694.

[5] Nakazawa T., Tomita H., Yamaguchi K., Sato Y., Shimura M., Kuwahara S., Tamai M., 2002. Neuroprotective effect of nipradilol on axotomized rat retinal ganglion cells. *Curr. Eye Res.* **24**, 114-122.

[6] McKinnon S.J., Lehman D.M., Kerrigan-Baumrind L.A., Merges C.A., Pease M.E., Kerrigan D.F., Ransom N.L., Tahzib N.G., Reitsamer H.A., Levkovitch-Verbin H., Quigley H.A., Zack D.J., 2002. Caspase activation and amyloid precursor protein cleavage in rat ocular hypertension. *Invest. Ophthal. Vis. Sci.* **43**, 1077-1087.

[7] Neufeld A.H., Hernandez M.R., Gonzalez M., 1997. Nitric oxide synthase in the human glaucomatous optic nerve head. *Arch. Ophthalmol.* **115**, 497-503.

[8] Osborne N.N., Barnett N.L., Herrera A.J., 1993. NADPH diaphorase localization and nitric oxide synthetase activity in the retina and anterior uvea of the rabbit eye. *Brain Res.* **610**, 194-198.

[9] Kim Y.M., Talanian R.V., Billiar T.R., 1997. Nitric oxide inhibits apoptosis by preventing increases in caspase-3-like activity via two distinct mechanisms. *J. Biol. Chem.* **272**, 31138-31148.

[10] Thippeswamy T., McKay J.S., Morris R., 2001. Bax and caspases are inhibited by endogenous nitric oxide in dorsal root ganglion neurons in vitro. *Eur. J. Neurosci.* **14**, 1229-1236.

[11] Farinelli S.E., Park D.S., Greene L.A., 1996. Nitric oxide delays the death of trophic factor-deprived PC12 cells and sympathetic neurons by a cGMP-mediated mechanism. *J. Neurosci.* **16**, 2325-2334.

[12] Estevez A.G., Spear N., Thompson J.A., Cornwell T.L., Radi R., Barbeito L., Beckman J.S., 1998. Nitric oxide-dependent production of cGMP supports the survival of rat embryonic motor neurons cultured with brain-derived neurotrophic factor. *J. Neurosci.* **18**, 3708-3714.

[13] Kim Y.M., Chung H.T., Kim S.S., Han J.A., Yoo Y.M., Kim K.M., Lee G.H., Yun H.Y., Green A., Li J., Simmons R.L., Billiar T.R., 1999. Nitric oxide protects PC12 cells from serum deprivation-induced apoptosis by cGMP-dependent inhibition of caspase signaling. *J. Neurosci.* **19**, 6740-6747.

[14] Kim Y.M., Kim T.H., Seol D.W., Talanian R.V., Billiar T.R., 1998. Nitric oxide suppression of apoptosis occurs in association with an inhibition of Bcl-2 cleavage and cytochrome c release. *J. Biol. Chem.* **273**, 31437-31441.

[15] Neufeld A.H., Kawai S., Das S., Vora S., Gachie E., Connor J.R., Manning P.T., 2002. Loss of retinal ganglion cells following retinal ischemia: the role of inducible nitric oxide synthase. *Exp. Eye Res.* **75**, 521-528.

[16] Chidlow G., Schmidt K.G., Wood J.P.M., Melena J., Osborne N.N., 2002. α-lipoic acid protects the retina against ischemia-reperfusion. *Neuropharmacology* **43**, 1015-1025.

[17] Lipton S.A., Choi Y.B., Pan Z.H., Lei S.Z., Chen H.S., Sucher N.J., Loscalzo J., Singel D.J., Stamler J.S., 1993. A redox-based mechanism for the neuroprotective and neurodestructive effects of nitric oxide and related nitroso-compounds. *Nature* **364**, 626-632.

[18] Kashii S., Mandai M., Kikuchi M., Honda Y., Tamura Y., Kaneda K., Akaike A., 1996. Dual actions of nitric oxide in N-methyl-D-aspartate receptor-mediated neurotoxicity in cultured retinal neurons. *Brain Res.* **711**, 93-101.

[19] Kim Y.M., Bombeck C.A., Billiar T.R., 1999. Nitric oxide as a bifunctional regulator of apoptosis. *Circ. Res.* **84**, 253-256.

[20] Uchida Y., Nakamura M., Shimizu S., Shirasawa Y., Fujii M., 1983. Vasoactive and β-adrenoceptor blocking properties of 3,4-dihydro-8-(2-hydroxy-3-isopropylamino) propoxy-3-nitroxy-2H-1-benzopyran (K-351), a new antihypertensive agent. *Arch. Int. Pharmacodyn. Ther.* **262**, 132-149.

[21] Ohira A., Wada Y., Fujii M., Nakamura M., Kasuya Y., Hamada Y., Shigenobu K., 1985. Effects of nipradilol (K-351) on alpha-adrenoceptor mediated responses in various isolated tissues. *Arch. Int. Pharmacodyn. Ther.* **278**, 61-71.

[22] Kanno M., Araie M., Tomita K., Sawanobori K., 1998. Effects of topical nipradilol, a beta-blocking agent with alpha-blocking and nitroglycerin-like activities, on aqueous humor dynamics and fundus circulation. *Invest. Ophthalmol. Vis. Sci.* **39**, 736-743.

[23] Kanno M., Araie M., Koibuchi H., Masuda K., 2000. Effects of topical nipradilol, a beta-blocking agent with alpha-blocking and nitroglycerin-like activities, on intraocular pressure and aqueous dynamics in humans. *Br. J. Ophthalmol.* **84**, 293-299.

[24] Kashiwagi K., Iizuka Y., Tsukahara S., 2002. Neuroprotective effects of nipradilol on purified cultured retinal ganglion cells. *J. Glaucoma* **11**, 231-238.

[25] Tomita H., Nakazawa T., Sugano E., Abe T., Tamai M., 2002. Nipradilol inhibits apoptosis by preventing the activation of caspase-3 via *S*-nitrosylation and the cGMP-dependent pathway. *Eur. J. Pharmacol.* **452**, 263-268.

[26] Adachi T., Hori S., Miyazaki K., Takahashi E., Nakagawa M., Udagawa A., Hayashi N., Aikawa N., Ogawa S., 1995. Rapid increase in plasma nitrite concentration following intravenous administration of nipradilol. *Eur. J. Pharmacol.* **286**, 201-204.

[27] Genaro A.M., Hortelano S., Alvarez A., Martinez C., Bosca L., 1995. Splenic B lymphocyte programmed cell death is prevented by nitric oxide release through mechanisms involving sustained Bcl-2 levels. *J. Clin. Invest.* **95**, 1884-1890.

[28] Wang C.Y., Mayo M.W., Baldwin Jr A.S., 1996. TNF- and cancer therapy-induced apoptosis: potentiation by inhibition of NF-kappaB. *Science* **274**, 784-787.

[29] Tamatani M., Mitsuda N., Matsuzaki H., Okado H., Miyake S., Vitek M.P., Yamaguchi A., Tohyama M., 2000. A pathway of neuronal apoptosis induced by hypoxia/reoxygenation: Roles of nuclear factor-κB and Bcl-2. *J. Neurochem.* **75**, 683-693.

[30] Bentires-Alj M., Dejardin E., Viatour P., Van Lint C., Froesch B., Reed J.C., Merville M.P., Bours V., 2001. Inhibition of the NF-κB transcription factor increases Bax expression in cancer cell lines. *Oncogene* **20**, 2805-2813.

[31] Poulaki V., Mitsiades C.S., Joussen A.M., Lappas A., Kirchhof B., Mitsiades N., 2002. Constitutive nuclear factor-κB activity is crucial for human retinoblastoma cell viability. *Am. J. Pathol.* **161**, 2229-2240.

[32] Ando, A et al: Cytoprotection by nipradilol, an anti-glaucomatous agent, via down-regulation of apoptosis related gene expression and activation of NF-kappaB. *Exp Eye Res* **80**: 501-507, 2005

[33] Schreiber E., Matthias P., Muller M.M., Schaffner W.,1989. Rapid detection of octamer binding proteins with 'mini-extracts', prepared from a small number of cells. *Nucleic. Acids Res.* **17**, 6419.

[34] Heid CA,

[35] Goldstein I.M., Ostwald P., Roth S., 1996. Nitric oxide: a review of its role in retinal function and disease. *Vision Res.* **36**, 2979-2994.

[36] Vorwerk C.K., Hyman B.T., Miller J.W., Husain D., Zurakowski D., Huang P.L., Fishman M.C., Dreyer E.B., 1997. The role of neuronal and endothelial nitric oxide synthase in retinal excitotoxicity. *Invest. Ophthalmol. Vis.* Sci. **38**, 2038-2044.

[37] Neufeld A.H., Sawada A., Becker B., 1999. Inhibition of nitric-oxide synthase 2 by aminoguanidine provides neuroprotection of retinal ganglion cells in a rat model of chronic glaucoma. *Proc. Natl. Acad. Sci. USA.* **96**, 9944-9948.

[38] Antonsson B., Conti F., Ciavatta A., Montessuit S., Lewis S., Martinou I., Bernasconi L., Bernard A., Mermod J.J., Mazzei G., Maundress K., Gambale F., Sadoul R., Martinou J.C., 1997. Inhibition of Bax channel-forming activity by BCL-2. *Science* **277**, 370-372.

[39] Slee E.A., Harte M.T., Kluck R.M., Wolf B.B., Casiano C.A., Newmeyer D.D., Wang H.G., Reed J.C., Nicholson D.W., Alnemri E.S., Green D.R., Martin S.J., 1999. Ordering the cytochrome c-initiated caspase cascade: hierarchical activation of caspases-2-3-6-7-8, and -10 in a caspase-9-dependent manner. *J. Cell Biol.* **144**, 281-292.

[40] Du C., Fang M., Li Y., Li L., Wang X., 2000. Smac, a mitochondrial protein that promotes cytochrome c-dependent caspase activation by eliminating IAP inhibition. *Cell.* **102**, 33-42.

[41] Verhagen A.M., Ekert P.G., Pakusch M., Silke J., Connolly L.M., Reid G.E., Moritz R.L., Simpson R.J., Vaux D.L., 2000. Identification of DIABLO, a mammalian protein that promotes apoptosis by binding to and antagonizing IAP proteins. *Cell.* **102**, 43-53.

In: Genetic Translocations and Other Chromosome Aberrations ISBN: 978-1-60456-344-3
Editor: Guy T. Leyden, pp. 161-172 © 2008 Nova Science Publishers, Inc.

Chapter 5

PRINS for Detection of Unique Chromosome Sequences and Single Gene Deletions

Stephen S. Wachtel* and Avirachan T. Tharapel
University of Tennessee, Memphis
Memphis, Tennessee, 38163, U.S.A.

Abstract

By *PRINS,* unlabeled oligonucleotide primers are annealed to denatured complementary sequences on chromosomes *in situ,* and extended in the presence of labeled nucleotides and Taq DNA polymerase. Chromosomal DNA acts as template for the extension, and the labeled nucleotides act as substrate for the DNA polymerase. The labeled extended target is visualized by fluorescence microscopy. Thus PRINS can be used to detect subtle chromosome rearrangements and microdeletions that are not amenable to analysis by standard analytical methods such as FISH. The PRINS method can also be used to identify single copy sequences and individual genes.

Keywords: Primed in situ labeling, PRINS, chromosome, gene mapping, single-copy gene, microdeletion

Introduction

Primed in situ labeling (PRINS) combines aspects of the polymerase chain reaction (PCR) and fluorescence in situ hybridization (FISH) for the localization of repetitive and unique nucleotide sequences on chromosomes. The PRINS signals, generated by annealing and subsequent extension of oligonucleotide primers, are identified and scored by

* E-mail address: swachtel@utmem.edu. Correspondence to: Prof. Stephen Wachtel, Department of Obstetrics and Gynecology, University of Tennessee 853 Jefferson Ave. Memphis TN 38163 USA.

fluorescence microscopy. Although the method was described in 1989 (Koch et al. 1989) it has not been used widely, ostensibly because of technical difficulties experienced in its application. Yet PRINS offers a unique, rapid approach to labeling, and could be used to identify any DNA sequence for which one or more oligonucleotide primers are available—i.e. any published sequence.

Anomalies of chromosome structure and number generate a variety of abnormal phenotypes including physical malformations, mental retardation and infertility. This is significant because some 0.7% of live-born children exhibit congenital malformations secondary to visible chromosome anomalies (Nussbaum et al., 2004). For that reason, fine point analysis of the chromosomes is a significant factor in the study of congenital diseases, with implications for clinical management and counseling. FISH has been an important and useful adjunct to the cytogenetic evaluation of chromosome rearrangements and gene deletions and duplications for more than a quarter century. FISH enables resolution of DNA sequences separated by only a few megabases in metaphase or prophase chromosomes (the low-end resolution is approximately 20 kb). Yet production of locus-specific FISH probes can be tedious and costly, and many of the probes are not commercially available. By contrast, PRINS is fast and relatively inexpensive, and the probes can be synthesized routinely.

Although most early applications of the PRINS method dealt with repetitive sequences such as the alpha-satellite, *Alu* and telomeric repeats (Pellestor et al., 1995; 1994; Gosden et al., 1991), PRINS can be used to identify single copy genes. In fact PRINS was used to identify the X-linked *Factor IX* gene more than 10 years ago (Cinti et al., 1993). That study was the first published example of identification of a single copy gene by PRINS, and it represents a significant biologic milestone, because every known gene sequence is a potential primer source. It follows that any gene or exon could be located in chromosome preparations *in situ,* provided that a unique nucleotide sequence is available.

In this chapter, we describe our experience with the PRINS method, with special emphasis on the detection of (i) unique sequences and single copy genes such as *SRY* and *SOX3* and (ii) microdeletions involving the *p53, RB1, SNRPN, GABRB3* and *DGCR2/TUPLE1* loci.

The Basic Method

In brief, the PRINS technique comprises four steps: (i) denaturation of chromosomal DNA, (ii) hybridization of primers *(annealing),* (iii) extension of DNA chains in the presence of labeled nucleotides and Taq DNA polymerase, and (iv) detection of the newly-synthesized labeled DNA by use of antibodies attached to fluorescent dyes. In our experience, a typical PRINS reaction mixture has a volume of 50 μl and contains: 200-500 pmol of primer, 0.2 mM dATP, dCTP and dGTP, 0.02 mM dTTP, 0.02 mM bio-16-dUTP, digoxygenin-11-dUTP, or fluorescein dUTP, 50 mM Tris-HCl (pH 8.3), 1.5 mM $MgCl_2$, 50 mM KCl, 0.01% bovine serum albumin and 2 units Taq polymerase. The reaction mixture is placed on the denatured chromosome preparation. Denaturation is accomplished by immersing slides in 70% formamide-2 x SSC for 2-3 minutes at 72°C, and dehydrating them through a series of cold ethanol washes. After the slides are air-dried, the reaction mixtures are layered onto the slides, and overlaid with coverslips that are sealed in place with rubber cement. The slides are put on a flat plate programmable thermal cycler and maintained at annealing temperatures for

5-10 minutes followed by extension for 20-30 minutes at 70-72°C. The reaction is stopped by immersing the slides in 500 mM NaCl-50 mM EDTA, pH8 for 5 minutes at 72°C. Labeled sequences are detected by immunocytochemistry and fluorescence microscopy. Biotin-complexed fragments can be visualized by use of fluoresceinated avidin, and digoxigenin, by use of fluoresceinated digoxigenin antibody. The chromosome preparations are counterstained with propidium iodide or DAPI, and scored under ultraviolet light.

Although our reactions are carried out on a programmable thermal cycler equipped with a flat plate for slides (HYBrite, Vysis Inc., Downers Grove IL), annealing and extension can readily be accomplished on thermoblocks or even in metal containers suspended in hot water baths (Koch et al. 1995). Of course temperature is a critical factor under these conditions, and must be carefully controlled. For detection of single copy genes and unique sequences, we used a modified PRINS method incorporating (i) multiple primers for the same locus, (ii) single-step annealing and extension, (iii) anti-Taq DNA polymerase antibody and (iv) stringent washing. A comprehensive description of the modified PRINS method is given in Tharapel and Wachtel (in press).

PRINS Primers

Oligonucleotide primers were synthesized and HPLC-purified at Research Genetics, Huntsville AL. The primers were stored at –20°C.

After counting the individual A, C, G and T nucleotide residues of the primers, annealing temperatures may be computed by use of either of the following formulas:

$$T_M = 69.3 + 0.41\,(\%\,G + C) - 650 / L$$

where L (length) = the total number of nucleotides in the primer.

$$T_M = 4\,(G + C) + 2\,(A + T)$$

When different temperatures are obtained, the results may be averaged. In general, satisfactory annealing occurs at temperatures between 55°C and 72°C. The higher temperatures increase annealing specificity.

The sequences of the PRINS primers are described in the various citations and in Table 1. When available, standard FISH probes were used to validate the data obtained with the PRINS primers.

Table 1. Primer sequences from cited studies

Locus	Sequence 5'—3'	Reference
RBM	GACCACCAGCTTCTTCGAGA	Kadandale et al. (2002)
	GGATGAACATTCCTCTAGAGG	
	GCCCTTCAGGAAGTCTGAGA	
	AGAGGCCATGCATACCGTGA	

Table 1. Continued

Locus	Sequence 5'—3'	Reference
DAZ	CTCTGCCTCTGGCTTTACCA GAGGAGGCATCTGGAAATCATT GGAAGCTGCTTTGGTAGATAC TAGGTTTCAGTGTTTGGATTCCG	Kadandale et al. (2002)
SNRP	ACGTACCAGAGGTGGAAGTC GTCTTCAGAAGCATCAAGTTT ATGACTGTGGAGGGGCCAC AGGCATTATGGCTCCTCCAC	Tharapel et al. (2002)
GABRB3	TTCTGCAGGCACCGTCGGG AGAGGTCCGAGTCCCGTG AGGAAGGCTTTTCGGCATCT TAAGTGGGCTGCGCGCGGC	Tharapel et al. (2002)
DGCR/TUPLE1	GCGGTGCAACCCTGGGCA CAGTATGTTATCACTGGCCGG CGCCTGGTCGTCAGCTGC CAACCACAATGGCAAGCCGA AAGTTCCCAGAAATTCTAGCTA GTCATCCATGAACTGTTTGAC	Tharapel et al. (2002)

PRINS in Detection of the *SRY* Gene

The *SRY (sex-determining region Y)* gene, discovered by Sinclair et al. (1990), is the switch that determines the sex of the mammalian embryo. In the presence of *SRY,* the indifferent embryonic gonad becomes a testis. The testis secretes testosterone and antimullerian hormone and the male body type is induced. In the absence of *SRY,* the gonad becomes an ovary. In the absence of testosterone and antimullerian hormone, the female body type is induced.

Errors in the expression of *SRY* accordingly generate errors in the development of the sex phenotype in humans and in mammals generally. For example translocation of *SRY* from the Y to the X chromosome in the male germ line can cause development of the male phenotype in the absence of the Y chromosome *(XX male syndrome).* And deletion or mutation of *SRY* can cause development of the female phenotype in the presence of the Y chromosome *(XY gonadal dysgenesis)* (XY gonadal dysgenesis can also occur in the presence of *SRY* and XX male syndrome can occur in the absence of *SRY)* (Wachtel, in press; and see below).

Study of the *SRY* gene can accordingly provide insights into the etiology and nature of various sex-reversed syndromes with implications for theory and clinical management. The *SRY* gene is located on the short arm of the human Y chromosome, in distal Yp11.2, close to the pseudoautosomal region. It is 2.1 kb in length and contains a single exon, including a conservative motif found among the high mobility group (HMG) proteins. The conservative motif spans 79 amino acids that are associated with DNA binding. This supports the presumptive role of the SRY protein as a transcription factor.

Using the PRINS method, we studied the *SRY* gene in two 46,XX males, in a woman with 46,XY gonadal dysgenesis, and in an azoospermic man with an Xp-Yp interchange (Kadandale et al., 2000). One of the XX males was a 30-year-old who had been referred for small testes and infertility—hallmarks of the XX male syndrome. The other XX male was a newborn referred for small phallus, third degree hypospadias, and bifid scrotum with labioscrotal folds. The scrotum contained abnormally small testes. These abnormalities are encountered in some 30% of XX males. The XY female was evaluated originally for primary amenorrhea. She was 19 years old, and 1.65 m tall with normal breast development but her uterus was hypoplastic.

The azoospermic male was 27 years old, and was first studied for infertility. His phenotype was unremarkable, but the karyotype in white blood cells was 45,der(X)t(X;Y)(p22.3;p11.2)/46,t(X;Y)(p22.3;p11.2) signifying the presence of mosaic cell lines, each containing a Y;X translocation.

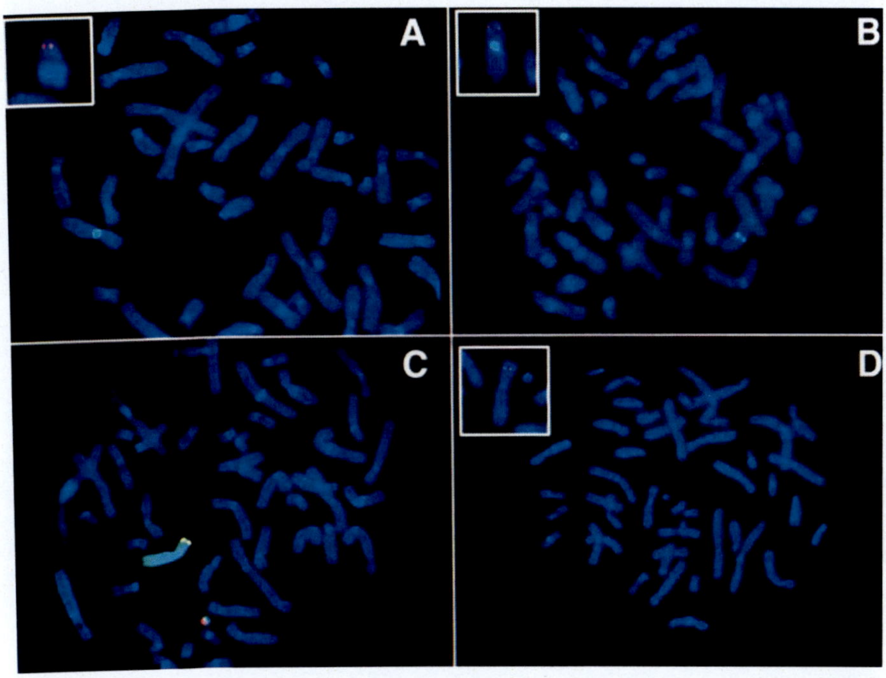

Figure 1. A. PRINS signals (red) for *SRY* in normal male control. The X centromere is stained green by FISH. Similar PRINS and FISH signals were observed in female with 46,XY gonadal dysgenesis B. PRINS signals (red) for SRY in der(X) of XX male. X centromeres are stained green by FISH C. The X chromosome is stained green by FISH in cells of the azoospermic male (see text). Y-derived material (yellow) is seen on Xp and the red-stained der(Y) D. SRY signals (red dots) identified by PRINS on Xp of the azoospermic male.

By use of the modified PRINS method described above, we detected SRY on band Yp11.2-p11.31 [ISCN, 2005] in normal male controls (Fig. 1A) and in cells from the XY woman. Presence of the *SRY* testis-determining gene in a woman with gonadal dysgenesis is not unusual, occurring in more than 80% of cases. It is attributed to perturbation of other genes in the sex-determining pathway or to the pleiotropic effects of genes such as *DAX1 (dosage sensitive sex reversal-adrenal hypoplasia congenital-critical region of the X chromosome, gene*

1), which block SRY-mediated testis differentiation when they are duplicated (Wachtel, 1998). Further application of PRINS could help to determine whether *DAX1* was in fact duplicated in cells from this patient, or whether failure of male development was due to some other lesion. Thus PRINS can be used to elucidate specific genetic defects.

We observed the SRY signal in Xp22 in the derivative X of the 30-year-old XX male (Fig. 1B) and in the derivative X chromosome of the azoospermic male (by FISH in Fig. 1C; by PRINS in Fig 1D). But the signal was absent in cells from the XX male infant with hypospadias. Absence of signal in the infant with hypospadias is not surprising. Approximately 70% of cases of XX male syndrome are due to X;Y crossover in the germ line of the father. These cases are notable for presence of *SRY.* The other 30%, often characterized by ambiguity of the external genitalia, are attributable to perturbations downstream of *SRY* in the sex-determining pathway. These cases are notable for absence of *SRY.* In that connection, duplication of *SOX9 (SRY-like HMG box 9)* on chromosome 17 has been described in some SRY-negative XX males, consistent with the idea that SRY-mediated induction of *SOX9* is a crucial step in differentiation of the male gonad (reviewed in Wachtel, in press). The question whether *SOX9* is duplicated in this case, could be addressed by application of the methods describe in this chapter.

The SOX Family of Genes: Detection of *SOX3*

The *SRY* testis-determining gene is one of a family of genes encoding high mobility group (HMG) transcription factors with a highly conservative DNA binding motif called the *HMG box.* The *SOX9* gene, discussed in the preceding paragraph, is another. Many of the *SOX* genes exhibit tissue-specific patterns of expression in early embryogenesis, consistent with the idea that they play significant roles in tissue-specific induction.

SOX3 (SRY-like HMG box 3)—yet another member of the SOX family—is located on the long arm of the human X chromosome at Xq26q27 (Stevanovic et al., 1993). *SOX3* is 2.4 kb in length and consists of a single exon. The corresponding *Sox3* gene of the mouse, also on the X, is required for neural development and gametogenesis (Raverot et al., 2005). Because the *SOX3* sequence is most closely related to that of *SRY,* and because *SOX3* is X-linked, it has been suggested that *SOX3* represents a primordial X-sequence from which the Y-linked *SRY* gene is derived (Foster and Graves, 1994).

In order to further evaluate the efficacy of the PRINS method in visualizing single copy genes, we tested the cells of a normal 46,XX female for presence of *SOX3.* The fluorescent gene signal was visualized as a doublet—two green dots—representing the two chromatids in the long arm on both X chromosomes. For these studies, identification of the X was facilitated by use of FISH with the DXZ1 probe, after the PRINS reaction, to highlight the X centromere.

Deletion Studies of RBM and DAZ in an Azoospermic Man

Genes governing spermatogenesis have been mapped to the long arm of the human Y chromosome at proximal Yq11.2 in a domain called the *azoospermia factor region (AZF).*

Thus deletions within the AZF region are correlated with various defects of sperm production. AZF has been subdivided into 4 subregions—AZFa, AZFb, AZFc and AZFd. Candidate genes for azoospermia include *RBM (RNA binding motif)* corresponding to AZFb, and *DAZ (deleted in azoospermia)* corresponding to AZFc. Both these genes are expressed in the testes, and both are situated in Yq11.2 (Reijo et al., 1996).

We evaluated cells from a 27-year-old man referred for infertility. He was 172 cm tall and weighed 68.5 kg. He appeared to be in good health but there was a paucity of body hair, and left and right testes were soft and measured 3 cm and 3.5 cm. No sperm cells were found in the seminal fluid (azoospermia). A mosaic karyotype was identified by GTG-banding: 45,der(X)t(X;Y)(p22.3;p11.2)/46,t(X;Y)(p22.3;p11.2). Studies by FISH with dual color paint probes confirmed reciprocal translocations between the short arms of the X and Y chromosomes, and showed absence of Yq11.2, i.e. the part of the Y that contains the AZF region.

PRINS was used to evaluate *SRY, RBM* and *DAZ* in the patient's lymphocytes. *SRY* was located on Xp22, confirming X;Y translocation. The *RBM* and *DAZ* candidate genes were absent, consistent with a role for those elements in spermatogenesis. In aggregate, the data obtained in this study led us to suspect that the derivative Y chromosome was in fact a ring chromosome, formed by Yp-Xp exchange and loss of material from Yq. This could explain loss of the ring in some cells of the patient, and the resulting mosaic karyotype (see discussion in Kadandale et al., 2002).

Gene Deletions in Cancer

The preceding data show too that the PRINS method can be employed gainfully in the detection of microdeletions and subtle chromosome rearrangements. This has been borne out by the use of PRINS in deletion studies of the *RB1 (retinoblastoma 1)* and *p53* tumor suppressor genes in the human (Tharapel and Kadandale, 2002). Rearrangements on chromosomes 13 and 14—in particular regions 13q14 and 17p13—occur frequently in leukemias and lymphomas. But they may not be observable by the standard methods of cytogenetics. In fact, deletions within these regions occur in >50% of cases, yet the karyotypes appear to be normal because the deletions are submicroscopic. In the past, the deletions have been detected by FISH.

Tharapel and Kadandale (2002) applied PRINS in patients with known deletions in these regions and compared their results with those obtained with FISH. They studied 5 patients with deletions in 13q14 (the *RB1* region) and 5 patients with deletions in 17p13 (the *p53* region). The patients in this trial had a normal chromosome 17 and an isochromosome of the 17 long arm, designated i(17q), and were thus monosomic for 17p.

FISH was performed before PRINS. The chromosome preparations from bone marrow cell pellets that were used for FISH were also used for PRINS: Normal controls were processed alongside the patient samples. For each patient and each control, 10 metaphase spreads and 40 interphase nuclei were evaluated in each procedure. FISH and PRINS methods yielded comparable results for normal and deletion patterns in all cases—for both *p53* and *RB1* genes. Although the PRINS signals were smaller than the FISH signals, the authors described the PRINS signals as "distinct and unambiguous."

In such cases, tumors presumably arise as a result of *RB1* or *p53* mutations in the *intact* chromosomes of hematopoietic cells. Having already lost a wild-type copy of the relevant gene due to microdeletion, the affected cells lose suppressor function, and the result is uncontrolled cell division leading to cancer. Neoplastic hematopoietic cells from the patients should accordingly exhibit only one PRINS signal, representing the remaining copy of the *RB1* or *p53* gene. Cells from normal controls should exhibit two signals.

Diagnostic Applications: Gene Deletions in Prader Willi and Angelman Syndromes

Whether a particular chromosome comes from the mother or father can effect significantly the expression of clinical syndromes resulting from genetic lesions on that chromosome. This phenomenon is called *parental imprinting*. An example is the differential appearance of Huntington disease, a degenerative disorder of the central nervous system with progressive dementia and spasmodic movements of the face and limbs, caused by a dominant gene on chromosome 4 (Nussbaum et al., 2004). Patients who inherit the gene from their father show symptoms in adolescence—*early onset* disease. Patients who inherit the gene from their mother show symptoms in middle age—*late onset* disease. A more dramatic example is provided by the differential response to deletions in the proximal long arm of chromosome 15 (q11.2-q13). When the affected chromosome is paternal in origin, the result is the Prader-Willi syndrome. When the affected chromosome is maternal, the result is Angelman syndrome.

Prader-Willi syndrome (PWS) occurs in 1/10,000 to 1/25,000 live births. It is characterized by mild mental retardation, obesity, short stature, small genitalia and small hands and feet. Affected children have a region 15 (q11.2-q13) with genetic information that comes only from the mother because the paternal chromosome 15 has been deleted for this region. The other side of the coin is the phenotypically distinct Angelman syndrome (AS), also called the "happy puppet" syndrome. Children with this condition are severely retarded and cannot speak. They are characterized by spastic "puppet-like" movements, easily provoked laughter, microcephaly and extruding tongue. Although the chromosomal region deleted in this condition is the same one that is deleted in PWS (15q11.2-q13), it is deleted here from the *maternal* copy of chromosome 15. Thus patients with AS have genetic information in 15 (q11.2-q13) deriving only from the father. In some cases, the disorder is found in children who have inherited both chromosomes 15 from the father and thereby lack a maternal 15 *(uniparental disomy)* (Nicholls, 1993).

For evaluation of patients with PWS or AS, we used primer sequences specific for the *SNRPN (small nuclear ribonucleoprotein polypeptide N)* and *GABRB3 (gamma-aminobutyric-acid receptor beta-3 subunit precursor)* genes, both thought to reside normally in the deleted regions of chromosome 15. PRINS primer sequences were derived from published sequences for these genes. We studied 4 males and 4 females with PWS, a male and a female with AS (Tharapel et al., 2002). Three normal males and 2 normal females served as controls. In all cases, venous blood was collected in sodium heparin.

Phytohemagglutinin-stimulated lymphocyte cultures were harvested, and slides prepared according to conventional protocols. Chromosome preparations of the patients were exposed to Wright stain after treatment with a dilute solution of pancreatin, for G-banding; 20 cells were studied for each patient. For FISH, we obtained *SNRPN* probes from Vysis Inc., Downer's Grove, IL. The *SNRPN* locus is generally deleted in patients with PWS and AS. We analyzed 20 metaphase spreads and 80 interphase nuclei from each patient and compared the results with those obtained after PRINS.

In all cases, our PRINS results were in full agreement with the results obtained by FISH (Table 1). Thus, among the 5 controls, a normal staining pattern was observed after PRINS in >90% cells studied for *SNRPN* and in >90% studied for *GABRB3.* And the corresponding pattern was observed after FISH with probes for the same genes. But among 4 patients with PWS and one with AS, a deletion was observed after PRINS in >80% of cells studied for *SNRPN* and *GABRB3.* And the corresponding pattern was observed for the same patients after FISH with a probe for *SNRPN* (the *GABRB3* locus was not studied with FISH in these patients). Normal staining patterns were observed by FISH and PRINS in the other 4 patients with PWS and one with AS (Table 2). Normal patterns could be attributed to uniparental disomy in these patients, or occasionally, to ambiguity of diagnosis.

Table 2. Analysis by PRINS and FISH of 15q11.2-q13 microdeletions

Subjects	N	Test	Probe/ locus	Labeling pattern	In percent cells **
Controls	5	PRINS	SNRPN	Normal	>90
			GABRB3	Normal	>90
PWS/AS	5	PRINS	SNRPN	Normal	>90
			GABRB3	Normal	>90
PWS/AS *	5	PRINS	SNRPN	Deleted	>80
			GABRB3	Deleted	>80
Controls	5	FISH	SNRPN	Normal	>95
PWS/AS	5	FISH	SNRPN	Normal	>95
PWS/AS *	5	FISH	SNRPN	Deleted	>90

* Four patients with PWS; one with AS

** Scores of less than 100% are attributed to poor nuclear or chromosomal morphology

100 cells tested for each subject

Diagnostic Applications: Gene Deletions in DiGeorge/Velocardiofacial Syndrome

The DiGeorge syndrome (DGS) is characterized by congenital absence of the thymus and parathyroid glands with multiple associated abnormalities including cardiac and aortic malformations, characteristic facial appearance with malformed ears, micrognathia, cleft palate, short philtrum etc. In the absence of an effective T-cell population, there is an increased susceptibility to viral and fungal infections. Most patients with DGS have a hemizygous deletion in chromosome 22(q11.2) which subsumes these loci: *DGCR2 (DiGeorge syndrome critical region gene 2)* and *TUPLE1 (TUP like enhancer of split gene 1) and N25.* The incidence of children with the deletion is about 1/4000 live births. The DiGeorge critical region spans 2 mb and is deleted in about 90% of DiGeorge patients. The

N25 locus (D22S75) is situated in the proximal end of the DiGeorge critical region, whereas the *TUPLE1* gene is located within a minimum critical region of 480-575 kb.

The Velocardiofacial syndrome (VCFS) is associated with deletions in the same region of chromosome 22 that DiGeorge syndrome (DGS) is. These patients sometimes show extended deletions: 22(q11.2-q12.3). Patients with this condition commonly exhibit cleft palate, cardiac malformations, abnormal facies, learning disabilities, short stature, immune defficiency, hypoparathyroidism and other malformations. The incidence has been reported as greater than 1/1000, although VCFS and DGS appear to have a similar etiology and thus may represent alternative molecular lesions within the same critical region of chromosome 22.

We studied 3 females and 4 males with DGS, all of whom had the characteristic malformations of derivatives of the third and fourth paryngeal pouches: hypoplastic thymus and parathyroid, diverse abnormalities of the aortic arch, and mild dysmorphic features. We also studied 2 females and 1 male with VCFS. Those patients were notable for short stature, elongated facies, submucosal cleft palate, tetralogy of Fallot, long thin hands and feet, and mild to moderate mental retardation.

As in the previous cases, we collected venous blood in heparinized syringes, and analyzed 20 metaphase spreads and 80 interphase nuclei from each patient. For FISH, we used the N25 cosmid probe (D22S75) which hybridizes with sequences in the proximal end of the DiGeorge critical region. Absence of N25 signal would be expected in cases of microdeletion within 22q11.2. In fact the deletions in VCFS and DGS are usually indistinguishable at the chromosomal level, consistent with a similar etiology in these conditions.

For PRINS, we use primers derived from published sequences of the *DGCR2* and *TUPLE1* genes (Table 1). The results of our tests in 10 patients with DGS/VCFS are shown in Table 3. By use of PRINS, we identified deletions in >80% of cells from 5 patients with DGS and 1 patient with VCFS. The results were confirmed by FISH. Deletions were not found in controls or in 4 of the patients. Analyses by PRINS and FISH gave similar results in all cases (Tharapel et al., 2002).

Table 3. Analysis by PRINS and FISH of 22q11.2 microdeletions

Subjects	N	Test	Probe/ locus	Labeling pattern	In percent cells **
Controls	5	PRINS	DGCR2/TUPLE1	Normal	>90
DGS/VCFS	4	PRINS	DGCR2/TUPLE1	Normal	>90
DGS/VCFS*	6	PRINS	DGCR2/TUPLE1	Deleted	>80
Controls	5	FISH	N25	Normal	>95
DGS/VCFS	4	FISH	N25	Normal	>95
DGS/VCFS*	6	FISH	N25	Deleted	>90

* Five patients with DGS; one with VCFS
** Scores of less than 100% are attributed to poor nuclear or chromosomal morphology
100 cells tested for each subject

Conclusion

The studies reviewed above suggest potential applications of PRINS in the evaluation of individual genes, gene duplications and microdeletions. In fact, with further development and adaptation, the PRINS method might have a fundamental impact on medical genetics; if the method could be applied routinely, physical mapping of individual loci would be possible in most laboratories concerned with the study of human chromosomes and the diagnosis of genetic disease.

Certain difficulties remain to be overcome. The PRINS signal is small compared with the FISH signal. This problem may be alleviated by the use of more than one primer for single loci, and by combining annealing and extension into a single step (we have used as many as 5 primers for a single locus; Tharapel and Kadandale, 2002). Moreover, primer concentrations must be optimized for optimal hybridization signal. This is a critical step in the PRINS reaction, as improper concentration of primer can compromise the result. Background staining can be minimized by increasing stringency of washes, and low signal can be reduced by optimizing concentration of other reagents and by increasing reaction times (see Notes in Tharapel and Wachtel, in press).

Despite these shortcomings, we look forward to more widespread use of the PRINS method and to broader application of PRINS in the study of individual loci and gene deletions in the diagnosis of various cancers and microdeletions, and of genetic disorders in general.

References

Cinti, C., Santi, S., and Maraldi, N.M. (1993) Localization of single copy gene by PRINS technique. *Nucleic Acids Res.* **21**, 5799-5800.

Foster, J.W., and Graves, J.A. (1994). An SRY-related sequence on the marsupial X chromosome: implications for the evolution of the mammalian testis-determining gene. *Proc. Nat. Acad. Sci. U.S.A.* **91**, 1927-1931.

Gosden, J., Hanratty, D., Starling, J., Fantes, J., Mitchell, A., and Porteous, D. (1991). Oligonucleotide-primed in situ DNA synthesis (PRINS): a method for chromosome mapping, banding, and investigation of sequence organization. *Cytogenet. Cell Genet.* **57**, 100-104.

ICSN 2005. *An International System for Human Cytogenetic Nomenclature* (2005). (2005). L.G.Shaffer, and N. Tommerup (Eds.). S. Karger, Basel.

Kadandale, J.S., Wachtel, S.S., Tunca, Y., Martens, P.R., Wilroy, R.S., and Tharapel, A.T. (2002). Deletion of RBM and DAZ in azoospermia: evaluation by PRINS. *Amer. J. Med. Genet.* **107**, 105-108.

Kadandale, J.S., Wachtel, S.S., Tunca, Y., Wilroy, R.S. Jr., Martens, P.R., and Tharapel, A.T. (2000). Location of SRY by primed in situ labeling in XX and XY sex reversal. *Amer. J. Med. Genet.* **95**, 71-74.

Koch, J., Hindkjaer, J., Kolvraa, S., and Blund, L. (1995). Construction of a panel of chromosome-specific oligonucleotide probes (PRINS-primers) useful for the identification of individual human chromosomes in situ. *Cytogenet. Cell Genet.* **71**, 142-147.

Koch, J.E., Kolvraa, S., Gregerson, N., Petersen, K.B., and Bolund, L. (1989). Oligonucleotide-priming methods for the chromosome-specific labeling of alpha satellite DNA in situ. *Chromosoma* **98**, 259-265.

Nicholls, R.D. (1993). Genomic imprinting and candidate genes in the Prader-Willi and Angelman syndromes. *Curr. Opin. Genet. Dev.* **3**, 445-446.

Nussbaum, R., Thompson, M.W., McInnes, R.K., and Willard, H.F. (2004). *Thompson and Thompson Genetics in Medicine.* Philadelphia: W.B. Saunders.

Pellestor, F., Girardet, A., Andreo, B., and Charlieu, J.P. (1994). A polymorphic alpha satellite sequence specific for human chromosomes 13 detected by oligonucleotide primed in situ labeling (PRINS). *Hum. Genet.* **94**, 346-348.

Pellestor, F., Girardet, A., Lefort, G., Andreo, B., and Charlieu, J.P. (1995). Use of primed in situ labeling (PRINS) technique for a rapid detection of chromosomes 13, 16, 18, 21, X and Y. *Hum. Genet.* **95**, 12-17.

Raverot, G., Weiss, J., Park, S.Y., Hurley, L., and Jameson, J.L. (2005). Sox3 expression in undifferentiated spermatogonia is required for the progression of spermatogenesis. *Dev. Biol.* **283**, 215-225.

Reijo, R., Alagappan, R.K., Patrizio, P., and Page, D.C. (1996). Severe oligospermia resulting from deletions of azoospermia factor gene on Y chromosome. *Lancet* **347**, 1290-1291.

Sinclair, A.H., Berta, P., Palmer, M.S., Hawkins, J.R., Griffiths, B.L., Smith, M.J., Foster, J.W., Frischauf, A.M., Lovell-Badge, R., and Goodfellow, P.N. (1990). A gene from the human sex-determining region encodes a protein with homology to a DNA binding motif. *Nature* **346**, 240-244.

Stevanovic, M., Lovell-Badge, R., Collignon, J., and Goodfellow, P.N. (1993). SOX3 is an X-linked gene related to SRY. *Hum. Molec. Genet.* **2**, 2013-2018.

Tharapel, A.T. and Wachtel, S.S. (In press). PRINS for detection of Gene deletions in cancer, In F. Pellestor (Ed.), *PRINS and in situ PCR.* Totowa, New Jersey: Humana Press. (In press).

Tharapel, A.T.,Kadandale, J.S., Martens, P.R., Wachtel, S.S., and Wilroy, R.S., Jr. (2002). Prader Willi/Angelman and DiGeorge/Velocardiofacial syndrome deletions: Diagnosis by primed in situ labeling (PRINS). *Amer. J. Med. Genet.* **107**, 119-122.

Tharapel S.A., and Kadandale, J.S. (2002). Primed in situ labeling (PRINS) for evaluation of gene deletions in cancer. *Amer. J. Med. Genet.* **107**, 123-126.

Wachtel, S.S. (1998). X-linked sex-reversing genes. *Cytogenet. Cell Genet.* **80**, 222-225.

Wachtel, S.S. SRY—sex reversal, In D. Ganten, and K. Ruckpaul (Eds.), *Encyclopedic Reference of Genomics and Proteomics in Molecular Medicine.* Berlin, Heidelberg, New York: Springer. (In press).

In: Genetic Translocations and Other Chromosome Aberrations
Editor: Guy T. Leyden, pp. 173-190
ISBN: 978-1-60456-344-3
© 2008 Nova Science Publishers, Inc.

Chapter 6

Efficiency of Intrachange Formation at Low Doses Is Determined by Condensation State of Interphase Chromosomes

S. G. Andreev and Yu. A. Eidelman
Institute of Biochemical Physics, Moscow, Russia

Abstract

A self-organization mechanism of establishment of interphase chromosome organization formerly proposed for large-scale interphase chromatin/chromosome self-organization is investigated here by computer simulation. Large-scale structure of interphase chromosome is suggested to be determined by volumetric interactions between chromosome subunits resulting in folding of subunit chain in a globular state. Structure units of interphase chromosome are studied as a spherical nonoverlapped structures consisting of a cluster of compact chromatin domains. These structure units/domain clusters were termed introduced as superdomains (SD) to emphasize their supradomain organization as well as domain cooperative behaviour within SDs. Unlike Gauss chain model for interphase chromosomes it is proposed the existence of globular state for interphase chromosomes. Globular state of interphase chromosomes is studied by dynamic Monte-Carlo approach. superdomain chain (SD chain) with excluded volume as well as together with phase transitions between condensed, globular and decondensed, coiled states Globular model/structural state is supported in agreement with by confocal microscopy data. It extends current view on interphase chromosome as a random ideal Gauss polymer chain. Radiation-induced intrachromosome exchange aberrations (intrachanges) are modeled on the basis of the contact hypothesis. It implies that Intrachanges are resulted with high frequency by misrejoining of damaged chromatin subunits (SDs) being in contacts. Calculated intrachange frequencies for X-rays agree well with experimental data. Results of calculations for globule provide biophysical modeling evidence that high frequency of localized multiple damaged sites per chromosome exists leading to enhanced interactions within the chromosome may be responsible for observed high frequency of interstitial deletions visible as a fragments at low doses of alpha particles.

Abbreviation

superdomains	SD;
chromosome aberrations	CA;
linear energy transfer	LET;

Introduction

Current experimental data and theoretical models (findings) obtained for inter- as well as intrachromosome aberrations, which show indicate an important role of structural chromosome organization in formation of CA.

The differential sensitivity of chromosomes and chromosome regions to ionizing radiation has been discovered described. More damage appears in euchromatin regions of chromosomes than in heterochromatin ones, and the damage reparation is also more fast and effective in euchromatin than in heterochromatin [1]. Many chromosomes participate in CA with frequencies unproportional to the DNA content in these chromosomes. Formation of X-induced dicentrics takes place mainly in euchromatin regions of chromosomes [1,2]. The distribution of breakpoints participating in exchange aberrations is nonrandom over the chromosome [3].

The degree of chromatin condensation in interphase nucleus influences yield of exchange aberrations. Frequency of dicentrics formation following γ-irradiation is higher in G1-stage than in G0, where chromatin in lymphocytes nuclei is more condensed. However, translocations frequencies are independent on the degree of chromatin condensation at the time of irradiation [2].

The data on intrachromosome aberrations show that the ratio between intra- and interarm exchanges is significantly more than 1 which was expected on the basis of the model of random damage interactions [4]. The experimental estimates of this ratio are as follows: paracentric inversions : centric rings = 10:1 for γ-irradiated human fibroblasts [5], acentric fragments (interstitial as well as terminal) : dicentrics : centric rings = 12:6.6:1 [6], deletions : centric rings = 6:1 for X-irradiated and 10:1 for α-irradiated human epithelial cells [7]. Intra- and interchromosome aberrations in mouse splenocytes were recently analyzed using the painting of whole chromosome (FISH) and simultaneously of telomeric probes [8]. The yield of interstitial fragments was comparable with the yield of dicentrics. Rejoining of open ends of interstitial fragments with one and two chromosome fragments was rare (8%). The majority of interstitial fragments were formed as small acentric rings or double minutes [8].

These and other data show that a knowledge of chromosome (structure) organization in interphase nucleus is important for mechanistic analysis of (formation) of radiation-induced chromosome aberrations (CA). Quantitative interpretation of the data on intra- and interchromosome exchanges on basis of theoretical models requires the knowledge of large-scale geometry of chromosomes in nuclei [4,9,10]. In the absence of systematic experimental information the modeling of the chromosome structure gives new opportunities for the estimating role of structure in formation of CA, for instance, of intrachromosome exchanges. The theoretical analysis of results of measurements of distances between probes with given quantity of DNA base pairs (bp) between them has been performed [11,12]. On this basis the idea model of interphase chromosome as a Gaussian polymer chain has been formulated [11,13]. After that The random walk/giant loop (RWGL) model of giant loops on the flexible backbone has been proposed [12,14]. On large scale (>10 million base pairs, Mbp) RWGL-model also exhibits Gaussian properties. The Gaussian polymer model of chromosome implies that each chromosome behaves like an ideal chain i.e. chain with not interacting units. Each element of chromosome chain may occupy the same element of volume, thus an excluded volume effects are ignored [15]. Moreover it implies that chromosome must be in the state of polymer coil [15,16]. However, applicability of these assumptions for interphase chromosomes is questionable. There is opportunity to propose a number of alternative models.

Thus the goal of this work was to look at the structural organization of chromosomes and intrachromosome exchange aberrations using the alternative concepts and models of chromosome organization (structure). We were based on our early ideas of principles of chromatin compactization in the nuclei of differentiated cells [17-19]. In this work chromosome structure as well as radiation-induced γ-induced intrachromosome aberrations are studied by using Monte-Carlo approach. Properties of chromosome structure are analyzed which may be important also for prediction of interchromosome exchange aberrations. We conclude that under certain circumstances structure of chromosome (chromosome arms) may be critical for prediction of (intrachromosome aberration) intrachange frequencies. As a consequence efficiency of high LET radiation in production of intraarm intrachanges At low doses may be very high.

Methods

Interphase chromosome structure was investigated using the methods of statistical polymer physics. The assumptions about Interphase chromosome structure: involved were the following: interphase chromosome consisting of one DNA molecule is the continuous chain of compact chromatin subunits. Loop chromatin domains (30 -150 thousands base pairs (kbp)) in an interphase nucleus were suggested to be associated into structures, in which they behave cooperatively rather than individually. Moreover, domains in a nucleus were suggested to be in a condensed, compact state rather than in a looped one.

These units of cooperativity were called superdomains to emphasize their supradomain organization. The estimated DNA content in a superdomain and its diameter [17,18] were the following. The diameter depended on the kind of domains association – either as a densely packed globule or on the surface of the inner structure (nuclear matrix) [18]. If the DNA content is 3.2-4.8 Mbp the minimal superdomain diameter is appr. 350-700 nm without nuclear matrix and approximately 620-1400 nm counting it. These are the lower estimates because the extremely dense packing of rigid spherical domain units into a globular structure of superdomain was considered. Individual chromosomes in interphase nuclei occupy spatially separated areas or "territories" [20]. It was proposed that higher-order chromatin organization into the separate interphase chromosomes in the nuclear interior is determined by volumetric attractive interactions between elements of chromatin fibers and/or chromatid chain. This implies that interactions with solvent are taken into account too, via renormalisation.

The structure of different chromosomes is a result of differentiated chromatin condensation being followed by the formation of heterogeneous globular structures. One can suppose that inside the "territory" chromosome structure is determined by equilibrium between attractive and repulsive interactions in the same chromosome. Condensation of a polymer chain consisting of N subunits (superdomains) is studied using the dynamic version of the Monte Carlo approach. We based our work on the algorithm developed by us earlier [21]. In this work it is modified, allowing us to study kinetics of formation of the condensed structure of chromosome and its arms. The non-lattice model of the polymer with unfixed ends is applicated. The chromosome is modeled as a flexible N-mer chain. The subunit is modeled as a rigid ball with diameter d. The chain elements interact between each other: they attract on distances, where $r \geq d$, and repulse on $r < d$ (excluded volume). Distance between adjacent balls is d.

Globularization modeling was described in details elsewhere [45]. It consists of two stages: modeling of the initial conformation and modeling of behavior of the chain with time. The initial conformation is being built as a random self-avoiding chain of subunits, not interacting with each other. At $t > 0$ the attraction between subunits begin. On each Monte-Carlo step (δt) there is one conformational reorganization of three neighbouring elements chosen randomly. At this time either the random rotation of this segment or random change of its inner angles takes place. The new conformation is being checked for the effect of excluded volume. If two or more subunits are found in the same place, the new conformation is being rejected. The energy E of interaction of all chain elements (except of pairs of the next elements) is being calculated. According to N. Metropolis *et al* [22], the energy of the new conformation E is being compared with the previous energy E_0. If $E < E_0$ then the new conformation is being accepted with the probability 1; otherwise it is being accepted in accordance with the Boltzman distribution [21].

If two subunits which are not neighbours in the chain approach the distance between their centers lower than the pre-given one, we call this a contact. One can define simple and complicated contacts. A simple contact is a contact, in which participate only two chain elements. A complicated contact, for example, of three balls is considered as two or three simple contacts. It depends on how many balls are close enough simultaneously. The chromosome arms overlap is defined as a part of balls belonging to different arms and contacting at the current moment. The chain behavior is considered over the time interval from 0 to T_{max} divided to small subintervals δt. The chosen value of T_{max} is large enough for the chain condensation to reach an equilibrium. The 2D projection $\langle r_{ij}^2 \rangle$ for globules was determined from calculated values of 3D mean-squared distance $\langle R_{ij}^2 \rangle_{3D}$: $\langle r_{ij}^2 \rangle = \frac{2}{3} \langle R_{ij}^2 \rangle_{3D}$ [13]. For each of the investigated structures the calculations was performed for 5000 - 10000 stories unlike low statistics calculations in previous paper [45]. The results didn't change significantly after this increasing the number of stories.

Alpha particle track (LET=100 keV/μm) intersected a chromosome structure. Chromosome is traversed by different number of alpha particle tracks/events at given dose according to Poisson distribution ($\langle N \rangle = D/Zf$). The distribution of specific energy imparted to an each SD is calculated. Then it is converted into the distribution of primary damage (DNA dsb) in SD. Energy deposition in SD result in induction of DNA dsb or several dsb in SD and hence in an altered SD structure (or state) but not in a break of SD chain. SD with one or more dsb i.e. with dsb cluster is determined as damaged SD. It means that this SD aquires a potential to exchange process with another damaged SD in parallel with exchange hypothesis of Revell. But unlike Revell's hypothesis our lesion of SD is stable. Thus a primary effect of radiation on the level of chromosome structure is altered state of whole chromosome chain which doesn't results in a real "break" or chromatid discontinuity and hence no free ends to interact. These proposals integrate (incorporate) features of two earlier ideas - contact-first hypothesis [27,28] and exchange hypothesis [29].

Information about a number and position of damaged SD as well as information about intrachromosomal contacts (between SD) is recorded to calculate and score a different types and probability of induction of intrachromosomal exchange aberrations - intraarm and interarm symmetrical and asymmetrical exchanges or intrachanges [45]. Two damaged SD are involved in a complete reciprocal exchange. Three and more damaged SD can also form complex intrachromosome exchange (see below). Damaged SD not participating in exchange doesn't result in any change of chromosome chain (chromatid) continuity. On the other hand, damaged SD not participating in exchange (if damaged SD has the features of an *unstable* lesion, accordingly to Revell's hypothesis) has to disappear or decay (accordingly to Revell's terminology [29]), without formation of real chromosome break. At present calculations we don't consider a finite life span of damaged SD and hence consider *stable (SD)* lesions. SDs in centromeric region of chromosome are not included in the exchange formation.

According to given proposals, probability of exchange between given pair of damaged SD is proportional to the probability of induction of two damaged SD P(Ndsb>=1) and probability of their contact (rate of collisions) within chromosome. Number and position of contacting SD is computed according with chromosome structure simulation code [45]. Complex intrachanges (with more than one contact of more than three damaged SD in single chromosome. A part of them, interarm intrachanges may be detected by different arm painting protocol [46]) are taken into account. They are formed by several intrachromosome contacts of damaged SD, two, three etc. and they are taken into account as sequential simple reciprocal exchanges. For example, three contact's exchange is scored as two, four contact's as three etc. The cutoff molecular weight of rearrangement (interstitial fragments) is introduced in calculations which allows reflect real restriction: if exchange aberration=interstitial deletion/fragment/acentric ring is small (less than fixed value, about (6)10-11 Mbp), this is not detected in experiment. For example, the exchange between neighbouring SD (deltaM=0) is recorded but is not scored in the yield of aberrations. All computations are repeated both for globular and coiled model of interphase chromosome and for X-rays and alpha particle tracks.

Results

A Globular State of Interphase Chromosome Consisting of Distinct Structural Subunits

A condensation of polymer chain (coil-to-globule phase transition) is accompanied by a change of macroscopic parameters as to inertia radius, index ν in formula $\langle R_{ij}^2 \rangle_{3D} \sim |i\text{-}j|^{\nu}$, where $\langle R_{ij}^2 \rangle_{3D}$ is three-dimensional mean square of distance between elements *i* and *j*, and

$|i\text{-}j|$ is the separation between i and j elements over the chain [15,23]. For a coil with excluded volume ν=1.2, for an ideal Gaussian coil ν=1, and for a dense globule ν=0.67 [16,23]. The results of Monte-Carlo simulation of 2D projection of mean square distance $\langle r_{ij}^2 \rangle = \frac{2}{3}\langle R_{ij}^2 \rangle_{3D}$ in globule formed as a result of collapse of homopolymer chain with excluded volume are presented on fig. 1. Chromosome 4 was modeled as a chain of N=64 subunits or superdomains (SDs). SD with DNA content of approximately 3 Mbp (N=64) had diameter of 950 nm based on (best fit) equality of maximum $\langle r_{ij}^2 \rangle$ value with experimental point.

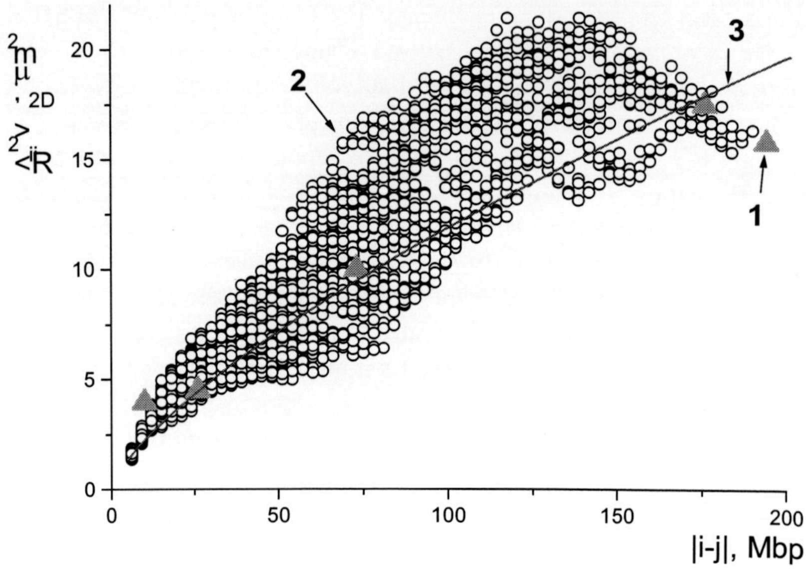

1 - experimental measurement of mean-square projection of interphase distance $\langle r_{ij}^2 \rangle$ for pair of probes i and j on chromosome as a function of genomic separation $|i\text{-}j|$. Data points for chromosome 4, human fibroblasts fixed with nonhypotonic conditions (paraformaldehyde) [23];
2 - Monte-Carlo simulation of $\langle r_{ij}^2 \rangle$ distances for globular model of interphase chromosome, N=64;
3 - approximation of $\langle r_{ij}^2 \rangle$ by scaling relationship $\sim |i\text{-}j|^{0.73}$.

Figure 1. Data and theory for the mean square interphase distance and genomic separation for human chromosome 4.

Independent estimation of DNA content and superdomain size based on different models [18] gives the following values. For DNA content per SD of 3-4.5 Mbp SD minimal diameter is 350-700 nm for compact structure without nuclear matrix and 620-1400 nm if nuclear matrix is taken into account. The description of Monte-Carlo results by analytical formula $\langle R_{ij}^2 \rangle_{3D} \sim |i\text{-}j|^{\nu}$ gives an assessment of ν equal ν=0.73. This is close to theoretical expectation for a dense globule (ν≈2/3=0.67) and significantly differs from ν for Gaussian coil and self-avoiding coil. The comparison of theoretical results with experimental measurements of the projection of mean square of distance $\langle r_{ij}^2 \rangle$ for chromosome 4 of human fibroblasts [14] is shown on fig. 1. The data have been obtained in special experiments on nuclei fixed in paraformaldehyde under nonhypotonic conditions [14]. Hypotonic buffer conventionally used in hybridization in situ (FISH technique) may artificially result in partial decondensation of structure. The $\langle r_{ij}^2 \rangle$ data from [14] for large scale genomic separation obtained under hypotonic conditions (not shown) in general don't differ significantly from nonhypotonic data except of specific "hook" at largest genomic separation, 170-200 Mbp region. Thus the agreement between theoretical results and experimental nonhypotonic measurement supports the conclusion about the globular state of at least some interphase chromosomes.

Chromosome Arms Overlap

Relative position as well as relative motion of chromosome arms can be manifested in the degree of chromosome arm overlap. An overlap parameter was determined as a relative fraction of SDs undergoing interarm contacts at $t=T_{max}$. The distribution of arms overlap calculated by Monte-Carlo method is shown on fig. 3. The calculations were performed for structure with the distinct condensed chromosome arms (fig. 2, b), except of the centromeric region, where the small overlap was observed.

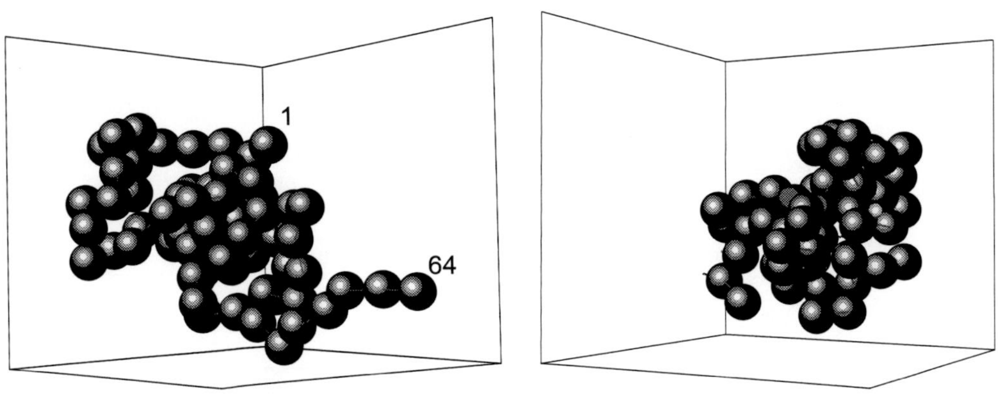

a - chain (N=64) is characterized by attraction of its elements to centromeres. Numbers 1 and 64 point on locations of the first and the last elements.
b – polymer chain (N=60) consisting of elements interacting inside arms.

Figure 2. Monte-Carlo simulated structures of interphase chromosomes.

The conclusion about small interarm overlap as well as the chromosome shape with distinct compact arms observed in the computer experiment (fig. 2, b), agree with the experimentally seen arm overlap and with recent alternative modelling approach [47] is shown on fig. 3.

Intrachromosome Exchange Aberrations at Low Doses for Different LET Radiation

Results of dose response predicted for different models of chromosome structure are presented in Tables1-IV.

Table I. Different types of intrachromosome aberrations irradiated with alpha particle and globular chromosome model

Dose, Gy	Exchange type	Cutoff, SD			
		1	2	3	4
0.1	Intraarm	6.3e-3	5.1e-3	4.6e-3	3.2e-3
	Interarm	1.1e-3	1.1e-3	1.0e-3	1.0e-3
	Multi-lesion	1e-4	<1e-4	<1e-4	<1e-4
1	Intraarm	8.90e-2	7.51e-2	6.83e-2	6.21e-2
	Interarm	1.3e-2	1.23e-2	1.17e-2	1.13e-2
	Multi-lesion	1.07e-2	6.3e-3	5.3e-3	4.4e-3
2	Intraarm	0.198	0.176	0.165	0.159
	Interarm	2.58e-2	2.57e-2	2.51e-2	2.52e-2
	Multi-lesion	3.93e-2	2.45e-2	1.74e-2	1.31e-2
3	Intraarm	0.294	0.268	0.267	0.259
	Interarm	3.36e-2	3.50e-2	3.46e-2	3.52e-2
	Multi-lesion	7.62e-2	4.93e-2	3.10e-2	2.48e-2

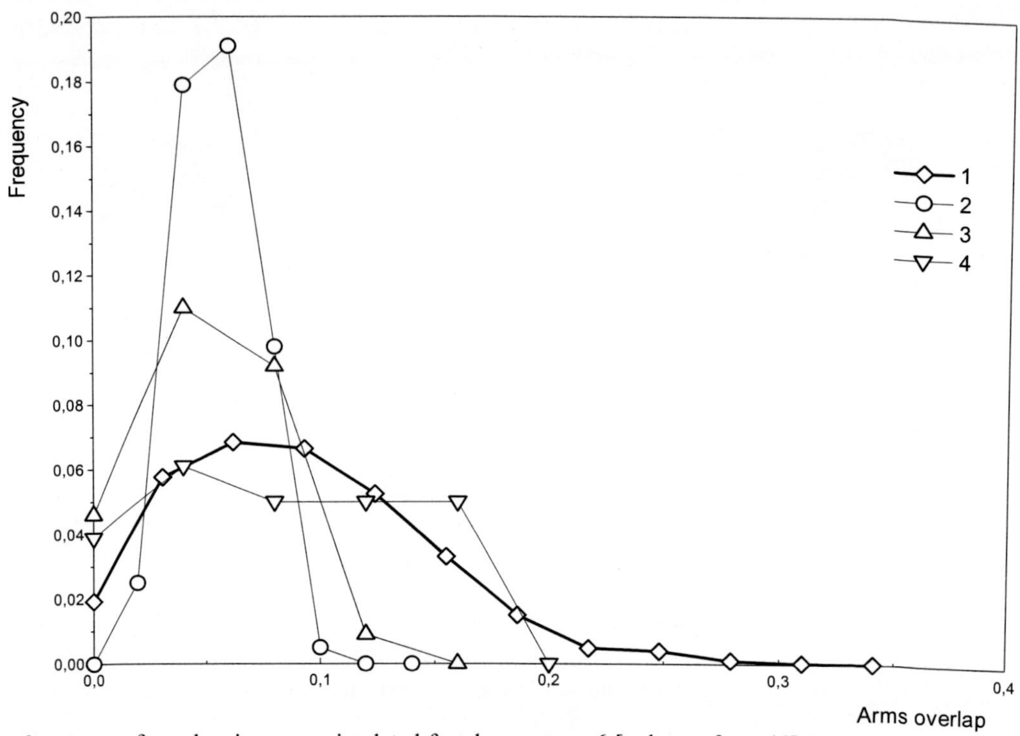

1 - frequency of overlapping arms simulated for chromosome 6 [redrawn from 45]. Relative arm overlap is calculated as the ratio of volume of contacted SD from different arm to whole chromosome volume

2- for comparison, overlap predicted by computer simulation approach developed in [47] is presented

3- observed overlap for chromosome 3 [redrawn from 47]

4- data for chromosome 6 [redrawn from 47]

Figure 3. Chromosome arms overlap.

Table II. Different types of intrachromosome aberrations irradiated with alpha particle and coiled chromosome model

Coiled chromosome structure (with excluded volume), alpha particles:

Dose, Gy	Exchange Type	Cutoff, SD			
		1	2	3	4
0.1	Intraarm	2.1e-3	5e-4	3e-4	3e-4
1	Intraarm	2.81e-2	1.41e-2	7.7e-3	5.3e-3
2	Intraarm	8.91e-2	5.83e-2	2.50e-2	1.08e-2

Table III. Different types of intrachromosome aberrations irradiated with X-rays and globular chromosome model

Dose, Gy	Exchange type	Cutoff, SD			
		1	2	3	4
0.1	Intraarm	1.3e-3	1.0e-3	9e-4	8e-4
	Interarm	2e-4	2e-4	2e-4	1e-4
	Multi-lesion	ND*	ND	ND	ND
1	Intraarm	0.106	8.60e-2	7.61e-2	6.94e-2
	Interarm	1.45e-2	1.36e-2	1.28e-2	1.21e-2
	Multi-lesion	6.7e-3	4.0e-3	2.8e-3	2.0e-3
3	Intraarm	0.346	0.342	0.337	0.331
	Interarm	4.85e-2	5.60e-2	5.90e-2	6.01e-2
	Multi-lesion	0.224	0.151	0.112	8.79e-2

* ND – not detected

Table IV. Different types of intrachromosome aberrations irradiated with X-rays and coiled chromosome model

Dose, Gy	Exchange type	Cutoff, SD			
		1	2	3	4
0.1	Intraarm	4e-4	2e-4	1e-4	<1e-4
	Interarm	<1e-4	<1e-4	<1e-4	<1e-4
	Multi-lesion	ND	ND	ND	ND
1	Intraarm	3.66e-2	1.63e-2	1.00e-2	7.2e-3
	Interarm	4.5e-3	3.5e-3	3.0e-3	2.7e-3
	Multi-lesion	7e-4	1e-4	<1e-4	<1e-4
2	Intraarm	0.124	5.94e-2	3.74e-2	2.67e-2
	Interarm	1.51e-2	1.24e-2	1.11e-2	1.01e-2
	Multi-lesion	9.1e-3	2.1e-3	8e-4	4e-4

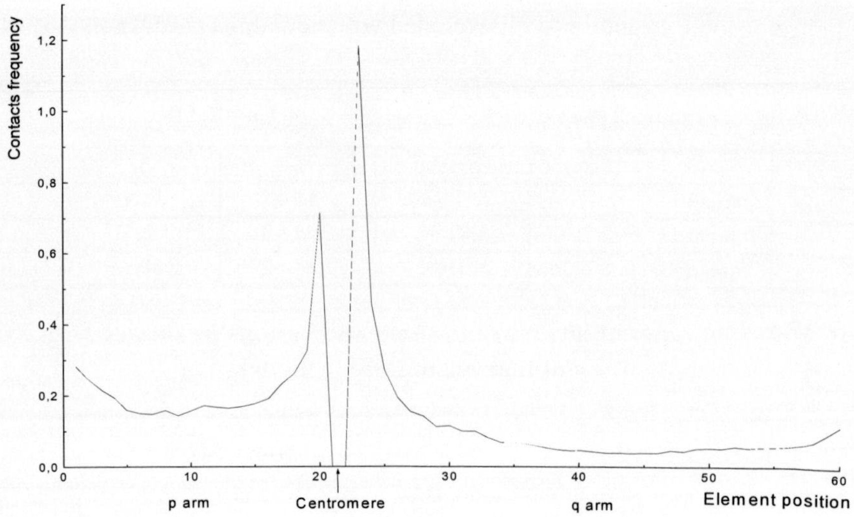

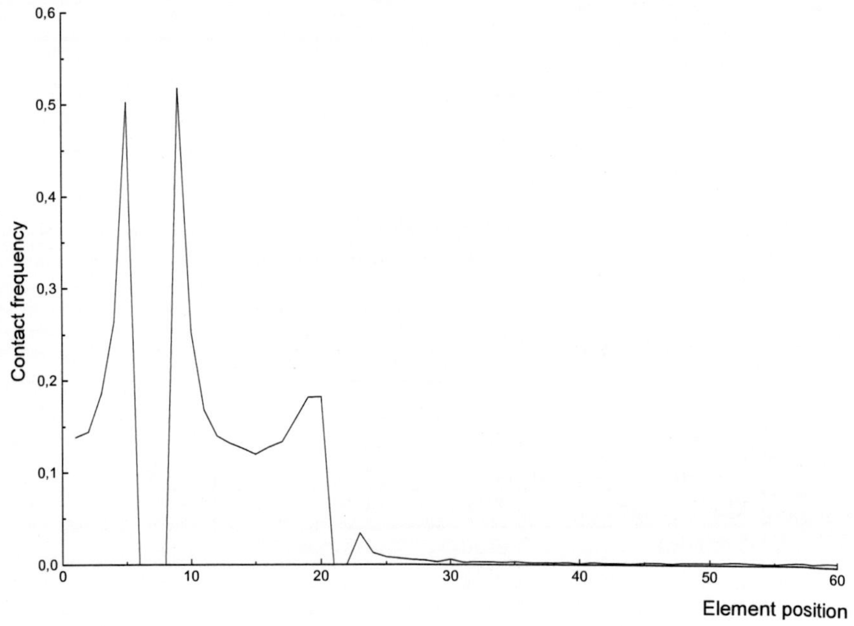

a - frequency of involvement of given (position, number on a chromosome) element or SD in contact between p and q arms.

b - frequency of contacts of element number 7 with any other one on p or q arm.

Figure 4. The simulated distribution of contacting elements along the chromosome.

The Distribution of Sites of Intrachromosome Exchange Aberrations along the Chromosome.

The Monte-Carlo simulation of intrachanges on the basis of globular chromosome model and contact hypothesis showes a nonrandom distribution of chromosomal sites (SDs) involved_ in intra-interarm intrachanges chromosome even if the distribution of primary damage is random. On fig. 4,a the frequency of contacts of an *i* element with any other one belonging to another arm of the same chromosome is shown. The more detailed distribution

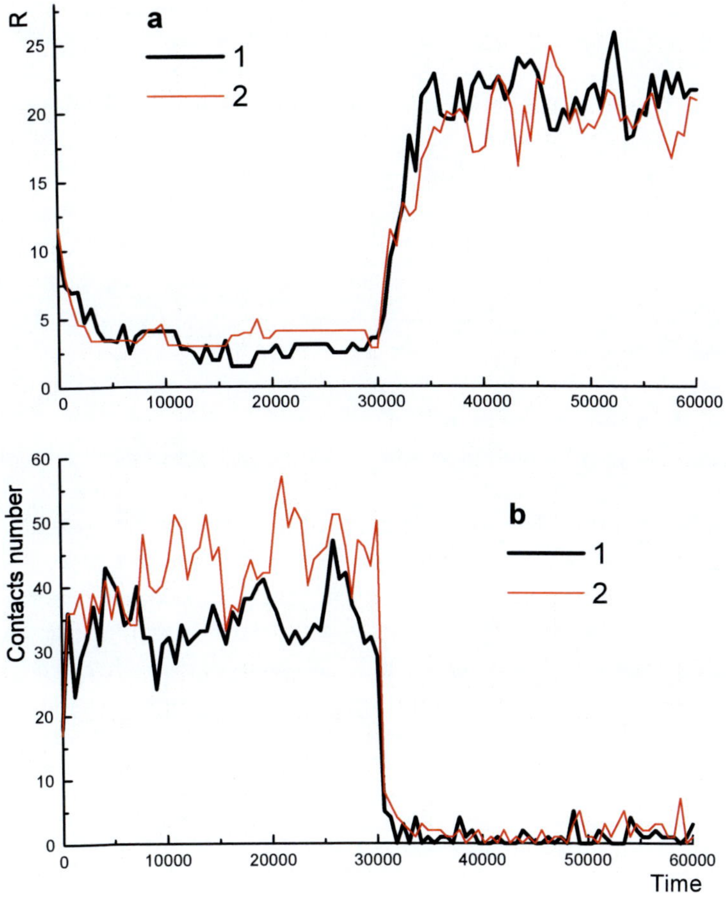

a - alteration of proximity for centrometic and telomeric regions of chromosome predicted by means of mean centromer-telomer distance

1 - for p arm

2 - for l arm

b - change of intraarm contacts in the course of condensation-decondensation of chromosome

1 - number of contact in p arm

2 - number of contact in q arm

Figure 5. Globule-to-coil (phase) transition of chromosome structure [45].

in the form of the frequency of contacts of the *i* element (7th element of short arm, p7) with any *j* element of any arm – is shown on fig. 4,b. These results as well as those shown on fig. 3 were obtained for the structure with spatially separated arms, fig. 2, b). The analysis of interaction frequency between elements p7 and p5, 9, 16-20 leads us to the following conclusions. 1) The high intra to interarm exchange ratio between different loci is ranged as 6 – 25. Such ratios were obtained in computer experiments with structures patterns being characterized by small arms overlap. 2) A high probability of intra-arm contacts of non-neighbouring elements of chain was observed which implies the high frequency of interstitial deletions. This conclusion also agrees with observations [8]. 3) The spatial proximity of free ends of interstitial fragment enables them to interact with high frequency with formation of acentric rings. Thus most interstitial deletions have to be formed as acentric rings.

Conformational Transitions of Chromosome Structure Following Irradiation

Under irradiation, structure of chromatin and chromosomes in an interphase nucleus can be changed [30-33]. The theoretical model of radiation-induced cooperative reorganization of high-order chromatin structures has been proposed [34]. According to this model, nuclear chromatin conformational reorganization between conditions with different macroparameters (e.g. density) can be considered as phase transitions [18,19]. In this work cooperative reorganization or phase transition is studied by means of Monte-Carlo approach. The chromosomal phase transition between different structure states (globule-coil type transition) is induced by the change of interaction volumeric potentials of structure elements. Phase transitions are accompanied by à change of contacts number, i.e. sites of intrachromosome exchanges (fig. 5,b). At t=0 chromosome is decondensed; during the time interval $0<t<30000$ the attractive interaction between subunits takes place; at t=30000 the repulsive interactions are turned on. The calculations were performed for structures with parameters similar to those described in the previous chapter. The change of contacts number is accompanied by motion of telomeric and centromeric chromosome parts regarding each other. The calculated time dependence of the distance between centromere and both telomeres is shown on fig. 5,a. The results were obtained on the model with free ends. The globule-to-coil phase transition leads to motion of chromosome locuses (centromeres and telomeres) regarding each other, and from different chromosomes too, over large distances. Phase transitions are accompanied by à change of a large number of SD contacts, which are potential locuses for chromosome exchanges (fig. 5, b).

Discussion

Globular Model of Interphase Chromosome

The results of Monte-Carlo modeling the chromosome structure, fig. 1, agree with recent experimental finding on DNA probes hybridization and laser confocal microscopy of cells [14] and support the globular model of large-scale interphase chromosome structure. Main features of this model and were proposed earlier [17-19] and modelled [45].

Model parameters generally reflect electrostatic interactions between different proteins and chromosomal DNA, intra-nuclear environment (ions, pH), attachment of chromosome ends to the nuclear matrix, to lamina, etc. Superdomain parameters used in presented work (DNA content ≈3 Mbp, diameter 950 nm) may vary depending on suggested fine structure of superdomain [17,18]. The variability of superdomain parameters as well as of distances between them over the chain may influence values of parameters of chromosome structure. The experimental data *in situ* obtained on chromosomes 3 and 4 (fig 1) agree well with results of computer modeling the globular structures [45].

However, one has to emphasize that the conclusion about the globular large-scale chromosome structure still has not been proved strictly. It is based on: a) comparison between theoretical and experimental values of mean squares of distances $\langle r_{ij}^2 \rangle$ between two probes on a chromosome [14]; b) comparison between interphase chromosomes 3D morphology, reconstructed – on one hand – after the laser microscopy in fixed cells [24,26], and – on the other hand – using the data of the computer modeling. Measurements of $\langle r_{ij}^2 \rangle$ have been performed for few probes on chromosome 4 in cells fixed under non-hypotonic conditions [14]. Chromosome structure depends on fixation conditions and is being experimentally observed to the resolution of the method of the laser confocal microscopy. [26]. Partial decondensation of high-order structures of chromatin and chromosomes as a result of fixation may appear even under "soft" conditions [14,26].

Chromosome structure in living cell is controversial, some experimental approaches were proposed but this is not very precise information. How one can explain some qualitative pecularities of chromosome organization observed for living cell comparing with fixed cells. According to the theoretical model [19], one can expect dynamic chromatin behavior/structure on the level of 30 nm fiber, and hence, on higher level of organization as to chromatin domains, superdomains, chromosome. In the living cell compact superdomain can undergo frequient transition between condensed and decondensed states. The number of superdomains in different states is determined from dynamic equilibrium of molecules (not specified) controlling the state and likely activity of given SD within the chromosome. Thus the state of SD chain (chromatid) the chromosome structure *in the nucleus in vivo* has to display dynamic features on the level of SD chain as well as on globular level or chromosome domains/territories.

RBE for Intrachromosome Exchange Aberrations at Low Doses

From Tables I-IV relative biological efficiency, RBE for every type of exchage aberration, can be derived, in particular, in the low dose range, RBE_{max}. For intraarm intrachanges or interstitial deletions RBE_{max}=4.5-5.5 and for interarm intrachanges or centric rings RBE_{max} =6.7 for globular structure for different paramaters M'=1,2,3. For coiled structure, RBE_{max} = 2.5-2.9 for intraarm and RBE_{max} <~1 for interarm intrachanges.

It should be pointed out that our estimates underestimate RBE due to several simplifying suggestions and paramaters. Calculations of RBE for different parameter of dsb induction with X-rays, 1.48 dsb per chromosome per Gy result in RBE_{max} =8.5-10.5 (instead of 4.5-5.5) for intraarm intrachanges.. Experimental estimates of RBE for intrachanges are ranged between RBE=4 for acentric fragments observed after irradiation of human lymphocytes with

3.45 MeV alpha particles and ~10 [48]. Thus only globular model predicts high RBE in agreement with experiment, coil model undersetimate this value.

The ratio of intra to interarm intrachanges or G ratio at low dose (0.1 Gy) is equal 3.2-8.8 for parameters M'=1-3 for globular chromosome irradiated with alpha particles. For X-rays and globule G=2.2-7.5 (M'=1-3). These values are about or some lower than observed in experiments: paracentric inversions : centric rings = 10:1 for γ-irradiated human fibroblasts [5], acentric fragments (interstitial as well as terminal) : dicentrics : centric rings = 12:6.6:1 [6], deletions : centric rings = 6:1 for X-irradiated and 10:1 for α-irradiated human epithelial cells [7]. G ratio at each dose is weakly dependent upon LET both for globulae and for coil. This is in contrast with theoretical predictions based on random walk chromosome model.

For alpha particle the yield of intrachanges is about 10 - 18 times higher for globular chromosome than for coiled in low doses (M=2,3). At high dose range (2 Gy) the difference is only 3 times (M'=2). Yield of interarm exchanges (=centric rings) for alpha particles at low doses (M'=2) is more that two orders of value more for globular than for coiled model of chromosome structure. For X-rays the difference between structures is less pronounced. Intraarm exchanges in globule are 5.6 times more frequent at 0.1 Gy and 6 times at 1 Gy (M'=2), for interarm exchanges the difference is 3.4 - 4.8 fold for M'=2,3 .

Thus the formation of both interstitial chromosome fragments/deletions, acentric rings and centric rings i.e. total intrachanges is frequent for globular and rare for coiled chromosome structure irrespectively of let although difference is more expressed for high lET radiation. Formation of interstitial chromosome fragments is more frequent than interarm exchanges.

Multi-lesion Exchange Aberrations

Multi-lesion intrachromosome SD interactions result in intrachromosome complex aberrations which are frequent for globular model and are rare for coiled model of chromosome. The incidence of complex, i.e. more than 2 contacts of more than 3 damaged SD per chromosome, intrachanges per chromosome for globule is 14% for M'=4 and 19% for M'=1of total aberrations at 3 Gy with X-rays. For coil the yield is 50 -100 times less.

High frequency of multiple lesions (damaged SD) per chromosome after alpha particle tracks suggests also that the chromosome may efficiently take part in complex interchromosome exchanges. If about one half of intrachromosomal lesions is involved in interchanges (as one can expect from the ratio 1:1 for interstitial deletions and dicentrics [49]) then the high frequency of alpha particles induced complex interchanges correlates well with the induction of high frequent multiple lesions per chromosome which is observed in our simulation.

If all exchanges are reciprocal then all excess acentric fragments predicted are acentric rings. To explain a high frequency of excess fragment observed a several theoretical models [9, 50-52] concluded that a high incopmletness of intrachromosome exchange has to exist. However recent data contradict this conclusion. Our calculations show a high yield of acentric fragments without any incompletness. It is in good agreement with recent findings that incompletness in rare event (<3%). Moreover, absolute yield of interarm exchanges measured by different arm paining [46] is in good agreemeent with model calculations for chromosome 3.

Cooperative Reorganization of Chromatin in Interphase Nucleus Following Irradiation

The idea of cooperative reorganization of higher-order chromatin structure in the whole interphase nucleus following irradiation was introduced [30-32] and modelled [17,18]. At present time several lines of experimental evidence support this proposal. We show, see also [45], by means of Monte-Carlo simulation that such a cooperative phenomenon, like a phase transition, is accompanied by a change of position of fast majority of SD in chromosome. Thus it results in a change of relative position of chromosomal loci, including damaged SD, in the single and many different chromosomes. If so, this effect must have an impotant implications for chromosome aberrations formation and DNA/chromosome damage repair mechanisms [45]. A movement of chromosome loci relative to each other, accompaining structure phase transition, may be also involved in the alteration of gene expression following irradiation. It was proposed that intrachromosome phase transitions could serve as a physical mechanism of position variegation effect [45].

The radiation-induced cooperative transition in the form of decondensation of chromosome arms (chromosome bands or subbands) could participate in a repair pathway, for example in repair of lesions inside the chromosome territory. If a DNA damage in condensed regions may be inaccessible for repair enzymes chromosome unfolding can be necessary for their efficient repair.

One of implication of phase transitions for intrachange aberration formation is that a dose rate effect has to exist for intrachanges. A second track will hit unfolded chromosome, or chromosome arm, with low number of intrachromosome contacts. Hence lower number of exchanges, especially multi-lesion or complex intrachanges, has to be produced with dose protraction.

This prediction is valid however under some circumstance. In this scenario the fast majority of intrachanges are to be produced in pre-existing contacts of SD within the globular chromosome. If decondensation occurs following each single track traversal then interaction of damaged SD after each track occurs in identical conditions or in coiled chromosome. Such a process would be insensitive to dose rate. But calculations show that in coiled structure a *random* interactions of SD are relatively rare.

If calculate the distance at which chromosome locus moves during cooperative decondensation process we get a large distances. Then widely separated loci of different chromosomes would reach (repositioned) each other and occur in close "contact" and thus would participate in complex exchanges aberrations. In this hypothetical process the movement of many different sites of several chromosomes, including damage SD, takes place.

Mechanisms of Cooperative Structure Rearrangements

Conformational chromosome reorganization by means of phase transitions can be induced by different factors, including irradiation, because of the change of one of three following parameters (or their combination): subunits interaction potentials, "solvent" properties, points of attachment to membrane and/or nuclear matrix. Radiation results in membrane damage and hence in a change of ion flux and intracellular pH [42], induces the expression of various genes [43,44], and causes plenty of other factors. They may change nonspecifically chromosome volumetric interactions and chromosome microenvironment and thus to induce phase transitions. One can speculate that some of gene products may act at points of anchorage of chromosome to nuclear lamina (nuclear matrix) or result to their detachment. Then free nonattached chromosomes will move for a long distance. The biological role for such a transition may be as follows. There is directed movement in the course of homological repair of chromosome damage. It implies that movement of damaged chromosomes to undamage homologues or to a pre-existing intranuclear sites where pairing of specific chromosomes has to occur.

Authors are grateful for valuable discussion with D.M. Spitkovsky, T.A. Talysina. Results were partially reported on 29th conference of European Society for Radiation Biology (Italy, 1998).

References

[1] Slijepcevic P., Natarajan A.T. // *Mutat. Res*. 1994. V.323. P.113-119.
[2] Natarajan A.T.,Balajee A.S., Boei J.J.W. et al. // *Mutat. Res*. 1996. V.372. P.247-258.
[3] Thucker J.D., Senft J.R. // *Radiat. Res*. 1994. V.372. P.247-258.
[4] Sachs R.K., Brenner D.J., Chen A.M.et al.//*Radiat. Res.* 1997. V.148. P.330-340.
[5] Muhlmann-Diaz M.C., Bedford J.S.. //*Radiat. Res*. 1995. V.143. P.175-180.
[6] Cornforth M.N., Bedford J.S. //*Radiat. Res*. 1987. V.111. P.385-405.
[7] Durante M., Grossi G.F, Gialanella G. et al. //*Radiat. Environ. Biophys*. 1995. V.34. P.195-204.
[8] Boei J.J.W.A., Natarajan A.T. //*Int. J. Radiat. Biol*. 1998. V.73. P.125-133.
[9] Wu H., Durante M., Sachs R.K., Yang T.C. //*Int. J. Radiat. Biol.* 1997. V.71. P.487-496.
[10] Ostashevsky J. //*Radiat. Res*. 1998. V.149. P.179-186.
[11] Van den Engh G., Sachs R., Trask B.J. //*Science*. 1992. V.257. P.1410-1412.
[12] Yokota Í., van den Engh G., Hearst J.E. et al. //*J. Cell Biol*. 1995. V.130. P.1239-1249.
[13] Hahnfeldt P., Hearst J.E., Brenner D.J. et al.//*Proc. Natl. Acad. Sci. USA*.1993. V.90. P.7854-7858.
[14] Sachs R.K., van den Engh G., Trask B.J.et al. //*Proc. Natl. Acad. Sci. USA*. 1995. V.92. P.2710-2714.
[15] Ostashevsky J.Y., Lange C.S.. //*J. Biomol. Struct. Dynam.* 1994. V.11. P.813-820.
[16] Lifshits I.M., Grosberg A.Yu., Khokhlov A.R. //*Rev. Mod. Phys*. 1978. V.50. P.683-702.
[17] Andreev S.G., Spitkovsky D.M., Talysina T.A. //*Microdosimetry*. Leningrad, "LIYaF",1986. P.109-142. (in Russian)
[18] Andreev S.G., Spitkovsky D.M., Talysina T.A. //*Proceedings of 5th All-USSR Meeting on Microdosimetry*. Ust-Narva, 5-12 March 1986. V.2. *Reviews and Problematics Reports*. *Ì*.,1987. P.4-60. (in Russian)
[19] Andreev S.G., Spitkovsky D.M. //*Reports of USSR AS*. 1983. V.269. P.1500-1502. (in Russian)
[20] Lichter P., Cremer T., Borden J., Manuelidis L. //*Hum. Gen*. 1988. V.80. P.224-234.
[21] Chepel V.Yu., Khvostunov I.K., Mirny L.A et al. //*Radiat. Prot. Dosim*. 1994. V.52. P.259-263.

[22] Metropolis N., Rosenbluth A.W., Rosenbluth M.N.et al. //*J. Chem. Phys*. 1953. V.21. P.1087-1092.
[23] Grosberg A.Yu., Khokhlov A.R.//*Statistical Physics of Macromolecules*. Moscow. "Science". 1989. 344p. (in Russian)
[24] Dietzel S., JauchA., Kienle D. et al. //*Chromosome Res*. 1998. V.6. P.25-33.
[25] Cremer T., Kurz A., Zirbel R et al. //*Cold Spring Harbor Symp. Quant. Biol*. 1993. V.58. P.777-792.
[26] Eils R., Dietzel S., Beritn E. et al. //*J. Cell Biol*. 1996. V.135. P.1169-1174.
[27] Serebrovsky A.S. //*Amer. Naturalist*. 1929. V63.¹687. P.374-378.
[28] Serebrovsky A.S., Dubinin N.P. //*Achievments of the Modern Biology*. 1929. V.8. P.235-238. (in Russian)
[29] Revell S.H. //*Proc. Roy. Soc. B*. 1959. V.150. P.563-589.
[30] Andreev S.G. //Stochastic and structure-temporary effects in the Physics of the Biological Action of Radiation: PhD. Theses. Moscow,"MEPhI". 1981. 16p. (in Russian)
[31] Spitkovsky D.M., Andreev S.G. //*Stud. Bioph*. 1981. V.86. P.83-84.
[32] Spitkovsky D.M., Andreev S.G. //The Theoretical Basis of Modification of Radiosensitivity. Alma-Ata, "Science". 1981. P.276-282. (in Russian)
[33] Andreev S.G., Karpukhin A.V., Lapidus I.L. *et al* //Proceedings of 5th All-USSR Meeting on Microdosimetry. Ust-Narva, 5-12 March 1986. V.2. *Reviews and Problematics Reports*. Ì.,1987. P.263-275. (in Russian)
[34] Andreev S.G., Spitkovsky D.M. // Proceedings of 5th All-USSR Meeting on Microdosimetry. Ust-Narva, 5-12 March 1986. V.2. *Reviews and Problematics Reports*. Ì.,1987. P.240-262. (in Russian)
[35] Spitkovsky. D.M., Ermakov A.V., Gorin A.I. *et al* //*Radiation Biology. Radioecology*.1994. V.34. P.23-31. (in Russian)
[36] Salimov A.G., Krpukhin A.V., Kuzmina I.V. *et al* //*3rd Congress on Radiation Research*. Moscow,14-17 Oct. 1997. Heads of reports. V.1. Puschino. P.40-41. (in Russian)
[37] Dolling J.A. Boreham D.R., Brown D.L. et al. //*Int. J. Radiat. Biol*. 1997. V. 72. P. 303-310.
[38] Belyaev I.Ya., Ivanischeva M.Yu, Edneral D.I., Akifiev A.P. //*Genetics*. 1990. V.26. P.971-975.
[39] Belyaev I.Ya., Akifiev A.P. //*Genetics*. 1988. V.24. P.1384-1392. (in Russian)
[40] Scink A.E., Henikoff S. // *Nature*. 1996. V.381. P.529-531.
[41] Simpson J.J., Savage J.R.K. // *Int. J. Radiat. Biol*. 1996. V.69. P429-436.
[42] Woloschak G., Paunesku T., Chang-Liu C_M., Grdina D.J. //*Proc. of Radiat. Res. Congr*. 1995. Germany. V.2. P.545-547.
[43] Al-Deiry. W.S.,Harper J.W., O'Connor P.M. et al. //*Cancer Res*. 1994. V.54. P.1169-1174.
[44] Spitkovsky D.M. // *Radiation Biology. Radioecology*. 1992. V.32. P.382-400.
[45] Andreev S.G., Eidelman Yu.A. //*Radiation Biology. Radioecology*. 1999. V.39.P.10-20.
[46] Natarajan A.T., Boei J.J.W.A., Vermeulen S., Balajee A.S. //*Mutat. Res*. 1996. V.372. P.1-7.
[47] Munkel C., Eils R., Dietzel S., Zink D., Mehring C., Wedemann G., Cremer T., Langowski J. //*J. Mol. Biol*. 1999. V.285. P.1053-1065.

[48] Durante M., Grossi G.F., Napolitano M., Pugliese M., Gialanella G. //*Int. J. Radiat. Biol.* 1992. V.62. P.571-580.

[49] Boei J.J.W.A., Vermeulen S., Fomina J., Natarajan A.T. // *Int. J. Radiat. Biol.* 1998. V.73. P.599-603.

[50] Chen A.M., Lucas J.N., Simpson P.J., Griffin C.S., Savage J.R.K., Brenner D.J., Hlatky L.R., Sachs R.K. //*Radiat. Res.* 1997. V.148. P.S93-S101

[51] Sachs R.K., Chen A.M., Simpson P.J., Hlatky L.R., Hahndfeldt P., Savage J.R.K. // *Int. J. Radiat. Biol.* 1999. V.75. P.657-672.

[52] Hahndfeldt P., Hlatky L.R., Brenner D.J., Sachs R.K. //*Radiat. Res.* 1995. V.141. P.136-152.

In: Genetic Translocations and Other Chromosome Aberrations ISBN: 978-1-60456-344-3
Editor: Guy T. Leyden, pp. 191-208 © 2008 Nova Science Publishers, Inc.

Chapter 7

Centromeric Breakage and Illegitimate Recombination Leads to Whole Arms Translocations in Squamous Cell Carcinoma

Marta Alonso Guervós[1], Gerrit Meijer[2], Evelin Schrock[3] and Mario Hermsen[1*]

[1]Dept of Otolaryngology, Instituto Universitario de Oncología del Principado de Asturias (IUOPA), Hospital Universitario Central de Asturias, Oviedo, Spain
[2]Dept of Pathology, VU University Medical Center, Amsterdam, The Netherlands
[3]Institute of Clinical Genetics, Medical Faculty Carl Gustav Carus, University of Technology, Dresden, Germany

Abstract

Structural chromosomal aberrations such as deletions, duplications, isochromosomes and translocations are common in epithelial tumors. However, the mechanisms leading to tumor type- and stage-specific chromosomal aberrations, as well as their biological role in tumor development and progression, are not well understood.

We discovered that in squamous cell carcinoma cell lines of the head and neck up to 40% of all chromosomal breakpoints occur in centromeric regions leading to whole arm translocations. When we studied epithelial tumors of other histological type and origin, such as colorectal adenocarcinomas, we found only 10% of all rearrangements to be whole arm translocations. More importantly, we demonstrated tissue-specific differences in the formation of these centromeric translocations. In head and neck squamous cell carcinoma, whole arm translocation chromosomes contained centromeric sequences of both contributing chromosomes, whereas in colorectal adenocarcinoma centromeric material of only one

* E-mail address: mhermsen@hca.es. Tel: +34 985 108000, ext. 36549. Fax: +34 985 10 62 76. Corresponding author: Dr. M.A.J.A Hermsen; Dept of Otolaryngology, Instituto Universitario de Oncología del Principado de Asturias (IUOPA), Unidad Administrativa del IUOPA, Edificio Santiago Gascón, Despacho 2.3, Campus El Cristo B, 33006 Oviedo (Asturias), SPAIN

chromosome was present. This suggests that the exact location of DNA breakage and refusion resulting in whole arm translocation chromosomes differs between squamous cell and adenocarcinomas. It occurs within centromeric repetitive sequences in head and neck squamous cell carcinomas and, in contrast, within pericentromeric sequences in colorectal adenocarcinoma.

The mechanisms leading to these types of chromosomal abnormalities are unknown, but clues can be found in two genetic instability syndromes. First, Fanconi Anemia (FA) predisposes to acute myeloid leukemia (AML) and head and neck squamous cell carcinoma. Centromeric translocations in head and neck squamous cell carcinoma from FA patients are as frequent as in sporadic head and neck squamous cell carcinoma and mutations in FA genes have been found in sporadic head and neck squamous cell carcinoma. FA cells are typically hypersensitive to DNA crosslinking agents, which is related to the role of FA genes in homologous repair. Second, centromeric instability of chromosomes 1, 9 and 16 is a characteristic of ICF syndrome (immunodeficiency, centromere instability, facial abnormalities) and is caused by hypomethylation of heterochromatic classical satellite DNA, due to mutations in a DNA methyltransferase protein.

Therefore it may be speculated that sporadic head and neck squamous cell carcinoma, or at least a subgroup of these tumors, have a defect in homologous repair leading to centromeric translocations upon exposure to crosslinking agents present in tobacco (a strong etiological factor in head and neck squamous cell carcinoma). Alternatively, sporadic head and neck squamous cell carcinoma may have hypomethylation of centromeric repeat sequences, making them vulnerable for DNA damaging compounds, or they may have a combination of centromeric hypomethylation and defective homologous repair which together lead to frequent centromeric breakage and illegitimate recombination.

1. Chromosome Aberrations in Epithelial Cancers

Epithelial cancers are the most frequent type of malignant tumors and have a large impact on human morbidity and mortality. Recurrent and stage-specific chromosomal aberrations have been observed in many tumor types ("NCI and NCBI's SKY/M-FISH and CGH Database, 2001; Albertson et al, 2003). How these chromosomal changes affect basic cellular mechanisms and drive cell transformation and tumor progression at the molecular level, has only been resolved to a limited extent. In addition, the mechanisms leading to tumor-specific aberrations are not well understood.

It has been recognized for some time, however, that tumors arising in individuals with mutations in genes that function in particular pathways that maintain genomic integrity such as BRCA1 (Israeli et al, 2003; Tirkkonen et al, 1997) or DNA mismatch repair (Nakao et al, 2004; Soulie et al, 1999; Eshleman et al, 1998; Schlegel et al, 1995) are associated with particular patterns of chromosomal aberrations. Analyses of a large series of tumor samples using comparative genomic hybridization (CGH) have revealed not only characteristic recurrent chromosomal gains and losses in adenocarcinomas (malignant tumors arising from glandular epithelium) compared to squamous cell carcinomas (malignant tumors arising from squamous epithelium) from different sites of origin (Hermsen et al, 2002, 2001a, 1998; Jin et al, 2000; Grieken et al, 2000; Meijer et al, 1998), but also a higher frequency of imbalances involving whole chromosome arms in squamous cell carcinomas compared to adenocarcinomas (Weiss et al, 2003; Varis et al, 2001; Pei et al, 2001; Luk et al, 2001; Petersen et al, 1997; Hermsen et al, 1996). Figure 1 gives an example of these differences, in this case between breast adenocarcinoma and larynx and pharynx squamous cell carcinoma, analysed by CGH.

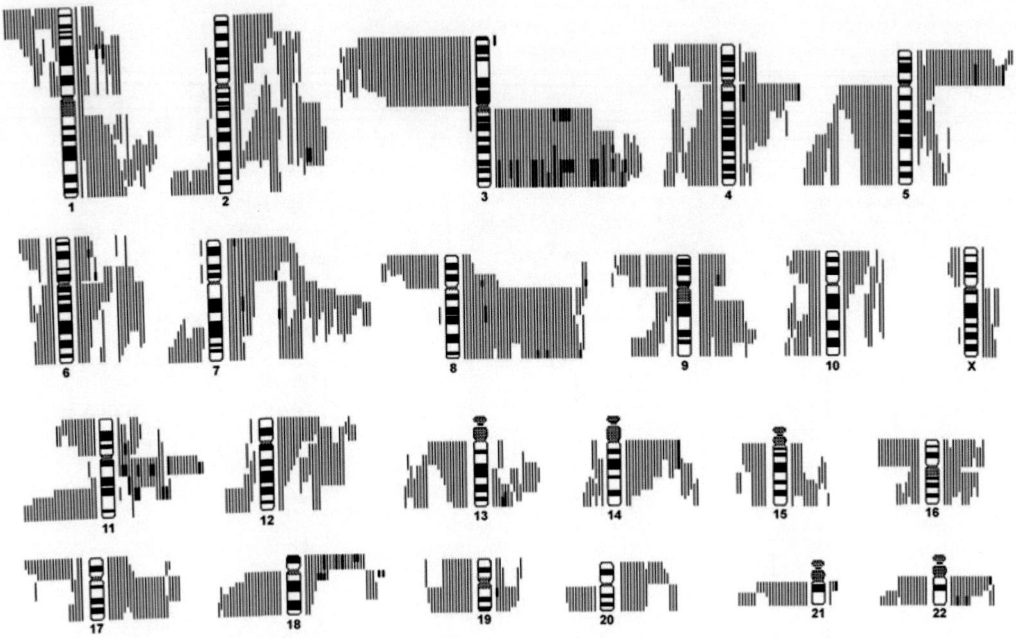

Figure 1. Overview of conventional CGH analyses on a series of 56 head and neck squamous cell carcinoma (A) and of 76 breast carcinomas (B). Bars to the right of the chromosomes represent gains and bars to the left losses. When comparing the two patterns of imbalances, it becomes clear that head and neck squamous cell carcinoma have many more whole arm gains and losses than breast carcinomas.

Whole chromosome arm gains and losses can be produced by isochromosome formation and whole arm translocation, both needing a break in pericentromeric or centromeric regions followed by illegitimate recombination. Why are the centromeres so frequently involved in translocations and why does this occur especially in head and neck squamous cell carcinoma? In this chapter, we will address these issues and speculate on possible mechanisms.

2. Centromeric Translocations Are a Key Feature of Head and Neck Squamous Cell Carcinoma

We performed a comprehensive analysis of chromosomal aberrations in two series of established adenocarcinoma (colorectal) and squamous cell carcinoma (oral cavity) cell lines using spectral karyotyping (SKY), array CGH and fluorescence in situ hybridization (FISH) analyses. In all cell lines, numerical and structural chromosomal aberrations were found by SKY, originally described by Schrock et al (1996). In adenocarcinoma cell lines, on average 11.1 (range 2-27) chromosomes with structural and/or numerical abnormalities were clonaly present compared to 19.4 (range 3-28) in the squamous cell carcinoma cell lines. In addition, on average 5.6 (range 0-12) non-clonal structural aberrant chromosomes (i.e. found in one cell only) were identified in the adenocarcinoma cell lines compared to 17.2 (range 0-33) in the squamous cell carcinoma cell lines.

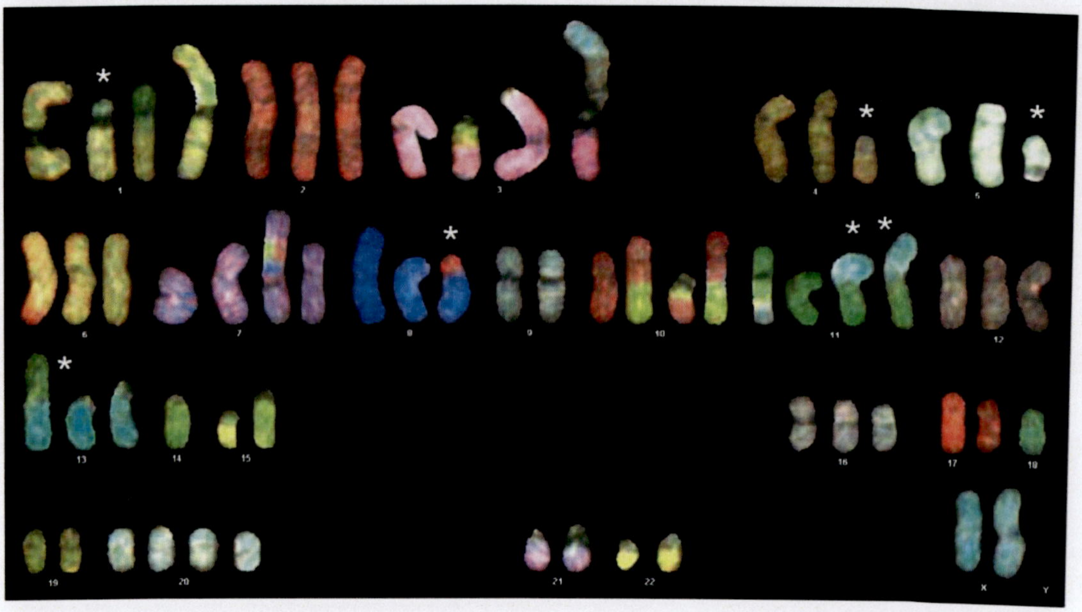

Figure 2a. Representative SKY-karyotype of cell line scc078, which is triploid and shows both numerical and structural chromosomal rearrangements. Amplification of chromosome 9 material can be seen at chromosome 3, between 3q and Xq. Whole chromosome arm translocations are indicated by asterisks: (1q;18p), (4p;12p), (5p;18p), (8q;10p), 2x(11q;Xq), and (11q;13q).

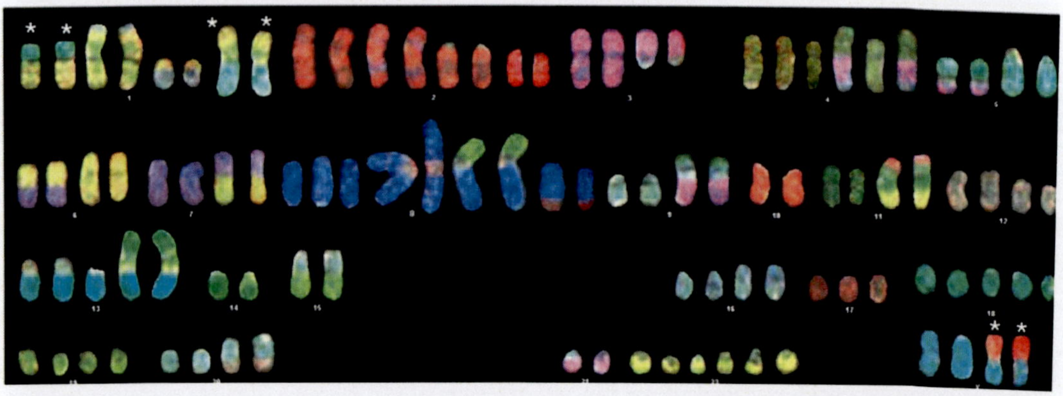

Figure 2b. Representative SKY-karyotype of cell line Colo320, which is tetraploid and shows both numerical and structural chromosomal rearrangements. This cell line contains two identical balanced whole chromosome arm translocation t(1q;5p), and two copies of an unbalanced whole chromosome arm translocation (Xq;10q), indicated by asterisks.

In both types of cell lines, adenocarcinoma and squamous cell carcinoma, the following types of rearrangements were found besides numerical aberrations: unbalanced translocations, whole arm translocations, deletions, duplications, isochromosomes, complex translocations and homogeneously staining regions. Reciprocal translocations, including inversions, occurred in all adenocarcinoma cell lines, but only once in a single squamous cell carcinoma cell line. Karyotype examples of a colorectal adenocarcinoma and an oral squamous cell carcinoma are shown in Figure 2.

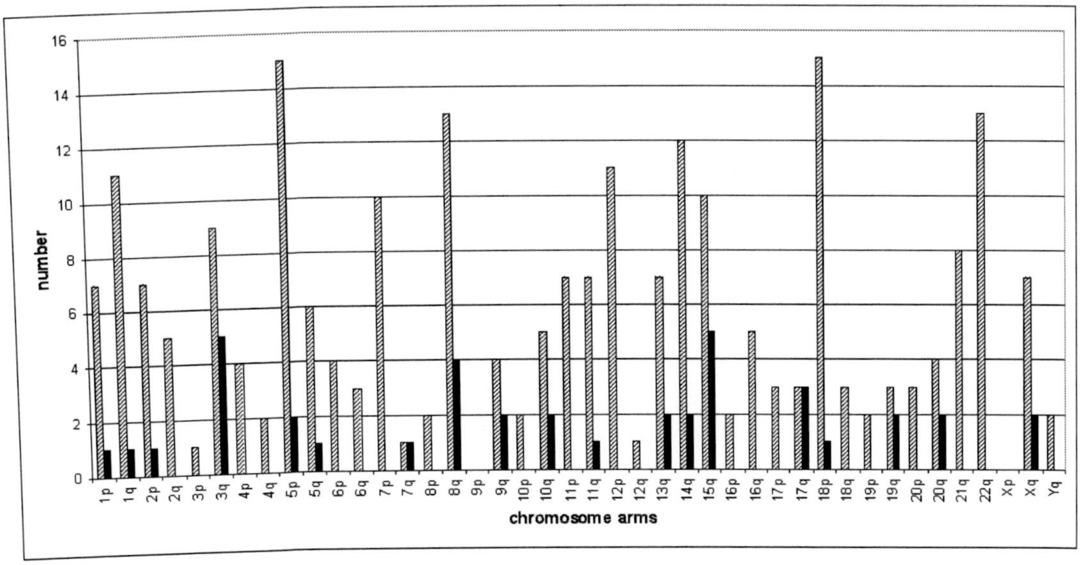

Figure 3. Frequencies of participation of chromosome arms in whole chromosome arm translocations or isochromosomes, as detected by SKY. Black bars represent the seven adenocarcinoma cell lines and hatched bars the nine squamous cell carcinomas.

Figure 3 demonstrates the frequency distribution of whole chromosome arm involvement in translocations between squamous cell carcinomas and adenocarcinomas indicating a clear difference. In squamous cell carcinomas, centromeres of all chromosomes were involved, while in adenocarcinomas this was only the case for 17 of 24 chromosomes. Moreover, in all squamous cell carcinomas, some chromosome arms were involved in two or more whole arm translocations with several partners.

Three classes of chromosomal rearrangements were categorized based on a combined interpretation of SKY and array CGH results: 1) band-band rearrangements, in which the breakpoints of both partner chromosomes are located on the chromosome arms; 2) band-centromere rearrangements, in which one of the partner chromosomes is broken and fused within the arm and the other partner within the centromere; and 3) centromere-centromere rearrangements, in which both partner chromosomes are broken and fused within the centromere, leading to isochromosomes and whole arm translocations. Only clonal aberrations were considered in this analysis. Band-band rearrangements were observed in 54% of the adenocarcinomas, as compared to 31% of the squamous cell carcinomas, whereas centromere-centromere rearrangements occurred more frequently in squamous cell carcinomas than in adenocarcinomas, 58% and 25% respectively (Table 1).

Table 1. Distribution of types of rearrangements detected by SKY, comparing cell lines of 7 colorectal adenocarcinomas and 9 squamous cell carcinomas. The data are expressed as percentages of the total number of clonal translocations. A marked difference can be seen in the frequency of centromere-centromere rearrangements, particularly those caused by whole arm translocations.

type of rearrangement	adenocarcinoma cell lines (n=7)	squamous cell carcinoma cell lines (n=9)
number of translocations	114	226
band-band (%)	54	31
cen-band (%)	21	11
cen-cen (%)	25	58
cen-cen: isochromosome (%)	14	17
cen-cen: whole arm (%)	11	41

Legend. band-band: rearrangements in which the breakpoints of both partner chromosomes are located on the chromosome arms; cen-band: rearrangements in which one of the partner chromosomes is broken and fused within the arm and the other partner within the centromere; cen-cen: rearrangements in which both partner chromosomes are broken and fused within the centromere, leading to isochromosomes and whole arm translocations. The percentages of the two types of cen-cen rearrangements are given in the lowest row.

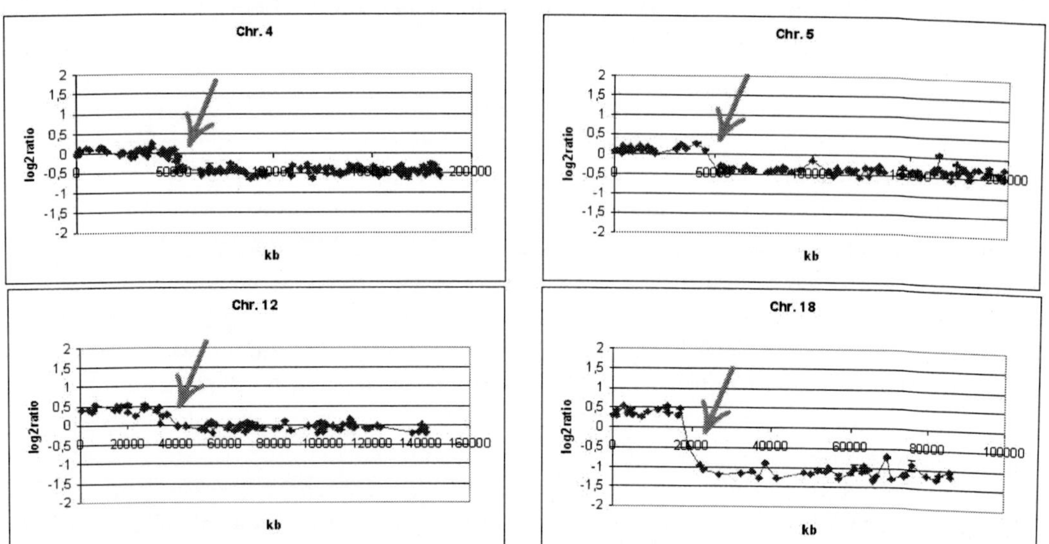

Figure 4. Whole arm translocations detected by SKY are in agreement with a change in copy number between the two BAC clones closest to the centromere, measured by array CGH analysis. The left panel shows results of chromosomes 4 and 12, the right panel shows results of chromosomes 5 and 18, both from cell line scc078 (see figure 2a). Red arrows indicate the position of the centromere, which in these cases coincide with the transition in copy number.

The comparison of array CGH and SKY analysis of these cell lines indicated that copy number transition points, i.e. loci with a significant change in copy number (Snijders et al, 2003) were consistent with the breakpoints of structural rearrangements detected by SKY. However, these breakpoints were mapped with a higher resolution using array CGH analysis (average resolution of 1.4 Mb). We were able to locate the breakpoints in between the two BAC clones flanking the centromeres (Figure 4) for unbalanced whole chromosome arm translocations both in squamous cell carcinomas and adenocarcinomas. Noteworthy, in one adenocarcinoma, the array CGH data allowed to redefine a translocation originally classified as whole arm der(5;17)(p10;q10) (cen-cen), into a band-band translocation der(5)t(5;17)(q11;q11.2), with chromosome 17 having the breakpoint close to, but not within the centromere.

Using FISH with centromeric probes for the chromosomes participating in whole chromosome arm translocations, we found differences between whole arm translocations comparing adenocarcinomas and squamous cell carcinomas. Seven of 9 clonal whole chromosome arm translocations detected in adenocarcinoma cell lines showed only one of the two possible signals, demonstrating that only the centromere of one chromosome was present and that the break occurred in the pericentromeric region. In strong contrast, all of 13 clonal whole chromosome arm translocations analyzed showed two signals (Figure 7) in squamous cell carcinoma cell lines. Thus, the break was located within the centromere and the aberrant chromosome contained a fused centromere with both translocated chromosomes contributing to it. This suggests that in adenocarcinoma the breakpoints generally are close to, but in squamous cell carcinoma they are within the centromeric sequences.

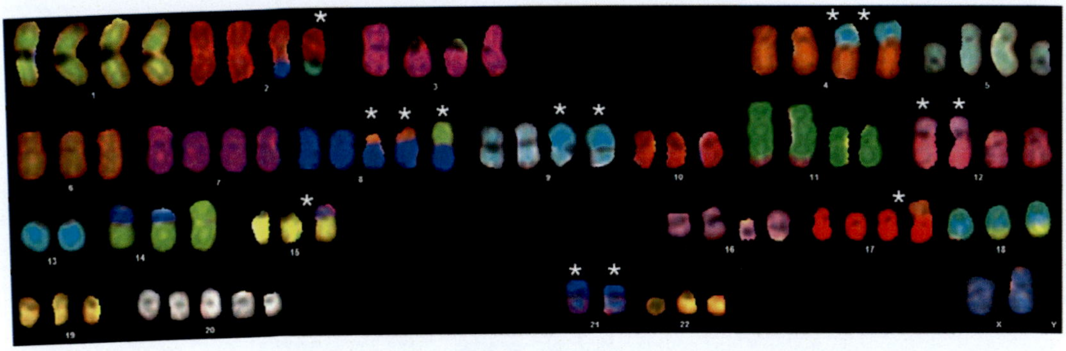

Figure 5. Representative SKY-karyotype of short term culture scc1365 showing both numerical and structural chromosomal rearrangements. Asterisks indicate the whole chromosome arm translocations: (2p;18p), 2x(4q;9p), 2x(8q;4p), (8q;14q), 2x(9q;13q), 2x(12q;16q), (15q;21q), (17q;4p), 2x(21q;8p).

In order to rule out the possibility that the observed difference in the structure of translocations in squamous cell carcinomas compared to adenocarcinomas had arisen during prolonged time in culture, we analyzed a short-term cell culture of a primary squamous carcinoma, scc1365. This tumor revealed a similar pattern of chromosomal abnormalities by SKY-analysis as the previously studied squamous carcinoma cell lines. Sixty-two percent (10/16) of all breaks were centromeric, mainly consisting of whole chromosome arm translocations (Figure 5). Also, both centromeres of the two translocated chromosomes were fused. In addition, we analyzed a series of primary tumor samples from head and neck squamous cell carcinomas by array CGH using a 30.000 oligo microarray which provides an average resolution as high as 140 kb (Carvalho et al, 2004). Again we found high frequencies of whole arm gains and losses, indicative of centromeric translocations. These high resolution data also showed that the copy number transition points were located between the two oligos flanking the centromeres (Figure 6). Thus, it is unlikely that the formation of centromeric rearrangements in the squamous cell carcinoma cell lines are artifacts of *in vitro* cell culture.

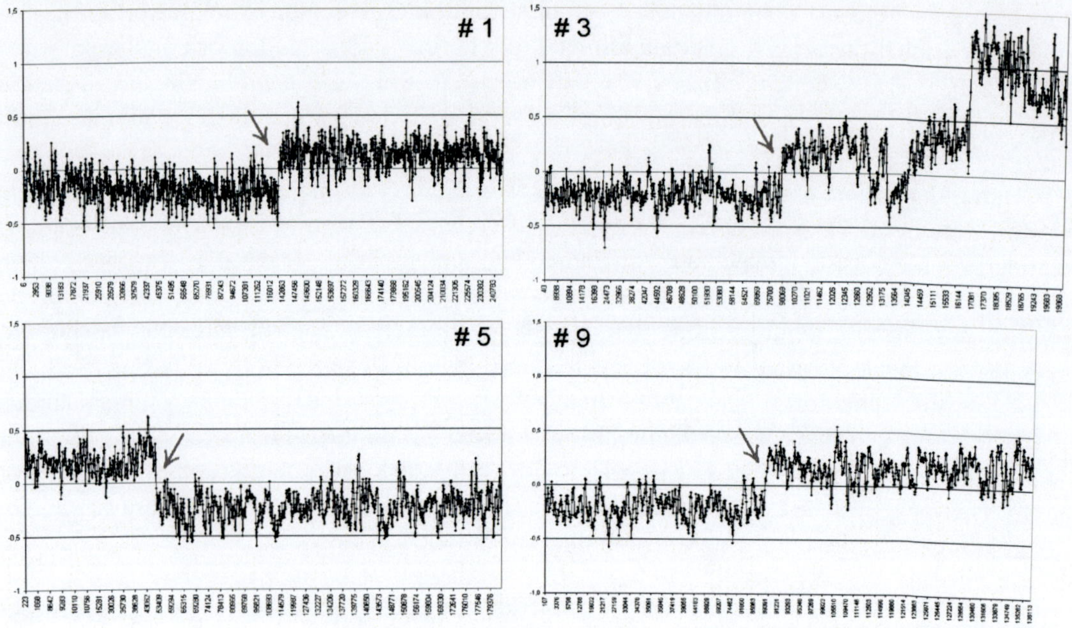

Figure 6. Part of the results of array CGH analysis on a primary, uncultured laryngeal SCC, using a 30.000 oligonucleotide chip representing all known genes in the human genome. The four panels show all oligos representing chromosomes 1, 3, 5 and 9. Arrows indicate the position of the centromeres.

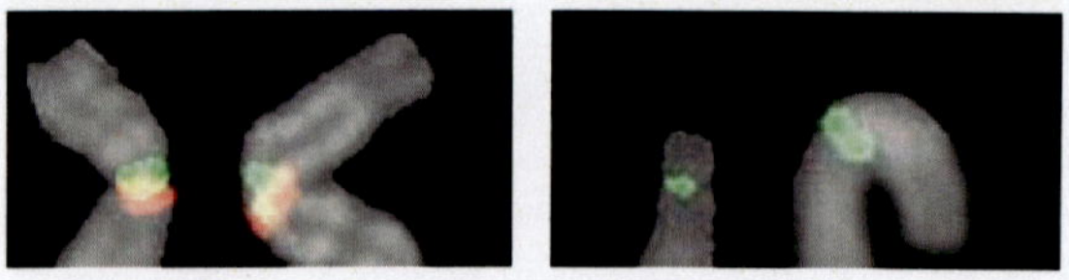

Figure 7. FISH with centromeric probes shows whole chromosome arm translocations in squamous cell carcinoma containing two signals (left panel), and in adenocarcinoma containing one signal. Left panel left marker: der(1;10)(q10;q10) from scc120 shows green for centromere 10 and red for centromere 1, and left panel right marker: der(1;14)(p10;q10) from scc094 shows green for centromere 1 and red for centromere 14. Right panel left marker: der(17;19)(q10;q10) from cell line WiDr shows only one FISH signal of chromosome 17. Right panel right marker: der(3;13)(q10;q10) from sw1398 shows only one FISH signal of chromosome 13.

In summary, squamous cell carcinomas exhibit up to 40% of all breakpoints within the centromeres, a phenomenon that may reflect a unique mechanism of chromosomal instability concerning centromeres (Hermsen et al, 2005). Its tumor specificity could be due to etiologic differences. Squamous cell carcinoma of the head and neck, lung and esophagus are all related to tobacco smoking whereas for adenocarcinomas from the same regions this relation is less obvious or absent. Further clues can be found in two genetic instability syndromes. First, head and neck squamous cell carcinoma from Fanconi Anemia (FA) patients show centromeric translocations as frequent as sporadic head and neck squamous cell carcinoma (Hermsen et al, 2001b) and mutations in FA genes have been found in sporadic head and neck squamous cell carcinoma (Marsit et al, 2004). FA cells are typically hypersensitive to DNA crosslinkers, compounds that are also present in tobacco. Second, centromeric instability of chromosomes 1, 9 and 16 is a characteristic feature of a syndrome termed ICF (immunodeficiency, centromere instability, facial abnormalities) and is caused by hypomethylation of heterochromatic classical satellite DNA, due to mutations in a DNA methyltransferase protein (Xu et al, 1999).

Therefore we speculate that (a subgroup of) sporadic head and neck squamous cell carcinomas have a defect in FA-related homologous repair which leads to centromeric translocations upon exposure to crosslinking agents present in tobacco. Alternatively, head and neck squamous cell carcinomas may have hypomethylation of centromeric repeat sequences present in all chromosomes, making the centromeres more vulnerable for DNA damaging compounds. Finally, head and neck squamous cell carcinoma may have a combination of centromeric hypomethylation and defective homologous repair which together lead to frequent centromeric breakage and illegitimate recombination. Below we will consider these possible causes of centromeric instability in more detail.

3a. Centromeric Translocations in Head and Neck Squamous Cell Carcinoma of FA Patients

Fanconi anemia (FA) is an autosomal recessive disease with a marked predisposition to specific malignancies, the second most frequent of which is squamous cell carcinoma (Joenje and Patel, 2001), in particular in the oral cavity (Kutler et al, 2003). Cells from FA patients are characterized by spontaneous chromosomal instability and are 10- to 100-fold more sensitive to intercalating agents, which may explain the early age of tumor onset (20-30 years in FA versus 60+ years in sporadic squamous cell carcinoma). Up to now, at least 11 genes related to FA have been discovered (*FANCA* to *FANCL*), many of which form a complex that activate FANCD2 that subsequently interacts with BRCA2, RAD51 and BRCA1 to perform homologous DNA repair (Levitus et al, 2004).

We established two head and neck squamous cell carcinoma cell lines from patients with mutations in either the *FANCA* gene or the *FANCC* gene and studied chromosomal aberrations as described before by conventional chromosome CGH (Hermsen et al, 2001), by array CGH and SKY analyses. Both tumors were negative for Human Papilloma as well as Epstein-Barr viral sequences and showed multiple alterations with a high proportion of whole-arm chromosomal gains and losses (Table 2).

Table 2. Summary of gains and losses of chromosomal regions found in two FA head and neck squamous cell carcinoma cell lines, detected by CGH. Both cell lines show a high percentage of whole arm gains or losses in relation to the total number of copy number changes.

	Gains		Losses		% Whole-arm*
	Sub-arm	Whole-arm	Sub-arm	Whole-arm	
Tumor 1	4p, 4q, 7q, 12q	3q, 6q, Xq	5q, 13q, 17q, 22q	18q	33
Tumor 2	1p, 6p, 6q, 10q	1q, 3q, 5p, 7p, 8q, 9q, 12p, 17q, 18p, 22q	1p, 6p, 6q, 9p	3p, 5q, 7q, 8p, 10p, 14q, 12q,18q	69

*Whole-arm gains or losses as percentage of all chromosomal imbalances involving chromosomal breakage

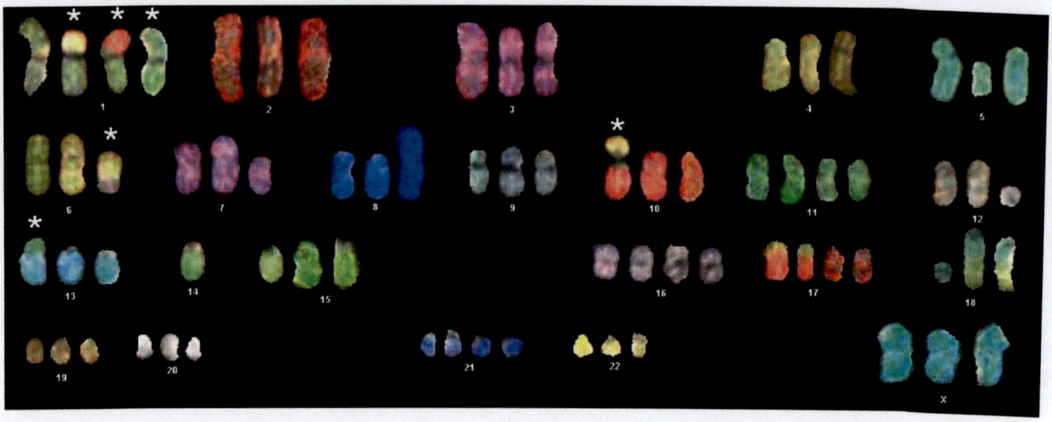

Figure 8. Representative SKY-karyotype of FA-head and neck squamous cell carcinoma cell line scc1131 showing both numerical and structural chromosomal rearrangements. Asterisks indicate the whole chromosome arm translocations: (1q;22q), (1q;10q), (1q;14q), (6p;12p), (10q;22q), (11p;13q).

In addition, SKY analysis revealed that these whole arm gains and losses were the result of centromeric translocations (Figure 8) and array CGH studies showed the transition points between the two BACs flanking the centromere (data not shown). These results suggest that the process leading to early occurrence of oral cancer in FA patients follows a similar pathway as in non-FA cancer patients. Since FA patients are uniquely hypersensitive to DNA cross-linking agents, while oral cancer in the general population is thought to be environmentally induced, these results also suggest that environmental DNA cross-linkers may be causally involved in the carcinogenesis of sporadic head and neck squamous cell carcinoma. Nevertheless, even though head and neck squamous cell carcinoma has been epidemiologically associated with smoking and alcohol consumption (Vokes, 1993; La Vecchia et al, 1997; Schmidt et al, 2004), these malignancies also occur in a significant proportion of individuals who do not smoke or consume alcohol. Such individuals might be people who have an increased sensitivity to other DNA cross-linkers found in the environment, in a comparable way as patients with FA.

3b. Defective FA Pathway in Sporadic Head and Neck Squamous Cell Carcinoma ?

Until now, little it is known on FA gene defects in sporadic cancers. Promotor methylation of the *FANCF* gene has been observed in lung, ovary as well as head and neck cancer (Taniguchi et al, 2003; Marsit et al, 2004). Mutations of *FANCC* and *FANCG* genes were found in pancreatic cancer (Rogers et al, 2004). BRCA 1, BRCA2 and Rad51 are other players in the FA mediated DNA repair pathway. In fact, the *BRCA2* gene was found to be identical with the *FANCD1* gene. BRCA 1 and BRCA2 are involved in the tumorigenesis of familial breast and ovarian cancer, but both have also been found to be mutated or silenced in various types of sporadic cancers (Turner et al, 2004).

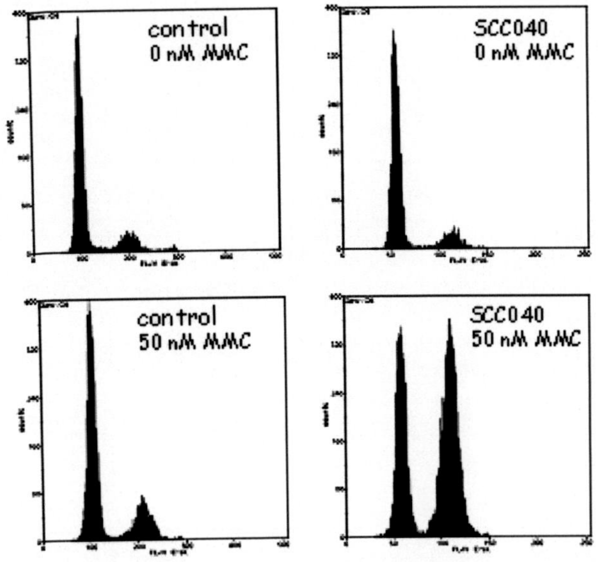

Figure 9a. Example of a G2 arrest experiment by flow cytometry. Here, a normal control is compared to a head and neck squamous cell carcinoma cell line. at 50 nm mmc exposure for 48 hours, the head and neck squamous cell carcinoma cell line shows a high percentage of cells in G2 as opposed to the normal control, meaning that this cell line is crosslinker hypersensitive.

FA is diagnosed by its hypersensitivity to crosslinking agents that cause interstrand crosslinks and block DNA replication, transcription and separation (Dronkert et al, 2001). Upon exposure, the cell cycle is arrested in the G2/M phase until the necessary repair has been performed. In FA cells, the percentage of cells in G2/M phase is high and becomes further enhanced after exposure to MMC because of its incapacity to repair the damage caused by it. Crosslinker hypersensitivity is tested by chromosome break analysis or by flow cytometric analysis of G2 arrest after MMC exposure. Using these tests, we found 4 of 11 sporadic head and neck squamous cell carcinoma cell lines to exhibit an elevated G2 arrest upon exposure to the crosslinker MMC (Figure 9), as compared to cultures of normal keratinocytes. Chromosome break analysis revealed high levels of damage of both the chromatid type (gaps) as of the chromosome type (triradials, quadriradials, acentric and dicentric chromosomes). Preliminary observations indicated frequent involvement of the centromere, but this needs to be followed by a combined quantitative and qualitative analysis. Finally, two of these 4 hypersensitive cell lines revealed a mutation either in the *BRCA2* or the *FANCD2* gene, respectively (unpublished data).

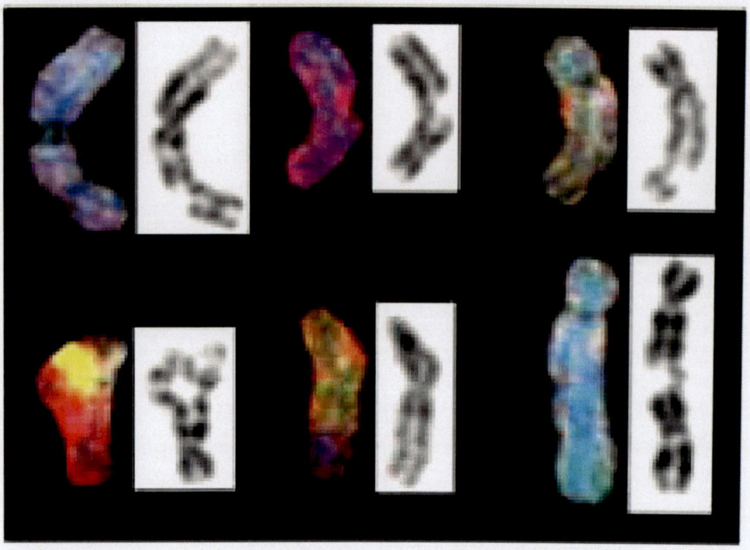

Figure 9b. Example of types of chromosomal damage after 48 hour 100 nM MMC exposure to FA lymphocytes, stained by DAPI and by SKY. In the top row, three examples of chromatid breaks are seen, respectively in chromosome 1, 3 and 5. In the bottom row exchange figures are shown, from left to right: a triradial formed by chromosome 10 and 22; a translocation between chromosomes 3 and 4, a marker chromosome formed by multiple translocations of chromosome 13 with itself (als showing a chromatid break).

In short, the available data suggest that mutations in the FA genes may also play a role in the tumorigenesis of sporadic head and neck squamous cell carcinoma. Furthermore, the disruption of the DNA repair pathway influenced by FA genes may be implicated in centromeric instability of squamous cell carcinomas.

4. DNA Repair Pathways

Two main types of double strand break (DSB) repair exists: non-homologous end-joining (NHEJ) and homologous recombination (HR). NHEJ is the predominant type of repair, active in all cell cycle phases, and is thought to restore the original sequence at the breakpoints. In the presence of multiple DSBs it has the potential to generate chromosomal rearrangements by incorrectly joining DNA ends. HR repairs reliably because it uses homologous sequences on the newly synthesized sister chromatid in S/G2 phase as a template, but can create sister chromatid exchanges.

Many known cancer-related genes are involved in HR, such as *ATM, BLM, NBS, BRCA1* and *BRCA2*, as well as the FA genes. Apart from DSBs, HR also repairs replication errors.

This explains why FA patients are hypersensitive to crosslinkers, known to cause replication stalling. Crosslinkers are present in tobacco, the main etiological factor in head and neck squamous cell carcinoma (Joenje and Patel, 2001). It may be speculated that tobacco causes the development and/or the progression of a subset of head and neck squamous cell carcinoma carrying impaired HR. It seems logical to assume that HR deficient cells have to use the error-prone NHEJ, which may trigger chromosomal rearrangments and cancer progression. However, recent work showed that two different forms of HR exist: Rad51 and BRCA2 dependent gene conversion (HR-GC), and Rad51 independent single strand annealing (HR-SSA). While HR-GC is considered to function error-free, HR-SSA can produce deletions, insertion and translocations, especially between repeat sequences (Richardson and Jasin, 2000). Elliot et al (2005) showed that HR-SSA repair of breaks near ALU repeats needed identical DNA sequences; when not identical, the break was repaired by NHEJ.

NHEJ is not impaired in FA cells, but both HR-GC and HR-SSA activity are reduced. However, in *BRCA2* (also named *FANCD1*) defective cells, HR-SSA is increased and HR-GC reduced. It has been speculated that FA cells have a mild reduction of HR-GC, leading to increased chromosome-abnormalities. FA proteins may therefore be involved in, but are not essential for both HR-GC and HR-SSA (Nakanishi et al, 2005). Cells with defective BRCA2/FANCD1 produce gross chromosomal rearrangements after MMC exposure, probably caused by HR-SSA repair, whereas in cells with functional BRCA2 mostly sister chromatid exchanges were observed (Tutt et al, 2001). Other reports showed that when more than one DSB is introduced, HR-SSA can lead to junctions between homologous repetitive sequences on heterologous chromosomes (Richardson and Jasin, 2000).

As much as 50% of all DNA is repetitive, the majority of which being ALU repeats in coding regions and satellite repeats in heterochromatic peri-centromeric regions (Murphy and Karpen, 1998). Rief and Lobrich (2002) have shown that DNA repair in centromeres is very effective, both in wild-type and in NHEJ-deficient cells and suggest a role for HR-SSA, which may also act in G1 phase. A possible confirmation was obtained by FISH studies, revealing that upon MMC exposure in G1 phase, paracentromeric sequences of chromosome 9 move together, whereas euchromatic sequences on chromosome 8 did not. In addition, chromatid interchanges were observed in metaphases between the two chromosome 9 homologues. (Abdel-Halim et al, 2005).

Finally, preliminary data from our own lab showed that human NHEJ-deficient cells (KU86) upon MTX exposure produced chromosomal rearrangements, which included frequent centromeric translocations. Non-reciprocal translocations were observed before in NHEJ-deficient cells (Ku70 and Lig4) mouse cells (Ferguson et al, 2000), but this study did not report on the status of centromeres.

In summary, several mechanisms have been proposed to be causally involved in gross chromosomal rearrangements, including NHEJ repair, but also a HR repair pathway: HR-SSA, which is equally error-prone. This repair acts in the vincinity of repeat sequences, but as yet it is unclear whether it could be part of the process of generating centromeric translocations in head and neck squamous cell carcinoma.

5. Epigenetic Modifications

Pericentromeric instability is also a characteristic of ICF syndrome (immunodeficiency, centromere instability, facial abnormalities). Patients with this syndrome show hypomethylation of juxtacentromeric heterochromatin on chromosomes 1, 9 and 16, in 60-70% of all cases caused by a defect in the DNA methyltransferase DNMT3B (Xu et al, 1999). Hypomethylation leads to de-condensation of the paracentromeric heterochromatin, which makes the regions more vulnerable to breakage. A similar process may cause breakage in the centromeric repeat sequences (followed by illegitimate repair) in squamous cell carcinomas.

It has been known for a long time that DNA of tumor cells is globally hypomethylated in comparison to normal cells and that this may contribute to genomic instability (Feinberg AP and Vogelstein, 1983). Hypomethylation of classical heterochromatic regions on chromosome 1 was found to be related with non-reciprocal translocations, e.g. in hepatocellular and prostate carcinomas (Wong et al, 2001). Also hypomethylation of centromeric alphasatellite repeats (present in all human centromeres) has been described and associated with chromosomal rearrangements (Ehrlich et al, 2003). Great advances have been made on the knowledge how hypermethylation of promotor regions affects gene expression in cancer (Esteller, 2003). In contrast, only little is known about how DNA methylation of specific chromosomal regions is achieved and regulated and how DNA methylation contributes to the development of tumors in specific tissues.

DNA methylated sequences such as in centromeres are packaged in nucleosomes that contain deacetylated H3 histones. Methylation of lysine 9 on H3 is closely associated with heterochromatin, whereas methylation of lysine 4 on H3 characterizes transcriptionally active euchromatin (Jones and Baylin, 2002).

DSB repair is also dependent on the chromatin state. Histone modifications correlate with DNA damage and might therefore serve as a code for repair. On the other hand, the chromatin state may influence the susceptibility of specific sequences to DNA damaging agents (Lydall and Whitehall 2005).

In summary, alterations in epigenetic modifications of centromeric heterochromatin, both in classical and in alpha satellites, have been described in association with centromeric translocations and cancer. A possible role for the tumorigenesis of head and neck squamous cell carcinoma still needs to be investigated.

6. Conclusion

Given its extraordinary high occurrence in this study and in previously reported cytogenetic studies on squamous cell carcinoma, irrespective of the organ of origin, we propose that centromeric breakage and illegitimate recombination should be considered a critical type of chromosomal instability in the development of this type of epithelial cancer, in contrast to adenocarcinoma. We demonstrated for the first time, to our knowledge, tissue-specific differences of centromeric breakage followed by illegitimate recombination leading to whole-arm translocations in epithelial cancer (Hermsen et al, 2005).

More insights into the mechanisms of DNA repair and classification of tumors according to possible DNA repair defects may be of great interest for clinical applications. For example,

HR impaired tumors may be extra sensitive to crosslinker-treatment. It has been proposed that MMC and Cis-Pt may be of particular merit in cancers with FA mutations (Lyakhovich and Surrallés, 2006). Kennedy et al (2004) suggested that sporadic tumors with loss of BRCA1 function would be sensitive to DNA damaging chemotherapy and perhaps resistant to spindle poisons. Alternatively, specific inhibitors of DNA repair or DNA repair checkpoints may sensitise tumors to therapy. Bryant and co-workers (2005) demonstrated that BRCA2 deficient cells are acutely sensitive to PARP1 inhibitors, which may kill a tumor, even in the absence of an exogenous DNA-damaging agent. Much effort is being put into finding specific inhibitors of the FA pathway. Recently, wortmannin was presented as an inhibitor of FANCD2 monoubiquitination (Surrallés et al, 2006), but its potential use in the clinic needs to be evaluated.

Understanding the mechanism of centromeric translocation may therefore help to develop new treatment modalities, especially tuned for tumors with centromeric instability, such as squamous cell carcinomas.

References

Abdel-Halim HI, Natarajan AT, Mullenders LH, Boei JJ. Mitomycin C-induced pairing of heterochromatin reflects initiation of DNA repair and chromatid exchange formation. *J Cell Sci*. 2005, 118:1757-67.

Albertson DG, Collins C, McCormick F, Gray JW. Chromosome aberrations in solid tumors. *Nat Genet* 2003, 34:369-376.

Bryant HE, Schultz N, Thomas HD, Parker KM, Flower D, Lopez E, Kyle S, Meuth M, Curtin NJ, Helleday T. Specific killing of BRCA2-deficient tumours with inhibitors of poly(ADP-ribose) polymerase. *Nature* 2005, 434:913-917.

Carvalho B, Ouwerkerk E, Meijer GA, Ylstra B. High resolution microarray comparative genomic hybridisation analysis using spotted oligonucleotides. *J Clin Pathol* 2004, 57:644-646.

Dronkert ML, Kanaar R. Repair of DNA interstrand cross-links. *Mutat Res* 2001, 486: 217-247.

Ehrlich M, Hopkins NE, Jiang G, Dome JS, Yu MC, Woods CB, Tomlinson GE, Chintagumpala M, Champagne M, Dillerg L, Parham DM, Sawyer J. Satellite DNA hypomethylation in karyotyped Wilms tumors. *Cancer Genet Cytogenet*. 2003, 141: 97-105.

Elliott B, Richardson C, Jasin M. Chromosomal translocation mechanisms at intronic alu elements in mammalian cells. *Mol Cell*. 2005, 17:885-94.

Eshleman JR, Casey G, Kochera ME, Sedwick WD, Swinler SE, Veigl ML, Willson JKV, Schwartz S and Markowitz SD. Chromosome number and structure both are markedly stable in RER colorectal cancers and are not destabilized by mutation of p53. *Oncogene* 1998, 17: 719–725.

Esteller M. Relevance of DNA methylation in the management of cancer. *Lancet Oncol.* 2003, 4:351-8

Feinberg AP, Vogelstein B. Hypomethylation distinguishes genes of some human cancers from their normal counterparts. *Nature*. 1983, 301:89-92.

Ferguson DO, Sekiguchi JM, Chang S, Frank KM, Gao Y, DePinho RA, Alt FW. The nonhomologous end-joining pathway of DNA repair is required for genomic stability and the suppression of translocations. *Proc Natl Acad Sci U S A.* 2000, 97:6630-3.

Grieken NC van, Weiss MM, Meijer GA, Hermsen MA, Scholte GH, Lindeman J, Craanen ME, Bloemena E, Meuwissen SG, Baak JP, Kuipers EJ. Helicobacter pylori-related and -non-related gastric cancers do not differ with respect to chromosomal aberrations. *J Pathol* 2000, 192:301-306.

Hermsen MAJA, Joenje H, Arwert F, Welters MJP, Braakhuis BJM, Bagnay M, Westerveld A, Slater R. Centromeric breakage as a major cause of cytogenetic abnormalities in oral squamous cell carcinoma. *Genes, Chromosomes and Cancer*, 1996, 15:1-9.

Hermsen MAJA, Baak JPA, Weiss J, Meijer GA, Weiss JM, Walboomers JJM, Snijders PJF, Diest PJ van. Genetic analysis of 53 lymph node-negative breast carcinomas by CGH and relation to clinical, pathological, morphometric, and DNA cytometric prognostic factors. *J Pathol* 1998;186:356-362.

Hermsen MAJA, Guervós MA, Meijer GA, Baak JPA, Álvarez-Marcos C, Salas A, Sampedro A, Van Diest PJ. New chromosomal regions with high level amplifications in larynx and pharynx squamous cell carcinoma identified by CGH. *J Pathol* 2001a, 194:177-182.

Hermsen MAJA, Xie Y, Rooimans MA, Meijer GA, Baak JPA, Plukker JTM, Arwert F, Joenje H. Cytogenetic characteristics of oral squamous cell carcinomas in Fanconi anemia. *Familial Cancer* 2001b, 1:39-43.

Hermsen MAJA, Postma C, Baak JPA, Weiss MM, Rapallo A, Sciutto A, Roemen G, Arends J-W, Richard Williams, Giaretti W, De Goeij A, Meijer GA. Colorectal adenoma to carcinoma progression follows multiple distinct pathways of chromosomal instability. *Gastroenterology* 2002, 123:1109-1119.

Hermsen MAJA, Snijders A, Alonso Guervós M, Taenzer S, Koerner U, Baak J, Pinkel P, Albertson D, Van Diest P, Meijer G, Schrock E. Centromeric chromosomal translocations show tissue-specific differences between squamous cell carcinomas and adenocarcinomas. *Oncogene* 2005, 24:1571-1579.

Israeli O, Gotlieb WH, Friedman E, Goldman B, Ben-Baruch G, Aviram-Goldring A, Rienstein S. Familial vs sporadic ovarian tumors: characteristic genomic alterations analyzed by CGH. *Gynecol Oncol* 2003, 90:629-636.

Jin C, Jin Y, Wennerberg J, Dictor M, Mertens F. Nonrandom pattern of cytogenetic abnormalities in squamous cell carcinoma of the larynx. *Genes Chromosomes Cancer* 2000, 28:66-76.

Joenje H, Patel KJ. The emerging genetic and molecular basis of Fanconi anaemia. *Nat Rev Genet* 2001, 2:446-457.

Jones PA, Baylin SB. The fundamental role of epigenetic events in cancer. *Nat Rev Genet.* 2002, 3:415-28.

Kutler DI, Auerbach AD, Satagopan J, Giampietro PF, Batish SD, Huvos AG, Goberdhan A, Shah JP, Singh B. High incidence of head and neck squamous cell carcinoma in patients with Fanconi anemia. *Arch Otolaryngol Head Neck Surg* 2003, 129:106-112.

La Vecchia C, Tavani A, Franceschi S et al. Epidemiology and prevention of oral cancer. *Oral Oncol* 1997; 33: 302–12.

Levitus M, Rooimans MA, Steltenpool J, Cool NF, Oostra AB, Mathew CG, Hoatlin ME, Waisfisz Q, Arwert F, de Winter JP, Joenje H. Heterogeneity in Fanconi anemia: evidence for 2 new genetic subtypes. *Blood* 2004, 103:2498-2503.

Luk C, Tsao MS, Bayani J, Shepherd F, Squire JA. Molecular cytogenetic analysis of non-small cell lung carcinoma by spectral karyotyping and comparative genomic hybridization. *Cancer Genet Cytogenet* 2001, 125:87-99.

Lydall D, Whitehall S. Chromatin and the DNA damage response. *DNA Repair (Amst).* 2005, 4:1195-207.

Lyakhovich A, Surralles J. Disruption of the Fanconi anemia/BRCA pathway in sporadic cancer. *Cancer Lett* 2006, 232:99-106.

Marsit CJ, Liu M, Nelson HH, Posner M, Suzuki M, Kelsey KT. Inactivation of the Fanconi anemia/BRCA pathway in lung and oral cancers: implications for treatment and survival. *Oncogene* 2004, 23:1000-1004.

Meijer GA, Hermsen MAJA, Baak JPA, Diest PJ van, et al. Progression from colorectal adenoma to carcinoma is associated with non-random chromosomal gains as detected by CGH. *J Clin Pathol* 1998, 51:901-909.

Murphy TD, Karpen GH. Centromeres Take Flight: Alpha Satellite and the Quest for the Human Centromere. *Cell* 1998, 93: 317-320.

Nakanishi K, Yang YG, Pierce AJ, Taniguchi T, Digweed M, D'Andrea AD, Wang ZQ, Jasin M. Human Fanconi anemia monoubiquitination pathway promotes homologous DNA repair. *Proc Natl Acad Sci U S A*. 2005, 102:1110-5.

Nakao K, Mehta KR, Fridlyand J, Moore DH, Jain AN, Lafuente A, Wiencke JW, Terdiman JP, Waldman FM. High-resolution analysis of DNA copy number alterations in colorectal cancer by array-based comparative genomic hybridization. *Carcinogenesis*. 2004 Aug;25(8):1345-57.

NCI and NCBI's SKY/M-FISH and CGH Database (2001), http://www.ncbi.nlm.nih.gov/sky/skyweb.cgi

Pei J, Balsara BR, Li W, Litwin S, Gabrielson E, Feder M, Jen J, Testa JR. Genomic imbalances in human lung adenocarcinomas and squamous cell carcinomas. *Genes Chromosomes Cancer* 2001, 31:282-287.

Petersen I, Langreck H, Wolf G, Schwendel A, Psille R, Vogt P, Reichel MB, Ried T, Dietel M. Small-cell lung cancer is characterized by a high incidence of deletions on chromosomes 3p, 4q, 5q, 10q, 13q and 17p. *Br J Cancer* 1997, 75:79-86.

Rogers CD, van der Heijden MS, Brune K, Yeo CJ, Hruban RH, Kern SE, Goggins M. The genetics of FANCC and FANCG in familial pancreatic cancer. *Cancer Biol Ther*. 2004, 3:167-169.

Richardson C, Jasin M. Frequent chromosomal translocations induced by DNA double-strand breaks. *Nature*. 2000, 405:697-700.

Rief N, Lobrich M. Efficient rejoining of radiation-induced DNA double-strand breaks in centromeric DNA of human cells. *J Biol Chem*. 2002, 277:20572-82.

Schlegel J, Stumm G, Scherthan H, Bocker T, Zirngibl HR, Rüschoff J and Hofstädter F. Comparative genomic in situ hybridization of colon carcinomas with replication error. *Cancer Res* 1995, 55 : 6002–6005.

Schmidt BL, Dierks EJ, Homer L, Potter B. Tobacco smoking history and presentation of oral squamous cell carcinoma. *J Oral Maxillofac Surg* 2004, 62:1055-1058.

Schrock E, Du Manoir S, Veldman T, Schoell B, Wienberg J, Ferguson-Smith MA, Ning Y, Ledbetter DH, Soenksen D, Garini Y, Ried T. Multicolor spectral karyotyping of human chromosomes. *Science* 1996, 273:494-497.

Snijders AM, Fridlyand J, Mans DA, Segraves R, Jain AN, Pinkel D, Albertson DG. Shaping of tumor and drug-resistant genomes by instability and selection. *Oncogene* 2003, 22: 4370–4379.

Soulie P, Fourme E, Hamelin R, Asselain B, Salmon R-J, Dutrillaux B and Muleris M. TP53 status and gene amplification in human colorectal carcinomas. *Cancer Genetics and Cytogenetics* 1999, 115: 118–122.

Surralles J, Jackson SP, Jasin M, Kastan MB, West SC, Joenje H. Molecular cross-talk among chromosome fragility syndromes. *Genes Dev* 2004, 18:1359-70.

Taniguchi T, Tischkowitz M, Ameziane N, Hodgson SV, Mathew CG, Joenje H, Mok SC, D'Andrea AD. Disruption of the Fanconi anemia-BRCA pathway in cisplatin-sensitive ovarian tumors. *Nat Med.* 2003, 9:568-74.

Tirkkonen M, Johannsson O, Agnarsson BA, Olsson H, Ingvarsson S, Karhu R, Tanner M, Isola J, Barkardottir RB, Borg A, Kallioniemi OP. Distinct somatic genetic changes associated with tumor progression in carriers of BRCA1 and BRCA2 germ-line mutations. *Cancer Res* 1997, 57:1222-1227.

Turner N, Tutt A, Ashworth A. Hallmarks of 'BRCAness' in sporadic cancers. *Nat Rev Cancer* 2004, 4:814-9.

Tutt A, Bertwistle D, Valentine J, Gabriel A, Swift S, Ross G, Griffin C, Thacker J, Ashworth A. Mutation in Brca2 stimulates error-prone homology-directed repair of DNA double-strand breaks occurring between repeated sequences. *EMBO J.* 2001, 20:4704-16.

Varis A, Puolakkainen P, Savolainen H, Kokkola A, Salo J, Nieminen O, Nordling S, Knuutila S. DNA copy number profiling in esophageal Barrett adenocarcinoma: comparison with gastric adenocarcinoma and esophageal squamous cell carcinoma. *Cancer Genet Cytogenet* 2001, 127:53-58.

Vokes EE, Head and neck cancer. *N Engl J Med* 1993, 328:184-193.

Weiss MM, Kuipers EJ, Hermsen MA, van Grieken NC, Offerhaus J, Baak JP, Meuwissen SG, Meijer GA. Barrett's adenocarcinomas resemble adenocarcinomas of the gastric cardia in terms of chromosomal copy number changes, but relate to squamous cell carcinomas of the distal oesophagus with respect to the presence of high-level amplifications. *J Pathol* 2003, 199:157-165.

Wong N, Lam WC, Lai PB, Pang E, Lau WY, Johnson PJ. Hypomethylation of chromosome 1 heterochromatin DNA correlates with q-arm copy gain in human hepatocellular carcinoma. *Am J Pathol.* 2001, 159:465-71.

Xu GL, Bestor TH, Bourc'his D, Hsieh CL, Tommerup N, Bugge M, Hulten M, Qu X, Russo JJ, Viegas-Pequignot E. Chromosome instability and immunodeficiency syndrome caused by mutations in a DNA methyltransferase gene. *Nature* 1999, 402:187-191.

In: Genetic Translocations and Other Chromosome Aberrations ISBN: 978-1-60456-344-3
Editor: Guy T. Leyden, pp. 209-222 © 2008 Nova Science Publishers, Inc.

Chapter 8

Fragile Sites on Human Chromosomes

Grant R. Sutherland*

Department of Cytogenetics and Molecular Genetics, Women's and Children's Hospital Adelaide, 5006, Australia;

Department of Paediatrics, University of Adelaide, Australia.

Introduction

Fragile sites on human chromosomes were first described in 1965 by Dekaban et al. In the early years of human cytogenetics these were rather enigmatic phenomena that were seen by some laboratories and not by others. It was not until 1977 that it was demonstrated that the type of culture medium in which lymphocytes were cultured to prepare chromosomes was a critical factor in determining whether fragile sites were seen (Sutherland 1977). It was subsequently shown that folic acid and thymidine needed to be absent from the culture medium for fragile sites to manifest themselves (Sutherland 1979).Some rare fragile sites were found not to be sensitive to folic acid but inducible by a range of other compounds.

In this chapter the classification of fragile sites will be given, the cytogenetics presented, the clinical significance of the rare fragile sites discussed and the inheritance and epidemiology of fragile sites presented.

Classification

Fragile sites are classified on the basis of their frequency in the population and the conditions of tissue culture under which they are expressed. Rare fragile sites are seen at a specific location on 1 in 40 chromosomes at most. Common fragile sites are probably present on all chromosomes and as such are normal components of chromosome structure.

* E-mail address: grant.sutherland@adelaide.edu.au. Correspondence concerning this article should be addressed to Grant R. Sutherland,

Most fragile sites are not expressed spontaneously but must be induced. When they are expressed spontaneously the proportion of metaphases in which they are seen can usually be greatly increased by appropriate conditions of induction. The classification, based upon population frequency and induction conditions, is shown in Table 1 and all known rare human fragile sites are listed in Table 2.

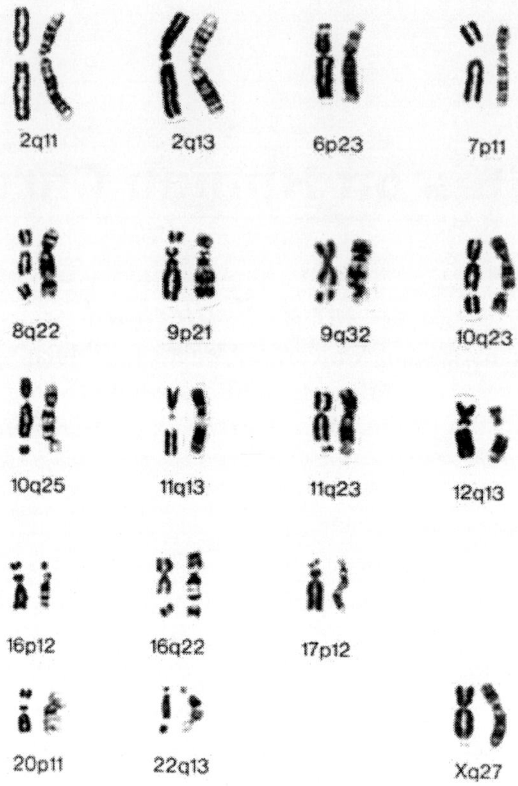

Figure 1. An assortment of rare fragile sites. The chromosome on the left of each pair is plain stained and the one on the right is G-banded. (Reproduced from Gardner and Sutherland, 1996).

Table 1. Classification of fragile sites

Rare	**Group 1**	folate sensitive fragile sites (n=22)
	Group 2	distamycin A inducible a) also inducible by BrdU (n=2, *FRA16B, FRA17A*) b) not inducible by BrdU, and only recorded in the Japanese (n=3, *FRA8E, FRA11I, FRA16E*)
	Group 3	BrdU requiring (n=2, *FRA10B, FRA12C*)
Common	**Group 4**	Aphidicolin inducible (n=76)
	Group 5	5-azacytidine inducible (n=4)
	Group 6	BrdU-inducible (n=7)

Table 2. All rare fragile sites

Group	Gene symbol	Chromosome Band	Reference*
1. Folate sensitive	FRA1M	1p21.3	Baker and Sutherland (1991)
	FRA2L	2p11.2	Schuffenhauer et al (1996)
	FRA2A	2q11.2	Ford and Madan (1973); Lejeune et al (1968); Quack et al (1978); Sutherland and Hecht (1985)
	FRA2B	2q13	Ford and Madan (1973); Lejeune et al (1968); Quack et al (1978); Sutherland and Hecht (1985)
	FRA2K	2q22.3	Mattei et al (1987
	FRA5G	5q35	Sutherland and Hecht (1985); Howell et al (1990)
	FRA6A	6p23	Sutherland et al (1983)
	FRA7A	7p11.2	Sutherland and Hecht (1985)
	FRA8A	8q22.3	Quack et al (1978)
	FRA9A	9p21	Guichaoua et al (1982); Sutherland et al (1983)
	FRA9B	9q32	Petit et al (1986); Sutherland et al (1983)
	FRA10A	10q23.3	Dutrillaux et al (1985); Giraud et al (1976); Sutherland (1979)
	FRA11A	11q13.3	Quack et al (1978); Sutherland (1979)
	FRA11B	11q23.3	Sutherland et al (1983); Voullaire et al (1987)
	FRA12A	12q13.1	Giraud et al (1976); Quack et al (1978); Sutherland and Hinton (1981)
	FRA12D	12q24.13	Amarose et al (1987); Sutherland and Baker (1993)
	FRA16A	16p13.11	Callen et al (1989); Day et al (1967); Sutherland (1979); Nancarrow et al (1994)
	FRA19B	19p13	Chodirker et al (1987); Tommerup et al (1985)
	FRA20A	20p11.23	Shabtai and Halbrecht (1981); Sutherland (1979)
	FRA22A	22q13.1	Kahkonen et al (1989), Webb and Thake (1984)
	FRAXA	Xq27.3	Lubs (1969), Sutherland (1977); Yu et al (1991); Oberle et al (1991);Verkerk et al (1991)
	FRAXE	Xq28	Sutherland and Baker (1992), Oberle et al (1992); Knight et al (1993)
	FRAXF	Xq28	Hirst et al (1993), Parrish et al (1994)
2. Distamycin A inducible (a)also BrdU inducible	FRA16B	16q22.1	Callen et al (1988); Magenis et al (1970); Schmid et al 1980, Simmers et al (1987); Yu et al (1997)
	FRA17A	17p12	Court Brown et al (1966); Sutherland et al (1984)
(b) not also BrdU inducible	FRA8E	8q24.1	Ochi et al (1988); Takahashi et al (1988)
	FRA11I	11p15.1	Hori et al (1988); Takahashi et al (1988)
	FRA16E	16p12.1	Hori et al (1988)
(c) BrdU requiring	FRA10B	10q25.2	Scheres and Hustinx (1980); Sutherland et al (1980); Hewett (1998)
	FRA12C	12q24.2	Sutherland et al (1985); Voiculescu et al (1988)

* References are to papers that either first specifically identified the fragile site or characterised it.

Cytogenetics

A fragile site is a region of chromatin that fails to compact normally for mitosis. When this happens the chromosome is attenuated at that point and fragile, both when prepared for microscopic examination and in vivo (as evidenced by configurations that must arise from breakage at the fragile site and nondisjunction of the acentric fragments (Figures 1-3). The fragile site is almost never seen in 100% of metaphases from an individual who has it, the proportion varying from only a few percent up to 50 – 60%. An exception is *FRA16B* which can be seen in close to 100% of metaphases when induced by berenil (Schmid et al, 1986).

While fragile sites can be seen on banded chromosomes they are easier to see and more distinctive when seen on unbanded chromosomes (Sutherland and Hecht, 1985).

Table 3. Features of the molecularly characterised rare fragile sites

Fragile site	Group	Repeat motif	Gene involved	Reference
FRA10A	1	CCG (perfect)	FRA10AC1	Sarafidou et al (2004)
FRA16A	1	CCG (complex)	? Yes	Nancarrow et al (1994), Nancarrow (1998)
FRA11B	1	CCG (perfect)	CBL2	Jones et al (1995)
FRA10B	2c	42bp AT-rich minisatellite	No	Yu et al (1997)
FRA16B	2a	33bp AT-rich minisatellite	No	Hewett et al (1998)
FRAXA	1	CCG (complex)	FMR1	Yu et al (1991), Oberle et al (1991)
FRAXE	1	CCG (perfect)	FMR2/FMR3	Gecz et al (1996), Gu et al (1996), Gecz (2000b)
FRAXF	1	CCG (complex)	FAM11A	Shaw et al (2002)

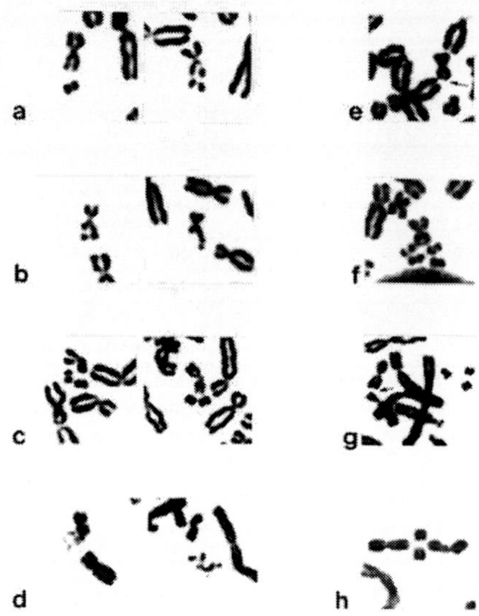

Figure 2. Forms of expression manifested by fragile sites exemplified by FRA16B. (a) isochromatid gap; (b) single chromatid gap; (c) duplication of chromatids distal to the fragile site arising from the mechanism in Fig. 3; (d) despiralisation of chromatin distal to the fragile site; (e) deletion of material distal to the fragile site (arrow); (f) quadruplication of chromatin distal to the fragile site; (g) deletion (arrow) and acentric fragments; (h) quadriradial figure. (Reproduced from Sutherland and Hecht 1985).

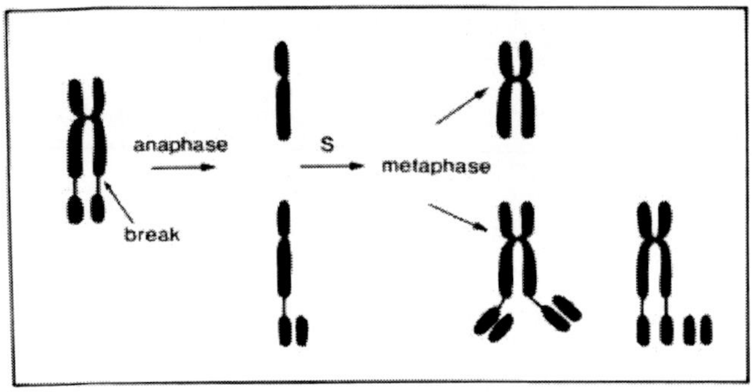

Figure 3. Mechanism of production of triradial figures and deleted chromosomes by breakage at the fragile site and nondisjunction of an acentric fragment during mitotic anaphase. (Reproduced from Sutherland, 1983).

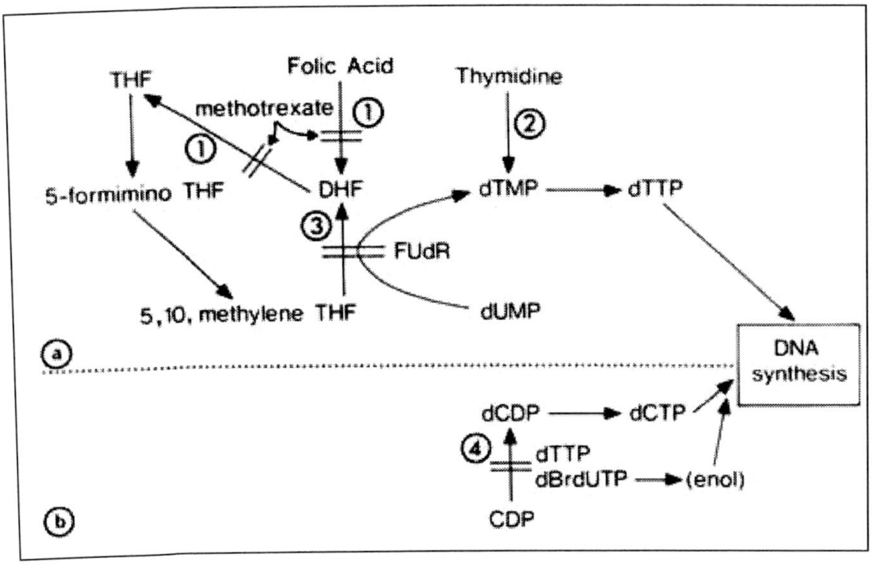

Figure 4. Areas of nucleotide metabolism involved in folate-sensitive fragile site expression. (a) pathway affected by restriction of folic acid and thymidine or induction by inhibitors of folate metabolism or inhibitors of thymidylate synthase: (b) pathway affected by high thymidine and BrdU concentrations. Enzymes crucial to this pathway are (1)dihydrofolate reductase, (2)thymidine kinase, (3)thymidylate synthase, (4) ribonucleotide reductase. Abbreviations: BrdU,bromodeoxyuridine; CDP, cytidine diphosphate; dCDP, deoxycytidine diphosphate; DHF dihydrofolate; FUdR, flurodeoxyuridine; dTMP, deoxythymidine monophosphate (thymidylate); dTTP, deoxythymidine triphosphate; THF, tetrahydrofolate; UMP, uridine monophosphate. (Reproduced from Sutherland, 1988).

Most studies of fragile sites have been on chromosomes prepared from short-term (3-4 day) cultures of lymphocytes. The lymphocytes must be cultured under specific conditions for the fragile sites to be expressed. Detailed protocols are given in Sutherland (1991). For the folate-sensitive fragile sites these conditions include absence of folic acid or presence of antifolates such as methotrexate; low concentrations of thymidine, and its analogs BrdU, BrdC, or presence of inhibitors of the enzyme thymidylate synthase (such as FudR); high concentrations of thymidine (but not its analogs). Any of these inducing conditions need to be present for at least 24 hours before chromosome preparation. All of these inducing conditions appear to lead to a relative deficiency of thymidine triphosphate at the time of DNA synthesis (Figure 4). The failure of high concentrations of BrdU to induce these fragile sites is unexplained, a possibility that the enol form of this compound is incorporated into the DNA under such conditions and inhibits fragile site expression is a barely satisfactory explanation.

The Distamycin A inducible fragile sites are induced by specific agents which bind the minor groove of the DNA helix. These agents include Hoechst 33258, netropsin and berenil. The group 2A fragile sites are also induced by BrdU (but not thymidine). The group 2B fragile sites, which have only been described in the Japanese, are not inducible by BrdU. The group 3 fragile sites are only inducible by BrdU (and BrdC).

An interesting observation that remains unexplained is that when there are two rare folate sensitive fragile sites on chromosomes in the one cell, or even two on the one chromosome, these tend to 'associate' in metaphase chromosome preparations (Sutherland and Baker, 2003). They give a somewhat similar appearance to acrocentric chromosome satellite stalk associations (Figure 5).

Figure 5. Three partial metaphases showing a chromosome 12 with two rare fragile sites adopting a U-shaped configuration to allow the fragile regions to associate. (From Sutherland and Baker 1993).

Fragile Site Expression

While much is known about the DNA sequences at fragile sites the mechanisms whereby these sequences give rise to the discontinuity in chromosome structure that is the visible manifestation of the fragile site remains conjectural. Almost certainly the fragile site results from a failure of chromatin compaction for normal mitosis. Many of the factors associated with induction of fragile sites, both chemical in terms of inducing agents, and physical in terms of timing of the action of such agents and of DNA replication, point to DNA at the fragile site not being completely replicated by the time a cell enters metaphase. Such under-replicated, perhaps even single-stranded, DNA is unlikely to package as would normal completely replicated DNA.

Alternatively, it has been shown that DNA comprising long tracts of CCG repeats, especially when CpG methylated, resist nucleosome formation thus providing a mechanism for packaging failure (Wang and Griffith, 1996). Furthermore, in vitro studies of the 33-bp sequence amplified in FRA16B suggest that the sequence itself, particularly in the presence of distamycin A, also excludes nucleosomes and the distamycin effect is not limited to single nucleosomes but also to nucleosome arrays (Hsu and Wang 2002). This could explain why *FRA16B* can be spontaneously expressed but that distamycin A greatly enhances its expression.

Replication Timing

Laird (1988) originally proposed that the fragile X (*FRAXA*) would be late replicating and he turned out to be correct, but almost certainly for the wrong reasons. Numerous studies have shown that *FRAXA, FRAXE* and some of the common fragile sites (reviewed by Arlt et al 2003) are later replicating than the same locus without either the repeat expansion (the rare sites) or the presence of specific fragile site inducers (the common sites). Data on *FRA10B* and *FRA16B* were less clear (Handt et al 2000) and indeed these authors concluded that while "... delayed replication may play a role in fragile site expression ... replication delay may not need to occur at the expanded repeat region itself ... to be permissive for fragility."

Clinical Significance

The Fragile X Syndromes

One fragile site on the X chromosome (*FRAXA*) is of major medical significance in that it is a visible manifestation of the full mutation at this locus which inactivates the *FMR*1 gene. This gives rise to fragile X syndrome, the most common familial form of intellectual handicap. This disease has been found in all population groups exposed to Western medicine but it is unknown whether it has the same frequency in all populations, however there is some evidence that it may be found less frequently in African Americans than in Caucasians (Crawford et al 2002). Approximately 1 in every 4000 boys has fragile X syndrome.

FRAXA permutations are now being recognised as clinically important and appear to be responsible for premature ovarian failure in a proportion of carriers and a late onset tremor/ataxia syndrome (FXTAS) affecting about one third of male carriers over the age of 50 years (Tassone et al, 2000; Leehey et al, 2003).

Another fragile site on the X chromosome (*FRAXE*) is associated with a mild form of mental retardation. These males have no physical anomalies and the associated gene, *FMR2*, was the first to be isolated for non-specific X-linked mental retardation (Gecz 2000a). More than 50 families have been recorded in which mental retardation associated with this fragile site is present.

Other Fragile Sites

Some mothers of children with a mental retardation/malformation condition known as Jacobsen syndrome have *FRA11B*. Jacobsen syndrome is due to a deletion of part of the long arm of chromosome 11 and the deletion breakpoints in some cases are at or very close to the fragile site on their mother's chromosome 11. This is the only autosomal fragile site with documented possible clinical significance. Even here, very few women with *FRA11B* will have a child with Jacobsen syndrome and most Jacobsen syndrome patients have their chromosome 11 breakpoint distal to the fragile site (Michaelis et al, 1998).

Pediatric chromosome studies are largely carried out on children with developmental problems, and from time to time rare autosomal fragile sites have been found. These are almost certainly unrelated to the developmental issues in the children in whom they are detected as family studies will almost always find the same fragile site in one parent and in half the normal siblings. This is an example of ascertainment bias and not 'cause and effect'.

The only rare autosomal fragile sites to have been recorded in homozygous form are *FRA10B*, *FRA16B* and *FRA17A* where they are without apparent phenotypic effect (Hocking et al, 1999). Group I autosomal fragile sites have never been seen in homozygotes. Any clinical effect in homozygotes would presumably depend upon the function of the gene in which the fragile site was located. It is likely that the presence of a fragile site detected cytogenetically means that a full mutation is present and any gene in which the fragile site is located is inactivated.

There has been controversy for many years about a possible role for fragile sites in oncogenesis. The hypothesis is that breakage or genomic instability at the fragile sites could give rise to chromosome rearrangements of the types seen in tumour tissue. Most studies had shown that when breaks in such rearrangements were carefully mapped in relation to rare fragile sites, the two did not coincide. However *FRA11B* is within the oncogene *CLB*2, and a number of the common (group 4) fragile sites have within them putative tumour suppressor genes. Nevertheless, there is no evidence that individuals with rare fragile sites (and everyone has all the common ones) are at any discernibly increased risk of developing malignant disease.

Inheritance

The fragile X (*FRAXA*) is the rare fragile site whose genetics and epidemiology has been best studied (Sherman et al 1984). Its segregation patterns departed greatly from regular Mendelian inheritance. Its most striking features were:

- Males could be asymptomatic carriers (transmitting males), a novel finding for an X-linked disease.
- The penetrance of the cytogenetic presence of the fragile site and the associated mental retardation syndrome was incomplete and depended upon the phenotype of the mother. Mothers who were intellectually affected showed complete penetrance in their offspring, whereas penetrance in the offspring of unaffected mothers was about 75%.

- The penetrance of mental retardation in the offspring of the mothers and daughters of transmitting males was different. This phenomenon became known as the Sherman paradox (Figure 6).

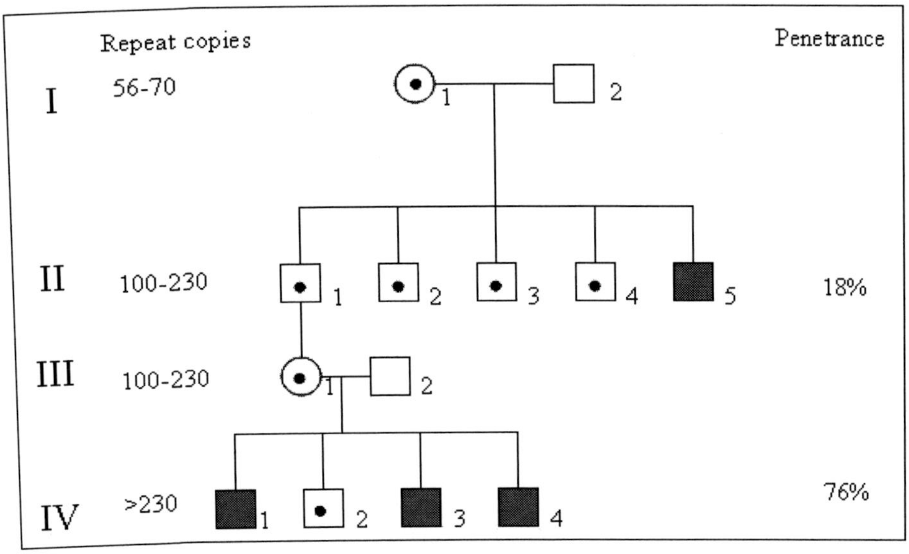

Figure 6. The Sherman paradox. A hypothetical pedigree is used to illustrate the different levels of penetrance of the fragile X mutation depending upon the position of the carrier in the pedigree. The fragile X carrier I-1 has an 18% chance (approximately one in five risk) that her fragile X sons are affected, while the fragile X carrier III-1 has a 76% chance (three in four risk) that her fragile X sons are affected. The size of the amplification of the fragile X p(CCG)n repeat is given as approximate number of copies. Copy numbers indicated are those expected hypothetically for individual 1 in each generation. Hatched squares indicate individuals with fragile X syndrome, black dots indicate fragile X premutations.

These anomalies in fragile XA inheritance were explained by the molecular characterisation of the fragile site (Yu et al 1992). The chance of a premutation expanding to a full mutation in a child is a function of the size of the mother's premutation. Full mutations only rarely regress to premutations when transmitted by women, but inevitably do so when transmitted by men.

The segregation of the rare autosomal fragile sites was studied by Sherman and Sutherland (1986). For the folate-sensitive fragile sites they found that there was a sex difference in penetrance (defined as the cytogenetic detection of the fragile site): when the fragile sites were transmitted by women they were fully penetrant but when transmitted by males they were only 50% penetrant. The great majority of probands in folate-sensitive fragile site families had, perhaps as a consequence of this sex difference, received their fragile site from their mother. The reasons for this sex difference have never been addressed, but sex differences in transmission are characteristic of most of the trinucleotide repeat expansions. It is known for example that *FRAXA* males never transmit a full mutation to their daughters (Reyniers et al, 1993) and it may be that the tendency for folate-sensitive fragile site repeat expansions to regress in size during spermatogenesis is a general phenomenon. Presumably the offspring of rare folate-sensitive fragile site males receive the fragile site carrying chromosome 50% of the time, but on about half of these occasions the size of its CCG repeat will have been reduced to the point where a fragile site is not manifested cytogenetically. This can only be tested extensively when more of this group of fragile sites has been characterised at the molecular level and their transmission studied. However, in three transmissions from a *FRA16A* male there were no discernible changes in CCG repeat copy number (Nancarrow et al. 1994).

While the three autosomal Group 1 sites that have been cloned so far (*FRA16A, FRA10A* and *FRA11B*) have similar molecular mechanisms to the fragile X, very few family studies at the molecular level have been performed. While in at least the case of *FRA16A* and *FRAXE* a male can transmit a full mutation, nothing is known about the likelihood of this happening.

FRA16B appears to follow a normal Mendelian segregation pattern. Although *FRA10B* appeared to show some of the features of the rare folate-sensitive fragile site segregation patterns the differences from normal Mendelian segregation did not reach statistical significance (Sherman and Sutherland, 1986).

References

Amarose AP, Huttenlocher PR, Sprudzs RM, Laitsch TJ, Pettanati MJ (1987) A heritable fragile 12q24.13 segregating in a family with the fragile X chromosome. *Human Genetics* **75**:4-6

Arlt MF, Casper AM, Glover TW (2003) Common Fragile Sites. *Cytogenetics and Genome Research* **100**:92-100.

Baker E, Sutherland GR (1991) A new folate sensitive fragile site at 1p21.3. *Journal of Medical Genetics* **28**:356-357

Callen DF, Hyland VJ, Baker EG, Fratini A, Gedeon AK, Mulley JC, Fernandez KE, Breuning MH, Sutherland GR (1989) Mapping the short arm of human chromosome 16. *Genomics* **4**:348-354

Callen DF, Hyland VJ, Bakar EG, Fratini A, Simmers RN, Mulley JC, Sutherland GR (1988) Fine mapping of gene probes and anonymous DNA fragments to the long arm of chromosome 16. *Genomics* **2**:144-153

Chodirker BN, Chudley AE, Ray M, Wickstrom DE, Riordan DL (1987) Fragile 19p13 in a family with mental illness. *Clinical Genetics* **31**:1-6

Court Brown WM, Buckton KE, Jacobs PA, Tough IM, Kuenssberg EV, Knox JD (1966) *Chromosome studies on Adults. Eugenics Laboratory Memoirs, Vol XLII.* Cambridge University Press, New York

Crawford DC, Meadows KL, Newman JL, Taft LF, Scott E, Leslie M, Shubek L, Holmgreen P, Yeargin-Allsopp M, Boyle C, Sherman SL. (2002) Prevalence of the fragile X syndrome in African-Americans. *American Journal of Medical Genetics* **110**:226-233

Day EJ, Marshall R, MacDonald PA, Davidson WM (1967) Deleted chromosome 18 with paternal mosaicism. *Lancet* **2**:1307

Dekaban (1965) Persisting clone of cells with an abnormal chromosome in a women previously irradiated. *Journal of Nuclear Medicine* **6**:740-746

Dutrillaux B, Muleris M, Prieur M (1985) Exact localization of several fragile sites remains uncertain. The example of fra(10) sensitive to folate. *Annales de Genetique* **28**:161-163

Ford CE, Madan K (1973) *Chromosome Identification*, Academic Press, New York and London, 98-103

Gardner RJM, Sutherland GR (1996) Chromosome abnormalities and genetic counselling. *Oxford monographs on medical genetics* No. **29**. Oxford University Press, New York.

Gecz J (2000a) The FMR2 gene, FRAXE and non-specific X-linked mental retardation: clinical and molecular aspects. *Annals of Human Genetics* **64**:95-106

Gecz J (2000b) FMR3 is a novel gene associated with FRAXE CpG island and transcriptionally silent in FRAXE full mutations. *Journal of Medical Genetics* **37**: 782-784

Gecz J, Gedeon AK, Sutherland GR, Mulley JC (1996) Cloning of FMR2: a gene associated with FRAXE mental retardation. *Nature Genetics* **13**:105-108

Giraud F, Ayme S, Mattei JF, Mattei MG (1976) Constitutional chromosomal breakage. *Human Genetics* **34**:125-136

Gu Y, Shen Y, Gibbs RA, Nelson DL (1996) Identification of FMR2, a novel gene associated with the FRAXE CCG repeat and CpG island. *Nature Genetics* **13**:109-113

Guichaoua M, Mattei MG, Mattei JF, Giraud F (1982) Genetic aspects of autosomal fragile sites. Apropos of 40 cases. *Journal de Genetique Humaine* **30**:183-197

Handt O, Baker E, Dayan S, Gartler SM, Woollatt E, Richards RI, Hansen RS (2000) Analysis of replication timing at the *FRA10B and FRA16B* fragile site loci. *Chromosome Research* **8**:677-688

Hewett DR, Handt O, Hobson L, Mangelsdorf M, Eyre H, Baker E, Sutherland GR, Schuffenhauer S, Mao J, Richards RI (1998) FRA10B structure reveals common elements in repeat expansion and chromosomal fragile site genesis. *Molecular Cell* **1**:773-781

Hirst MC, Barnicoat A, Flynn G, Wang Q, Daker M, Buckle VJ, Davies KE, Bobrow M (1993) The identification of a third fragile site, FRAXF, in Xq27-q28 distal to both FRAXA and FRAXE. *Human Molecular Genetics* **2**:197-200

Hocking T, Feichtinger W, Schmid M, Haan EA, Baker E, Sutherland GR (1999) Homozygotes for FRA16B are normal. *Chromosome Research* **7**:553-556

Hori T, Takahashi E, Ishihara T, Minamihisamatsu M, Kaneko Y, Murata M (1988) Distamycin A-inducible fragile sites and cancer proneness. *Cancer Genetics and Cytogenetics* **34**:177-187

Howell RT, McDermott A, Evans JL (1990) A new apparently folate sensitive fragile site, 5q35. *Journal of Medical Genetics* **27**:527-528

Hsu YY, Wang Y-H (2002) Human fragile site FRA16B DNA excludes nucleosomes in the presence of distamycin. *The Journal of Biological Chemistry* **277**:17315-17319

Jones C, Penny L, Mattina T, Yu S, Baker E, Voullaire L, Langdon WY, Sutherland GR, Richards RI, Tunnacliffe A (1995) Association of a chromosome deletion syndrome with a fragile site within the proto-oncogene CBL2. *Nature* **376**:145-149

Kahkonen M, Tengstrom C, Alitalo T, Matilainen R, Kaski M, Airaksinen E (1989) Population cytogenetics of folate-sensitive fragile sites. II. Autosomal rare fragile sites. *Human Genetics* **82**:3-8

Knight SJL, Flannery AV, Hirst MC, Campbell L, Christodoulou Z, Phelps SR, Pointon J, Middleton-Price HR, Barnicoat A, Pembrey ME, Holland J, Oostra BA, Bobrow M, Davies KE (1993) Trinucleotide repeat amplification and hypermethylation of a CpG island in *FRAXE* mental retardation. *Cell* **74**:127-134

Laird CD (1988) Fragile X mutation proposed to block complete reactivation in females of an inactive X chromosome. *American Journal of Medical Genetics* **30**:693-696

Leehey MA, Munhoz RP, Lang AE, Brunberg JA, Grigsby,J, Greco C, Jacquemont S, Tassone F, Lozano AM, Hagerman PJ, Hagerman RJ (2003) The fragile X permutation presenting as essential tremor. *Archives of Neurology* **60**:117-121

Lejeune J, Dutrillaux B, Lafourcade J, Berger R, Abonyi D, Rethore MO (1968) Selective endoreduplication of the long arm of the 2 chromosome in a woman and her daughter. *CR Acad Sci Hebd Seances Acad Sci D* **266**:24-26

Lubs HA (1969) A marker X chromosome. *American Journal of Human Genetics* **21**:231-244

Magenis RE, Hecht F, Lovrien EW (1970) Heritable fragile site on chromosome 16: probable localization of haptoglobin locus in man. *Science* **170**:85-87

Mattei MG, Philip N, Pellissier MC, Piquet C, Mattei JF, Giraud F (1987) A new rare fragile site on human chromosome 2. *Cytogenetics and Cell Genetics* **46**: 658

Michaelis RC, Velagaleti GV, Jones C, Pivnick EK, Phelan MC, Boyd E, Tarleton J, Wilroy RS, Tunnacliffe A, Tharapel AT (1998) Most Jacobsen syndrome deletion breakpoints occur distal to FRA11B. *American Journal of Medical Genetics* **76**: 222-228

Nancarrow JK, Kremer E, Holman K, Eyre H, Doggett NA, Le Paslier D, Callen DF, Sutherland GR, Richards RI (1994) Implications of FRA16A structure for the mechanism of chromosomal fragile site genesis. *Science* **264**:1938-1941

Nancarrow JK, Holman K, Mangelsdorf M, Hori T, Denton M, Sutherland GR, Richards RI (1995) Molecular basis of p(CCG)n repeat instability at the *FRA16A* fragile site locus. *Human Molecular Genetics* **4**:367-372

Nancarrow J (1998) *Studies of fragile sites on human chromosome 16. PhD Thesis*, University of Adelaide

Oberle I, Rousseau F, Heitz D, Kretz C, Devys D, Hanauer A, Boue J, Bertheas MF, Mandel JL (1991) Instability of a 550-base pair DNA segment and abnormal methylation in fragile X syndrome. *Science* **252**:1097-1102

Oberle I, Boue J, Croquette MF, Voelckel MA, Mattei MG, Mandel JL (1992) Three families with high expression of a fragile site at Xq27.3, lack of anomalies at the FMR-1 CpG island, and no clear phenotypic association. *American Journal of Medical Genetics* **43**:224-231

Ochi H, Watanabe S, Yamamoto H (1988) New heritable fragile site on chromosome 8 induced by distamycin A. *Japanese Journal of Cancer Research* **79**:145-177

Parrish JE, Oostra BA, Verkerk AJMH, Richards CS, Reynolds J, Spikes AS, Shaffer LG, Nelson DL (1994) Isolation of a GCC repeat showing expansion in FRAXF, a fragile site distal to FRAXA and FRAXE. *Nature Genetics* **8**:229-235

Petit P, Fryns JP, van den Berghe H, Hecht F (1986) Population cytogenetics of autosomal fragile sites. *Clinical Genetics* **29**:96-100

Quack B, Nantois Y, Mottet J, Noel B (1978) Constitutional stereotyped gap in human chromosomes. *Journal de Genetique Humaine* **26**:55-67

Reyniers E, Vits L, DeBoulle K, Van Roy B, Van Velzen D, de Graaff E, Verkerk AJ, Jorens HZ, Darby JK, Oostra B etal (1993) The full mutation in the FMR-1 gene of male fragileX patients is absent in their sperm. *Nature Genetics* **4**:107-108

Sarafidou T, Kahl C, Martinez-Garay I, Mangelsdorf M, Gesk S, Baker E, Kokkinaki M, Talley P, Maltby EL, French L, Harder L, Hinzmann B, Nobile C, Richkind K, Finnis M, Deloukas P, Sutherland GR Kutsche K, Moschonas NK, Siebert R, Gécz J. Folate-sensitive fragile site FRA10A is due to an expansion of a CGG repeat in a novel gene FRA10AC1, encoding a nuclear protein. *Genomics* (2004) 84: 69-81

Scheres JMJC, Hustinx TWJ (1980) Heritable fragile sites and lymphocyte culture medium containing BrdU. *American Journal of Human Genetics* **32**:628-629

Schmid M, Klett C, Niederhofer A (1980) Demonstration of a heritable fragile site in human chromosome 16 with distamycin A. *Cytogenetics and Cell Genetics* **28**:87-94

Schmid M, Feichtinger W, Jessberger A, Kohler J, Lange R (1986) The fragile site (16)(q22). I. Induction by AT-specific DNA-ligands and population frequency. *Human Genetics* **74**: 67-73

Schuffenhauer S, Lederer G, Murken J (1996) A heritable folate-sensitive fragile site on chromosome 2p11.2 (FRA2L). *Chromosome Research* **4**:252-254

Shabtai F, Halbrecht I (1981) Fragile sites on human chromosomes. *Clinical Genetics* **19**:536

Shaw MA, Chiurazzi P, Romain DR, Neri G, Gecz J (2002) A novel gene, FAM11A, associated with the FRAXF CpG island is transcriptionally silent in FRAXF full mutation. *European Journal of Human Genetics* **10**:767-772

Sherman SL, Morton NE, Jacobs PA, Turner G (1984) The marker (X) syndrome: a cytogenetic and genetic analysis. *Annals of Human Genetics* **48**:21-37

Sherman SL, Sutherland GR (1986) Segregation analysis of rare autosomal fragile sites. *Human Genetics* **72**:123-128

Simmers RN, Sutherland GR, West A, Richards RI (1987) Fragile sites at 16q22 are not at the breakpoint of the chromosomal rearrangement in AMMoL. *Science* **236**:92-94

Sutherland GR (1977) Fragile sites on human chromosomes: Demonstration of their dependence on the type of tissue culture medium. *Science* **197**:265-266

Sutherland GR (1979) Heritable fragile sites on human chromosomes. I. Factors affecting expression in lymphocyte culture. *American Journal of Human Genetics* **31**:125-135

Sutherland GR (1983) The fragile X chromosome. *International Review of Cytology* **81**: 107-143

Sutherland GR (1988) The role of nucleotides in human fragile site expression. *Mutation Research* **200**:207-213

Sutherland GR (1991) Chromosomal fragile sites. *Genetic Analysis Techniques and Applications* **8**:161-167

Sutherland GR, Baker E (1992) Characterisation of a new rare fragile site easily confused with the fragile X. *Human Molecular Genetics* **1**:111-113

Sutherland GR, Baker E (1993) Unusual behaviour of a human autosome having two rare folate sensitive fragile sites. *Annales de Génétiques* **36**:159-162.

Sutherland GR, Baker E (2000) The clinical significance of fragile sites on human chromosomes. *Clinical Genetics* **58**: 157-161

Sutherland GR, Baker (2003) Forgotten fragile sites and related phenomena. *Cytogenetics and Genome Research* **100**:89-91

Sutherland GR, Hecht F (1985) *Fragile sites on human chromosomes*. Oxford University Press, New York

Sutherland GR, Hinton L (1981) Heritable fragile sites on human chromosomes. VI, Characterization of the fragile site at 12q13. *Human Genetics* **57**:217-219

Sutherland GR, Baker E, Seshadri R (1980) Heritable Fragile Sites on Human Chromosomes. V. A New Class of Fragile Site Requiring BrdU for Expression. *American Journal of Human Genetics* **32**:542-548

Sutherland GR, Jacky PB, Baker E, Manuel A (1983) Heritable Fragile Sites on Human Chromosomes. X. New Folate Sensitive Fragile Sites: 6p23, 9p21, 9q32, 11q23. *American Journal of Human Genetics* **35**: 432-437

Sutherland GR, Jacky PB, Baker E (1984) Heritable fragile sites on human chromosomes. XI. Factors affecting expression of fragile sites at 10q25, 16q22 and 17p12. *American Journal of Human Genetics* **36**:110-122

Sutherland GR, Parslow MI, Baker E (1985) New classes of fragile sites induced by 5-azacytidine and BrdU. *Human Genetics* **69**:233-237

Takahashi E, Kaneko Y, Ishihara T, Minamihisamatsu M, Murata M, Hori T (1988) A new rare distamycin A-inducible fragile site, fra(11)(p15.1), found in two acute nonlymphocytic leukemia (ANLL) patients with t(7;11)(p15-p13;p15). *Human Genetics* **80**:124-126

Tassone F, Hagerman RJ, Taylor AK, Gane LW, Godfrey TE, Hagerman PJ (2000) Elevated levels of *FMR***1** mRNA in carrier males: A new mechanism of involvement in the Fragile-X syndrome. *American Journal of Human Genetics* **66**:6-15

Tommerup N, Nielsen J, Mikkelsen M (1985) A folate sensitive heritable fragile site at 19p13. *Clinical Genetics* **27**:510-514

Verkerk AJMH, Pieretti M, Sutcliffe JS, Fu Y-H, Kuhl DPA, Pizzuti A, Reiner O, Richards S, Victoria MF, Zhang F, Eussen BE, van Ommen G-JB, Blonden LAJ, Riggins GJ, Chastain JL, Kunst CB, Galjaard H, Caskey CT, Nelson DL, Oostra BA, Warren ST (1991) Identification of a gene *(FMR-***1***)* containing a CGG repeat coincident with a breakpoint cluster region exhibiting length variation in fragile X syndrome. *Cell* **65**: 905-914

Voiculescu I, Hausmann C, Wolff G, Back E (1988) A BrdU-requiring fragile site on chromosome 12. *Human Genetics* **78**:183-185

Voullaire LE, Webb GC, Leversha MA (1987) Chromosome deletion at 11q23 in an abnormal child from a family with inherited fragility at 11q23. *Human Genetics* **76**: 202-204

Wang YH, Griffith J. (1996) Methylation of expanded CCG triplet repeat DNA from fragile X syndrome patients enhances nucleosome exclusion. *Journal of Biological Chemistry* **271**:22937-22940

Webb T, Thake A (1984) Fragile 22q13 segregating in a family. *Clinical Genetics* **26**: 125-128

Yu S, Pritchard M, Kremer E, Lynch M, Nancarrow J, Baker E, Holman K, Mulley JC, Warren ST, Schlessinger D, Sutherland GR, Richards RI (1991) Fragile X genotype characterised by an unstable region of DNA. *Science* **252**:1179-1181

Yu, S., Mulley, J., Loesch, D., Turner, G., Donnelly, A., Gedeon, A., Hillen, D., Kremer, E., Lynch, M., Pritchard, M., Sutherland, G.R., Richards, R.I. (1992) Fragile-X syndrome: Unique genetics of the heritable unstable element. *American Journal of Human Genetics* **50**:968-980

Yu S, Mangelsdorf M, Hewett D, Hobson L, Baker E, Eyre H, Lapsys N, Le Paslier D, Doggett NA, Sutherland GR, Richards RI (1997). Human chromosomal fragile site FRA16B is an amplified AT-rich minisatellite repeat *Cell* **88**:367-374

Zhong N, Ju W, Curley D, Wang D, Pietrofesa J, Wu G, Shen Y, Pang C, Poon P, Liu X, Gou S, Kajanoja E, Ryynanen M, Dobkin C, Brown WT (1996) A survey of FRAXE allele sizes in three populations. *American Journal of Medical Genetics* **64**:415-419

In: Genetic Translocations and Other Chromosome Aberrations ISBN: 978-1-60456-344-3
Editor: Guy T. Leyden, pp. 223-252 © 2008 Nova Science Publishers, Inc.

Chapter 9

Novel and Uncommon Chromosome Aberrations in Chronic Lymphocytic Leukemia: Cytogenetic, FISH and Clinical Evaluation

Roxana Cerretini*[*1,2], Christian Chena[2] and Irma Slavutsky[2]

[1] Departamento de Genética Experimental, Centro Nacional de Genética Médica, Buenos Aires, Argentina.

[2] Departamento de Genética, Instituto de Investigaciones Hematológicas "Mariano R. Castex", Academia Nacional de Medicina, Buenos Aires, Argentina.

Abstract

The identification of the non-random genetic alterations associated with cancer has great importance not only to basic research but also to disease management as diagnostic and prognostic markers. In contrast to other types of lymphomas, translocations are rare events in chronic lymphocytic leukemia/small lymphocytic lymphoma (CLL/SLL) and not any CLL/SLL-specific translocation has been found yet. Moreover, translocations have been reported to be a prognostic marker for an unfavorable clinical outcome. The purpose of our study was to contribute to the understanding of the chromosome alterations associated with this group of lymphoid malignancies. The combination of conventional cytogenetics with FISH (fluorescence *in situ* hybridization) analysis allowed us a more accurate definition of the genomic aberrations involved in CLL/SLL. Novel and uncommon chromosome rearrangements were found in our series. In the present work, we describe the cytogenetic and FISH analyses of 12 CLL/SLL patients out of 150 cases with successful cytogenetic study, that interestingly showed in their karyotypes new or infrequent clonal chromosome rearrangements, all of them associated to adverse clinical course with partial or null response

* E-mail address: rcerretini@argentina.com; rcerretini@hotmail.com. Correspondence to: Roxana Cerretini; Departamento de Genética, Instituto de Investigaciones Hematológicas "Mariano R. Castex", Academia Nacional de Medicina; Pacheco de Melo 3081, Ciudad Autónoma de Buenos Aires, Argentina, CP: 1425. FAX: (5411) 4803-9475

to different chemotherapeutic agents. We identified 16 different chromosome rearrangements and translocations were the most frequent aberration (7/16; 43,8%). Some of these findings were recently reported by us as a new alterations in SLL/CLL patients: t(2;17)(p21;q23); der(17)t(12;17)(q13;q25); der(14)(t(11;14)(q13;q24), t(5;12;19)(q15;p11;q13); psu dic (17;2)(p11.2;p21), psu dic (12;21)(q24;q10), t(10;14)(q22;q32), del(3)(p11) and del(5)(q15). Translocation t(11;13)(p15;q22) and del(18)(p11) were not previously described, meanwhile, dic(17;18)(p11.2;p11.2), dup(12)(q13q24), del(7)(q34), t(2;14)(p11;q32) and del(7)(q11) were infrequently reported in the literature. Translocation t(2;17)(p21;q23) is recurrent in our series and t(2;14)(p11;q32) became recurrent from our data. Chromosomes 17, 2, 12 and 14 were the most frequently involved. Twenty one different breakpoints were found and 38.1% (8/21) of them, such as 2p21, 5q15, 17p11.2, 17q23, 12q13, 12q24, 14q32 and 18p11 were recurrent in our series. All but one patient of our series showed an unfavorable clinical outcome. A detailed analysis and a review of the cytogenetic and molecular cytogenetics findings in CLL/SLL were performed. Our findings may facilitate the understanding of genetic events associated with development and progression of CLL/SLL and may have implications in the clinical management of patients with this disease.

1. Introduction

Chronic lymphocytic leukemia (CLL) is a monoclonal disease of mature-appearing lymphocytes that are functionally incompetent and accumulate in bone marrow, blood, lymph nodes, spleen, liver and other organs, primarily of the lymphatic system. A diagnosis of small lymphocytic lymphoma (SLL) can be made histologically in the absence of bone marrow or blood involvement. Most cases (>95%) are characterized by monoclonal lymphocytes expressing normal B-cell surface proteins including immunoglobulin (Ig), CD19 and CD20 and aberrantly expressing CD5, a protein normally found on T-cells [1].

CLL is the most frequent type of leukemia among adults in the western world, accounting for about 30% of all leukemias [2-5]. However, its incidence is very low in Asia [6, 7]. Particularly in Japan, it accounts for just 3-4% of all leukemias and interestingly, CLL is quite rare not only among Japanese living in Japan, but also among Japanese immigrants and their descendants living in USA [8-10]. This suggests that the rareness of CLL in Japanese is caused by genetic rather than environmental factors. The basis for this ethnic variation is unknown.

This disease affects mainly people of advanced age, with a median age of 65 years, but about 20% of patients are younger than 55 years. The gender ratio varies from approximately 1.5 to 2.1 males to 1 female [11]. The clinical diagnosis of CLL requires an absolute lymphocytosis (more than 5000 mature-appearing lymphocytes/μL). Nevertheless, it is common to find small percentages of larger or atypical cells, cleaved cells, or prolymphocytes [12].

This pathology follows an extremely variable clinical course with overall survival times ranging from months to decades. Some patients have no or minimal signs and symptoms during their entire disease course and have a survival time similar to age-matched controls. However, other patients progress rapidly to symptomatic disease, needing immediate treatment [13].

A number of clinical and biologic markers of prognostic relevance have been identified. These include clinical characteristic, such as age, stage, gender, and performance status, as well as laboratory parameters reflecting the tumor burden or disease activity (i.e. lymphocyte count, higher lactate dehydrogenase [LDH] levels, bone marrow infiltration pattern, or lymphocyte doubling time) [14-16].

More recently, prognostic markers related to the biology of the disease have been identified: serum parameters such as soluble CD23 (sCD23), β_2-microglobuline (β_2-MG) or thymidine kinase (TK) [17, 18] and genetic markers of the CLL cells, such as genomic aberrations, the mutation status of the variable segment of immunoglobulin heavy chain genes (IgV_H), or surrogate markers for these factors [19, 16, 20-22].

Over the recent years the focus of research has moved towards the molecular genetic and epigenetic levels, which may not only provide insight into the biology and transforming events of CLL but may also allow the definition of mechanisms directly responsible for the course of the disease with regard to clinical progression, response to treatment, and overall survival.

Among epigenetic modifications, DNA methylation and post-translational modifications of histones tail proteins, in the form of acetylation, methylation, ubiquitination, and phosphorylation are the best documented epigenetic changes. These modifications are heritable and influence the chromatin structure on a genome wide scale. Many novel genes that are epigenetically silenced in CLL have been identified. At present, the field of epigenetics is at an evolving stage but there is no doubt that further unraveling of its cause and effects in transformed cells will bring a new revolution in cancer therapeutics [23].

One of the most important molecular genetic parameters defining pathogenic and prognostic subgroups of CLL is the mutation status of the IgV_H genes. The presence or absence of somatic mutations in IgV_H genes can be used to stratify CLL patients into two groups [24, 25]. In CLL with mutated IgV_H genes, the proliferating B cell may have transited through germinal centers, the physiologic site of hypermutation, whereas in CLL with unmutated IgV_H genes the malignant B cell may derive from a pre-germinal center naïve B cell [12]. Mutated and unmutated CLL patients clearly differ in terms of prognosis and may also differ with respect to oncogenic mechanisms. Patients whose CLL cells express unmutated IgV_H have a shorter median survival than those expressing mutated IgV_H [24].

In despite of these clinical and molecular differences, recent studies on gene expression profiling of CLL cells showed that this pathology is characterized by a common gene expression signature which is independent of IgV_H mutational status. This profile of gene expression differs from other lymphoid cancers and normal lymphoid subpopulations, suggesting that CLL cells share a common mechanism of transformation and/or cell of origin [26, 27].

However, in despite of sharing a common signature, CLL expressing mutated and unmutated IgV_H genes differentially express more than 100 genes. Among these differentially expressed genes, overexpression of *ZAP-70*, lipoprotein lipase (*LPL*), *BCL-7a*, dystrophin and gravin are observed in the aggressive unmutated cases, while stable mutated cases overexpress *Wnt3*, *CTLA-4*, *NRIPI* nuclear receptor gene, *ADAM* and the transcription factor *TCF7* [26-29].

ZAP-70, a receptor associated protein tyrosine kinase expressed by T lymphocytes, natural killer (NK) cells, activated tonsillar and splenic B cells and cases of CLL with unmutated IgV_H is the best surrogate considering IgV_H mutational profiles. The mode of action of *ZAP-70* may be to enhance B-cell receptor signaling, promoting the survival and/or proliferation of CLL [30]. Recently, Catherwood et al [31] confirmed that IgV_H unmutated CLL cells have a high expression of *ZAP-70* in comparison with IgV_H mutated-CLL B cells.

ZAP-70 is differentially methylated in CLL and a close association exits between *ZAP-70* methylation and expression and IgV_H mutational status [32]. Expressions of *ZAP-70* and CD38 have been identified by several groups as predictors of disease progression and overall survival in CLL [33]. The prognosis information given by *ZAP-70* and CD38 expression is complementary and these two markers have allowed for the separation of three CLL patient subgroups with good, intermediate and poor prognosis [34].

2. Genomic Aberrations

Genomic aberrations are the other genetic parameters shown to be of pathogenic and clinical relevance in this disease. The study of the genomic aberrations provides insights into the pathogenesis of the disease since they point to loci of candidate genes and identify subgroups of patients with distinct clinical pictures.

Cytogenetic analysis is a routine part of the diagnosis and management of a significant number of lymphoid malignancies. Particularly, in CLL, this study is problematic as malignant B cells tend to have a low spontaneous mitotic index, usually a poor response to most common mitogens, and in addition, the growth advantages of normal T cells in the cell culture [35]. Early studies reported normal karyotypes for most cases of CLL, almost certainly as the metaphase spreads obtained in these cases originated from normal cells present in the blood or bone marrow culture.

The addition of B-cell mitogens such as 12-0-tetradecanoylphorbol-13-acetate (TPA) and Pokeweed have been shown to be effective in stimulating malignant B cells to divide [36] and enabled the detection of cytogenetic abnormalities in approximately 40-50% of CLL cases using chromosome banding [37-39]. Peripheral blood is generally the most appropriate sample for cytogenetic analysis of CLL.

From patients with abnormal karyotype by standard cytogenetic, as many as 65% have one chromosome abnormality, 25% have two abnormalities and the reminder ones have more complex alterations [40].

The development of molecular cytogenetic techniques, such as fluorescence *in situ* hybridization (FISH), has enabled sensitive detection of specific changes in interphase and metaphase cells and has shown a higher incidence of abnormalities than cytogenetics. The incidence of FISH detection of genome aberrations is estimated to be 77-80% [19,22]. In CLL cases with normal karyotype by banding analysis but clonal aberrations by interphase FISH, the metaphase cells are derived either from non-clonal T-cells or clonal cells with cryptic interstitial deletions, and therefore do not reflect the karyotype of the malignant B-CLL cells [41, 42].

The most common cytogenetic and/or FISH anomalies in B-CLL involve chromosome 6, 11, 12, 13, 14 and 17 [19, 22, 36, 41, 43-47]. Meanwhile by interphase FISH, the most frequent abnormality is 13q14 deletion, followed by 11q22-23 deletion, trisomy 12, 17p13 deletion, 6q deletion and t(14q32) [19]. These anomalies have been associated with different prognoses.

13q Deletion: del(13)(q14.3)

The del(13)(q14.3), is the most common isolated chromosome abnormality in CLL, and is found in about 10-20% of the cases by cytogenetics or 55-64% of the cases by FISH [19, 22, 48]. CLL metaphases with 13q- are not always detected because these deletions are often cryptic and not apparent by conventional G-banding [41].

The 13q deletion in CLL uniformly involves 13q14. Commonly used FISH probes to detect the deletion in the lymphoid disorders are chromosome 13q14 specific, including *RB1*, D13S319 and D13S25. The most commonly deleted marker in CLL initially reported was D13S319, between the retinoblastoma and D13S25 loci. The deletion of an allele of *RB1* is detected by molecular cytogenetics in one-fourth of all CLL cases [49, 50]. The 13q14 deletions usually do not lead to inactivation of the *RB1* gene, and del(13q) is associated with the presence of one intact *RB1* gene on the homologous chromosome, implying the role of an adjacent locus that may harbor important tumor suppressor gene(s) [19, 49, 51].

A critically deleted region (CDR) of approximately 350 kb was identified at 13q14 with the proximal border around 120 kb centromeric from D13S319, encompassing an area rich in expressed sequence tagged sites and containing *DLEU1*, *DLEU2*, for which a pathogenic role has not yet been established, and *BCMS* (B-cell neoplasia associated gene with multiple splicing). Sequence analysis of the RNA variants suggest that *BCMS* transcripts belong to the group of non-coding RNAs (ncRNAs). The alignment of the gene with all critical subregions provides a strong argument for *BCMS* being a very good candidate for the tumor suppressor gene in 13q14 involved in the leukemogenesis of CLL. Due to the limited understanding of functional RNAs, however, the pathogenic role of *BCMS* has not yet been proven [52, 53]. Moreover, during the past few years, molecular studies have discovered hundred of genes that encode small RNA molecules, named microRNAs (miRNAs). More recently it has been shown the presence of two microRNA genes, *miR15* and *miR16*, located at 13q14 within the minimal deleted region (MDR) for CLL patients. Both genes are deleted or down-regulated in the majority of CLL cases [54]. In addition, a unique miRNA signature was associated with prognostic factors and disease progression [55], suggesting an important role for these genes in this pathology.

11q Deletion: del(11)(q22-q23)

Loss of genomic material on the long arm of chromosome 11 is frequently observed in various types of lymphoproliferative disorders, most frequently involving 11q23 [56, 57]. The recurring loss of material from chromosome bands 11q22-q24 suggests that this region harbors a tumor suppressor gene and many candidate genes. All 11q deletions affect a minimal consensus region of 2-3 Mb in size in bands 11q22.3-q23.1 [56, 58]. The ataxia telangiectasia mutated (*ATM*) gene, which is responsible for the autosomal recessive disorder ataxia telangiectasia, is located within this minimal consensus region in bands 11q22.3-q23.1 and is implicated in the pathogenesis of different lymphoid malignancies, mainly mantle cell lymphoma [59].

Patients with inherited inactivation of both alleles of the *ATM* gene by deletion and/or mutation are highly predisposed to the development of leukemia and lymphoma. However, mutation of the remaining *ATM* allele is found in about 15-23% of CLL cases, which indicates a possible involvement of additional genes in this region in CLL [19, 60]. By contrast, in mantle cell lymphoma, in which the 11q22-q23 deletion occurs in 50% of the cases, mutation of the remaining allele is observed in all cases [59, 61, 62].

ATM is a pleiotropic protein, which activities are induced by double-stranded chromosome breaks that occur spontaneously or after exposure to DNA-damaging agents. ATM protects the integrity of the genome by arresting the cell cycle and activating DNA-repair pathways [63]. If the DNA damage is more severe, ATM induces apoptosis. After DNA damage, the serine-threonine kinase activity of ATM increases, leading to the phosphorylation of downstream effectors involved in the regulation of cell-cycle checkpoints, DNA repair and apoptosis [64]. The most important of these effectors seems to be the *TP53* tumor suppressor gene [65]. Constitutional ATM deficiency in cells of ataxia telangiectasia patients and in atm-l- mice leads to the deregulation of apoptosis and cell-cycle checkpoint control and causes defects in DNA recombination [66].

Trisomy 12

Trisomy 12 is found in one third of cytogenetically abnormal CLLs by conventional karyotype [40, 67, 68] and in about 11-42% of cases when interphase FISH is used [19, 48, 69-71].

Using restriction fragment length polymorphism analysis, the extra chromosome 12 has been shown to derive from duplication of one chromosome 12, with retention of the other homolog rather than from triplication of one homolog. Trisomy 12 is thought to be a secondary change, associated with disease progression, since combined immunological and cytogenetic studies showed that it is present in only a part of the neoplastic B-cells [72-74].

The role of genes potentially involved in the pathogenesis of CLL with trisomy 12 is not established. While the critical region of most aberrations in CLL could be narrowed down to a set of candidate genes, this has been difficult in the case of trisomy 12. Cases with partial duplication 12q were analyzed by FISH, and a "minimal duplicated region" could be defined in segment 12q13-12q15 [75, 76]. Bands 12q13-q22 were the smallest duplicated segment in four CLL cases in the series of Dierlamm et al [75] and by DNA microarray technology it was identified the smallest replicated genomic regions in bands 12q13-q21 in this disorder [77, 78].

Recent RNA expression analyses showed an overexpression of genes located on chromosome 12 in CLL cases with trisomy 12 as compared to cases without this aberration [79, 80]. Genes of oncogenic potential, like *CDK4*, *GLI* and *MDM2* are localized in this genomic region, but they have not shown pathogenetic relevance for CLL to date. Recently, it has been identified a novel gene, CLL up-regulated gene 1 (*CLLU1*), in band 12q22, uniquely overexpressed in CLL but not in normal tissue or in other hematological disease [81].

6q Deletion: del (6)(q21)

Lymphoid malignancies are characterized by recurrent chromosome changes, involving deletion in bands 6q21-q23. The frequency of the deletions is difficult to estimate by conventional cytogenetic analysis because small interstitial deletions are beyond the sensitivity of the technique; furthermore, many studies have reported conflicting data on the putative region of overlap and the number of regions involved. The breaks occur predominantly in 6q21, but 6q15 is also often described; overall, del(6q) cases encompassed the 6q21 band [82, 83].

Comparison of the extent of the deletions among different cases by FISH allowed the identification of two distinct regions of minimal deletion (RMD) at 6q25 to 6q27 (RMD-1) and at 6q21 to 6q23 (RMD-2). These data support a role for 6q deletions in B-cell non-Hodgkin´s lymphomas (NHL) pathogenesis and provide a basis for identifying the corresponding tumor suppressor genes [84-86], defined the deletion breakpoints exactly and narrowed this region further. They found 3-cM (4-5 Mb) commonly deleted region in 6q21, involved in the pathogenesis of both low and high grade NHL and acute lymphoblastic leukemias (ALL).

Deletion in band 6q21 is often observed in malignant transformation and development of both T- and precursor B-ALL and it is the band most frequently loss in CLL. This lack of specificity for a particular immunophenotype may imply that the gene or genes affected by 6q abnormalities are broadly active in the multistep process of lymphoid leukemogenesis [87].

17p Deletion: del(17)(p13)

TP53 gene is located near the telomere of the chromosome 17 p-arm at 17p13, its absence can be detected by FISH in interphase nuclei of 8-10% patients with CLL. Results of chromosome studies suggest that loss of *TP53* is often caused by unspecified structural anomalies of chromosome 17 rather than by monosomy 17 [40, 88]. However, this type of chromosome 17 anomalies and their mechanisms of origin in CLL have not been well characterized.

Isochromosome 17q, or "i(17q)," is one of the most common structural abnormalities observed in human neoplasms. A breakpoint cluster region for i(17q) in 17p11.2 has been identified and hypothesized that genome architectural features could be responsible for this clustering. A complex genomic architecture in the i(17q) breakpoint cluster region, characterized by large (approximately 38-49-kb), palindromic, low-copy repeats, strongly suggest that somatic rearrangements are not random events but rather reflect susceptibilities due to the genomic structure [89]. Fink et al. [42] showed that loss of *TP53* results from the absence of nearly the entire chromosome 17 p-arm rather than to monosomy 17 or deletions of *TP53*. Translocation or isochromosome formations at sites of low copy DNA repeats (LCR) in 17p10 to 17p12 appear to be the mechanism for the loss of *TP53* in CLL. These results show that 17p10-p12 is a hot spot for structural anomalies in this pathology.

Further studies will be useful to identify if regions on other chromosomes have LCR that are homologous to 17p10-p12 and mediate translocation with chromosome 17. This cytogenetic mechanism involving LCR is responsible for most, if not all, instances for loss of *TP53* in CLL.

In addition, the high rate of *TP53* mutations in the remaining allele with 17p13 deletions suggests that *TP53* is the tumor suppressor gene affected by the aberrations.

Rearrangement of the IgH locus in Band 14q32

Although 14q32 involvement was frequent in early CLL studies, currently t(11;14)(q13;q32) with *CCND1/IgH* rearrangement is considered to be the hallmark of mantle cell lymphoma (MCL). Thus, lymphoproliferative disorders exhibiting the translocation t(11;14) should therefore be carefully examined for morphological and immunophenotypical characteristics of MCL. Other rare 14q32 rearrangements, such as t(14;19)(q32;q13), is not restricted to CLL but it has also been described in low and high grade NHLs [90].

A pathogenetic role was also suggested for *BCL2* at 18q21. While the breakpoint of t(14;18) in follicular lymphoma occur in the major breakpoint region (MBR) or in the minor cluster region (mcr) within the 3´portion of *BCL2*, the breakpoints in CLL tumors have been localized at the 5´end with preferential juxtaposition of *BCL2* to the immunoglobulin light chains. However, subsequent studies show that *BCL2* rearrangements are rare events in CLL [19, 91], more often as a secondary event occurring late during disease progression [92].

Thus, translocations involving 14q32/*IgH* locus are either rare events or likely associated with other types of lymphoproliferative disorders. Disregarding t(11;14)(q13;q32) and t(14;18)(q32;q21), 14q32 rearrangements are found in 4% of cases studied by FISH [19].

3. Clinical Impact of Genomic Aberrations in CLL

In general, complex karyotypes and a high percentage of abnormal metaphases appear to confer a worse prognosis than single or dual chromosomal abnormalities. The median survival in cases with diploid normal karyotype is about 15 years and about 6-7.7 years for those with complex karyotypes [40]. However, multivariate analysis confirmed neither the presence of clonal aberrations nor the number of clonal aberrations to be independent prognostic variables.

13q Deletions

Patients that have monoallelic deletion 13q14 (13q14x1) as a single anomaly have good prognosis. They are characterized by slow disease progression, initial stages of the disease at diagnosis and survival times similar to those with a normal karyotype, with a median survival of 133 months [19, 22, 88]. This chromosome anomaly predominates among patients displaying mutated profile of IgV_H genes [12, 16, 93].

By FISH, recent studies have observed about 10% of cases carrying biallelic deletion 13q14 (13q14x2) into MDR region. This abnormality may result from chromosomal evolution and would represent a more aggressive FISH anomaly than 13q14x1 [22, 47].

Deletion 11q22-Q23

Presence of the 11q deletion in CLL patients is associated with the diagnosis at younger age, more advanced clinical stages, extensive lymphadenopathy, and more rapid disease progression with a shorter treatment-free interval (9 months versus 43 months). The prognostic effect of 11q deletion on survival depends on the age such as patients under 55 years, median survival is significantly shorter in the deletion group (64 months versus 209 months) while there is not difference in those aged 55 years or older (94 months versus 111 months) [16]. So, deletion of 11q is considered an independent prognostic factor and is an important predictor of survival in younger CLL patients [16, 19, 94]. This chromosome anomaly predominates among patients displaying unmutated IgV_H genes [12, 16, 93].

Trisomy 12

Some studies found that CLL patients with trisomy 12 by banding analysis had both shorter treatment-free interval and shorter overall survival, but such adverse prognostic impact has not been confirmed by other groups [43, 67, 68, 95, 96].

However, when the data of conventional chromosome banding analysis were included together with FISH studies, median survival in patients with trisomy 12 was significantly shorter than that observed in patients with a normal karyotype (7.8 versus 14.4 years). In other series, however, the prognosis in the group with trisomy 12 (median survival time 114 months) was intermediate to that 13q deletion, 11q22-23 or 17p13 [19, 97].

At the present time it is known that trisomy 12 is associated with an atypical morphology and immunophenotype, a high proliferative rate, advanced clinical stage and intermediate outcome but remain unresolved its pathogenic impact. This chromosome anomaly shows a similar distribution among patients displaying unmutated and mutated IgV_H genes, suggesting that CLL patients with trisomy 12 have considerable heterogeneity [16, 19, 74, 93]

Deletion 6q

By cytogenetic studies, deletion 6q was observed as a product of clonal evolution, mainly affecting region 6q21-q25 [44, 98]. Recently, an association was found between del(6)(q21) by FISH and a high white blood cell count, more extensive adenopathy at presentation, atypical morphology, classical inmunophenotype with CD38+, and a therapy-demanding disease. They show an intermediate incidence of IgV_H somatic hypermutation between the patients with 11q-/17q- and the patients with isolated 13q-, suggesting that the 6q- patients may be biologically heterogeneous as is the case with CLL with trisomy 12. Clinico-biological features and outcome show that this cytogenetic subset of CLL should be allocated in an intermediate risk group [99].

Deletion 17p

CLL patients with a *TP53* deletion have aggressive clinical course, advanced stages of the disease, shorter survival than those without deletion and poor response to treatment with purine nucleoside analogues and alkylating agents [19, 22].

Multivariate analysis identified *TP53* gene deletion as the strongest independent prognostic factor for survival [16, 83, 100]. The presence of a 17p deletion predominates among patients displaying unmutated IgV_H genes [12, 16, 93].

Conclusion

The cytogenetic status correlates with the biological behavior of CLL in vitro. Poor cytogenetic prognosis was associated with more rapid spontaneous apoptosis in vitro and lower inmunogenicity. Good cytogenetic prognosis have lower levels of spontaneous apoptosis, but higher basal expression of molecules involved in co-stimulation, cellular adhesion and antigen presentation, and induced significantly T-cell proliferation in mixed lymphocyte cultures [22, 101].

Different genomic aberration allow to identify distinct risk subgroups being 17p- and 11q- the most aggressive, +12 and 6q- as intermediate risk and normal karyotype and 13q14x1 as the less aggressive ones. Recent data from the literature [22] and from our laboratory (unpublished data) suggest that 13q14x2 could represent a new aggressive risk category. These data reveal the importance of FISH panel to monitor genomic changes with different clinical outcome.

Based on different studies, it appears that IgV_H mutation status and genomic aberrations are among the strongest currently available parameters and are of independent value to predict outcome in CLL. The IgV_H mutational profile has the advantage that it remains constant during clonal evolution, which contrasts with genomic aberrations [12, 93].

In contrast to other types of lymphoproliferative disorders, translocations are rare events in CLL and not any CLL-specific translocation has been found yet. Moreover, translocations have been reported to be a prognostic marker for an unfavorable clinical outcome [102]. In this context, the purpose of our study was to contribute to the understanding of the genomic aberrations associated with this group of lymphoid malignancies. The combination of conventional cytogenetics with FISH analysis allowed us a more accurate definition of the genomic aberrations involved in CLL/SLL. Novel and uncommon chromosome rearrangements were found in our series. Interestingly, among them, translocations were the most frequent aberration.

Material and Methods

Patients

Between 1993 and 2006, 191 patients with CLL/SLL were referred to our institution and cytogenetic analysis was successfully performed on 150 (78.5%) of them (103 males and 88 females; mean age 63.8 yr; range 39-89 yr). Twelve cases, (8%, 12/150) (8 males and 4 females; mean age 57.6 yr; range 46-72 yr) (table 1) were selected for inclusion in the present work based on the presence of non frequent clonal structural chromosome aberrations.

All patients were histologically reclassified according to World Health Organization classification of lymphoid neoplasias [103, 104]. Three patients (2, 5 and 9) were diagnosed as SLL by clinical, immunophenotypic, and histopathologic analysis. In all cases the peripheral white blood counts were normal. The remaining 9 patients were diagnosed as CLL, based on the National Cancer Institute Working Group guidelines for diagnosis of CLL [105]. According to the morphologic evaluation, six patients were classified as typical (cases 1, 3, 4, 6, 10 and 12) and three as atypical CLL (cases 7, 8 and 11). Clinical stages were established according to the Ann Arbor classification for SLL [106] and to the Rai classification for CLL

[107]. Immunophenotypic analysis showed expression of pan-B antigen (CD19, CD20, CD22) with coexpression of CD5 and CD23. All of them were CD5+ and CD19+. Case 1 showed Hodgkin´s disease (HD) of the mixed cellularity subtype after the diagnosis of CLL [108] and Case 7 showed a transformation to a diffuse large B-cell lymphoma (Ritcher´syndrome) after the second cycle of treatment [92].

Table 1. Clinical characteristics of CLL/SLL patients

Case	Sex/ Age	Lymph ($x10^9$/l)	Clinical stage[1]	Treatment	Response	Survival (month)	
						treatment-free	total
1	F/50	22.4	I[2]/HL[3]	CLB/P[2]	PR	7	
				ABVD[3]	Null		
				PrVCLBP[3]	PR		45[4]
2	M/70	2.3	IV[2]	CHOP[2]	PR	1	
				F/D[3]	PR		
				Anti CD20[3]	CR		56
3	M/56	9.1	0[2]	F/C[3]		2	
				Anti CD20[3]	CR		
				AT			106[4]
4	M/61	94.6	II[2]	CAPVE[2]/			
				CLB/P[2]/F	Null		24[4]
5	M/58	5	IV[2]	CHOP-MINE[2]	Null	5	9[4]
6	F/72		0[2]	No Treatment			28
7	M/54	82	III[2]	CHOP[2]	Null	2	
				F/C/D[3]	Null		
				ESHAP[3]	PR		
				Hyper-CVAD[3]	Null		9[4]
8	F/51	104	II[2]	CLB[3]	PR	48	
				F[3]	null		
				F/C/ Anti CD20[3]			155
9	M/46	1.5	II[2]	F/C[3]		4	73
10	M/50	140	I[2]	CLB/P[3]	PR	2	
				F/C[3]	CR		
				CHOP/Anti CD20[3]			51
11	F/66	68	II[2]	CLB/P[2]	PR	20	
				CLB/P[2]	Null		62[4]
12	M/57	21	I[2]	CLB[2]	PR	28	
				CLB[3]	Null		
				F/C[3]	Null		
				CHOP[3]			72

[1] CLL: Rai., SLL: Ann Harbor, [2] initial stage, [3] progression stage, [4] deceased.

PR, partial response; Null, without response; CR, complete response; HL, Hodgkin´s lymphoma; CLB, chlorambucil; P, prednisone; CAVPE, cyclophosphamide, adryamicine, vincristine, prednisone, etoposide; F, fludarabine; C, cyclophosphamide; D, dexamethasone; ABVD, doxorrubicin, bleomycin, vinblastine, dacarbazine; Pr: procarbazine. CHOP: cyclophosphamide, adryamicine, vincristine, prednisone. MINE: uromitexan, ifosfamide, navelbine, etoposide; ESHAP, etoposide, cisplatin, cytosine, arabinoside; CVAD, cyclophosphamide, vincristina, adryamicine, dexamethasone; AT, autologous transplantation.

Clinical features, treatment and survival information are shown in table 1. This series included patients whose karyotypes were performed at diagnosis or at variable points during disease evolution. Six patients had sequential cytogenetic studies. Cytogenetic and FISH analysis are detailed in table 2. All but one (case 6) patients had an adverse clinical course, with partial or null response to different chemotherapeutic regimens. Seven patients had died at the time of this study (table 1).

Cytogenetic Analysis

Conventional cytogenetic studies were performed on peripheral blood, bone marrow and/or lymph nodes samples. Cells were cultured in F-10 medium supplemented with 15% fetal calf serum by direct method (24hs), synchronized cultures (48hs) and with stimulation of different mitogens such as lipopolysaccharide, pokeweed and phytohemagglutinin A for 72-96hs. G and C-banding techniques were used. Clonal chromosome abnormalities were described according to the International System for Human Cytogenetic Nomenclature (ISCN) [109].

FISH Analysis

For FISH analysis #2, #5, #12, #17, #18 and #21 whole chromosome painting probes (CAMBIO, Cambridge, UK), chromosomes #2, # 12 and #17 specific ∝-satellite DNA probes (VYSIS, Downer Grove, IL) and the locus specific DNA probes *Cyclin D1* at 11q13 (LSI Cyclin D1 Spectrum Orange/Cep 11 Spectrum Green, dual color), D13S319 at 13q14 band, *TP53* at 17p13.1 (Spectrum Orange), IgH Spectrum Green/BCL2 Spectrum Orange (VYSIS, Downer Grove, IL) and *ATM* (ataxia telangiectasia mutated) locus at 11q23 Spectrum Green (Q BIO gene) were used. FISH was performed according to manufacturer´s protocols. Briefly, probe mix was denatured for 2 minutes at 70°C and maintains at 45-50°C till use. Simultaneously, fresh spread slides were denatured for 2 minutes in 70% formamide 2 x SSC at 70°C and dehydrated in graded ethanol series. The probe mix was then applied to the slides and hybridized overnight at 37°C in a moist chamber. After hybridization, the slides were washed in 0.4 x SSC/ 0.3% NP40 for 2 minutes at 73°C, rinsed in 2 x SSC/ 0.1% NP40 for 1 minute at room temperature. The slides were counterstained with 4',6-diamidino-2-fenilindol diclorhidrato (DAPI). Four hundred interphase nuclei from the patients and 10 controls were scored for each probe. The cut-off for positive values (mean of normal controls+ 3SD) was 1.0, 3.0, 10.2, 5.1 and 7.7% for *IgH/BCL-2*, trisomy 12 and monosomies of D13S319, *TP53* and *ATM*, respectively.

Results

Table 2 shows the cytogenetic and FISH studies of 12 CLL/SLL patients of our series that presented non-frequent clonal structural chromosome aberrations in their karyotypes. In all cases, near diploid karyotypes (39-48 chromosomes) were observed.

Table 2. Cytogenetic and FISH results in the CLL/SLL patients

Case	Sample	Combined karyotype (GTG and FISH)	Interphase (FISH (%) +12	13q-	17p-	11q-
1	BM LNB	46,XX[20] 45,XX,t(2;17)(p21;p11.2)[11].ish psu dic(17;2)(p11.2;p21)(wcp2+,D2Z1+,wcp17+,D17Z1+)/46,XX[2].	0.8 0.4	ND 7.7	7.0 85.0	ND 7.1
2	BM	47,XY,+12[4]/47,XY,t(2;17)(p21;q23),+12[9]/ 47,XY, t(2;17)(p21;q23),-10,+12,+17[2]/ 46,XX[16].	60.0	2.0	0.9	1.8
3	BM BM	47,XY,+12[3],/ 46,XY[19] 46,XY,t(2;17)(p21;q23),del(11)(q23)[cp4]/46,XY[8].	11.0 2.8	7.0 8.0	3.5 3.0	9.0 13.0
4	BM	39-45,XY,del(3)(p11),-10,der(14)t(11;14)(q13;q24), der(17)t(12;17)(q13;q25),-13,+r[cp23].	0.5	ND	ND	ND
5	BM	48,XY,+3,t(5;12)(q15;p11),+del(5)(q15),del(7)(q11), der(19)t(19;?)(q13;?)[13].ish.t(5;12;19)(q15;p11;q13)(wcp5+,wcp12++),del(5)(wcp5+)/46,XX[20].	ND	ND	ND	ND
6	BM	46-48,XX,t(2;14)(p11;q32),+10,del(11)(q23),+14[cp11]/ 46,XY[14]	1.7	22.4	4.7	2.5
7	BM BM	43,XY,psu-dic(12;21)(q24;q10),dic(17;18)(p11.2;p11.2)[6]/46;XY[16]. 43,XY,psu-dic(12;21)(q24;q10),dic(17;18)(p11.2;p11.2)[6]/46;XY[12]	1.0 0.7	18.0 11.7	43.8 77.3	ND ND
8	BM PB	46,XX[11] 47,XX,t(10;14)(q22;q32),+12[8]/ 46,XX[5]	39 36.3	ND 4.8	ND 2.9	ND ND
9	BM	46,XX[10] 47,XX,del(7)(q34),+12[3]/ 46,XX [30]	3.3 ND	ND 16.0	ND ND	ND ND
10	BM	ND 46,XY,t(11;13)(p15;q22)[5]/ 46,XY[6]	0.7 0.2	50.0 27.3	1.7 1.9	ND 0.9
11	PB	46,XX,dup(12)(q13q24)[9]/ 46,XX [9]	2.4	20.6	4.4	ND
12	PB	46 ,XY,del(18)(p11)[4]/ 46,XY [11]	ND	ND	ND	ND

Bold chromosome rearragements correspond to unpublished or infrequent anomalies ; bold numbers correspond to abnormal clones; GTG: G-bands by Trypsin using Giemsa; ND: not determined; BM: bone marrow; LNB: lymph node biopsy; PB: peripheral blood.

As a whole, we identified 16 different novel or infrequent chromosome rearrangements by cytogenetic analysis: 7 translocations altogether, 4 of which were balanced, 2 unbalanced and 1 complex translocation; and 3 dicentric or psu dicentric chromosomes, 1 partial chromosome duplication and 5 deletions. Some of these findings were recently reported by us as a new alterations in SLL/CLL patients: t(2;17)(p21;q23); der(17)t(12;17)(q13;q25); der(14)(t(11;14)(q13;q24), t(5;12;19) (q15;p11;q13); psu dic (17;2)(p11.2;p21), del(3)(p11), del(5)(q15) [108], psu dic (12;21) (q24;q10) [92] and t(10;14)(q22;q32) [110]. Translocation t(11;13)(p15;q22) and del(18)(p11) were not previously described, meanwhile, dic(17;18)(p11.2;p11.2), dup(12)(q13q24), del(7)(q34), t(2;14)(p11;q32) and del(7)(q11) were infrequently reported in the literature [111]. In addition, the novel translocation t(2;17)(p21;q23)) was recurrent in our series. Chromosomes 17 (five cases), 2 and 12 (four cases each) and 14 (three cases each) were the most frequently involved in this series.

Twenty one different breakpoints were found in these infrequent structural rearrangements, and all of them were previously reported though with different partner chromosomes or in a different type of chromosome rearrangements. However, 38.1% (8/21) breakpoints, such as 2p21, 5q15, 17p11.2, 17q23, 12q13, 12q24, 14q32 and 18p11, were recurrent in our series.

Among these aberrations, psu dic(17;2) (p11.2;p21) (case 1), t(11;13)(p15;q22) (case 10), dup(12)(q13q24) (case 11) and del(18)(p11) (case 12), were observed as the sole cytogenetic abnormality. Associated clonal alterations were identified in the remaining patients. Several aberrations previously described in this neoplasm were noted, namely a 11q23 deletion in cases 3 and 6 and trisomies 3, 10, 12, 14 and 17, and monosomies 10 y 13.

In Case 1 [108] cytogenetic and FISH studies at the lymph node biopsy showed the presence of a psu dic (17;2)(p11.2;p21) (figure 1). Interphase FISH showed monoallelic loss of TP53 in 85% of cells. This chromosome rearrangement was seen at the moment in which the patient develop a mixed cellularity HD coexisting with the SLL in the same lymph node. A previous study on BM cells, at CLL diagnosis, had a low number of metaphases with normal karyotype but interphase FISH also showed a small clone (7% of cells) with *TP53* gene deletion, indicating the presence of the marker chromosome in pre-HD diagnostic tissue. In addition, EBV (Epstein Barr Virus) was detected in the Reed Sternberg (RS) cells, while the CLL/SLL tumour cells were EBV negative.

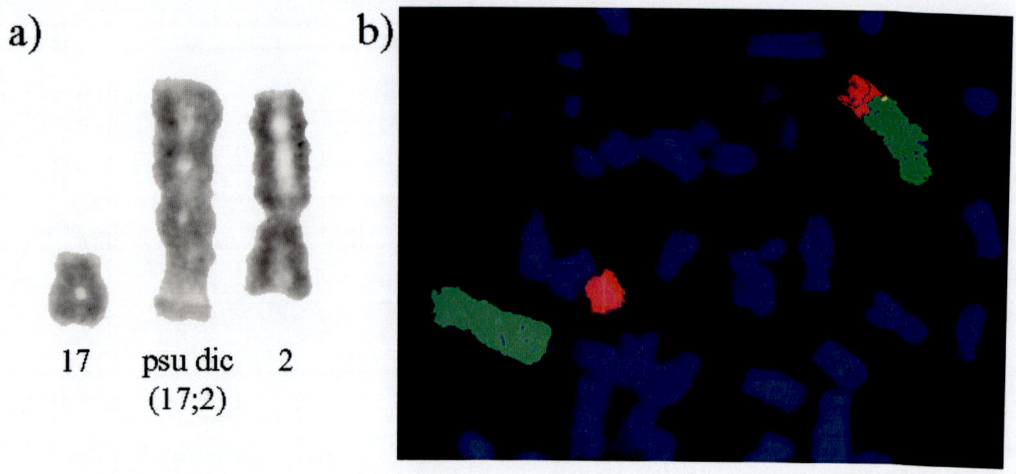

Figure 1. a) Partial karyotype showing psu dic(17;2)(p11.2;p21) from Case 1; b) Partial metaphase hybridized with chromosomes 2 and 17 painting probes showing normal chromosomes 2 (green) and 17 (red) and psu dic(17;2)(p11.2;p21) (arrow).

Cases 2 and 3, previously reported [108] showed the same translocation t(2;17)(p21;q23) (figure 2). In Case 2, a clear clonal evolution that included trisomies 12 and 17 and monosomy 10 was observed. Case 3 had two cytogenetic studies, the first one before treatment, showed a clone with trisomy 12 (13% of cells) and only one cell with t(2;17). A second study post-treatment (37 months later), showed a composite karyotype with t(2;17) and del(11)(q23), confirming the initial finding. Interphase FISH using ATM probe showed monoallelic loss of this locus with frequencies of 9% and 13% for the first and second study, respectively (table 2).

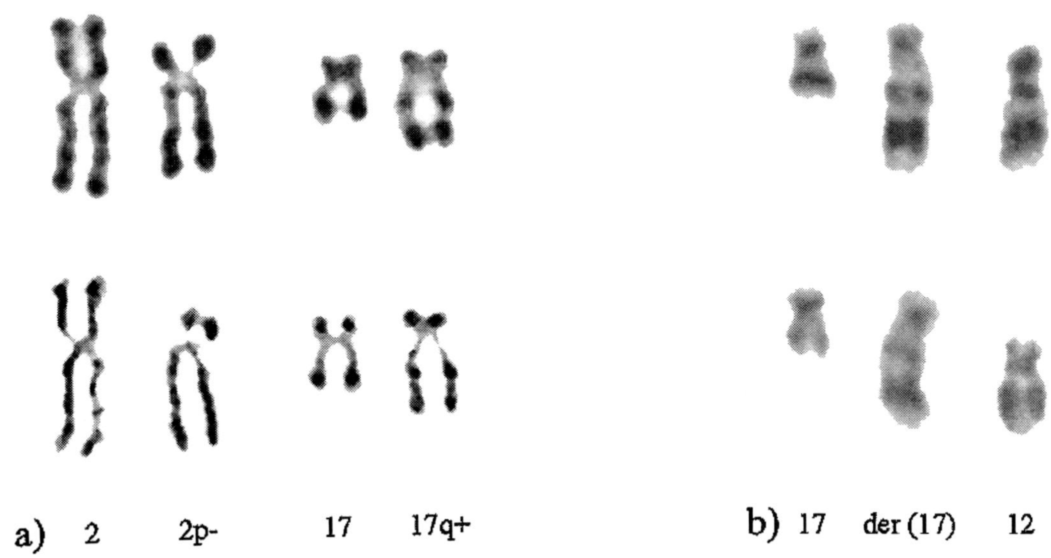

Figure 2. a) Partial karyotypes showing t(2;17)(p21;q23) from Cases 2 and 3; b) Partial karyotypes showing der(17)t(12;17)(q13;q25) from Case 4.

Cases 4, 5 and 6 had complex karyotypes including: del(3)(p11), der(14)t(11;14)(13;q24) and der(17)t(12;17)(q13;q25) (figure 2); t(5;12;19) (q15;p11;q13), del(5)(q15) and del(7)(q11); and t(2;14)(p11;q32), respectively. In addition of t(2;14), the karyotype of case 6 included trisomies of chromosomes 10 and 14, and del(11)(q23). Interestingly, FISH analysis did not showed ATM deletion (table 2), indicating that the breakpoint was telomeric to this gene. After 24 months from diagnosis, progression to Rai´s stage I without requirement of treatment was observed.

Case 7 [92], was studied in two opportunities, at diagnosis and during Richter´s transformation showing the same karyotype: 44,XY,der(12)t(12;21)(q24;q11), der(17;18)(q10;q10). Both rearrangements were confirmed by FISH using painting probes. C-banding technique revealed the presence of the inactivated centromere of chromosome 21, indicating the pseudo dicentric nature of der(12): psu dic(12;21)(q24;q10). FISH using α-satellite DNA probes of chromosomes 17 and 18 showed a signal of each probe on the der(17;18), confirming the dicentric condition of this chromosome: ish dic (17;18)(p11.2;p11.2).

Cases 8 and 9 were first studied at diagnosis showing normal karyotypes but trisomy 12 by FISH (39% and 3.3%, respectively). Second cytogenetic study, during disease progression, showed t(10;14)(q22;q32) and del(7)(q34), respectively, and trisomy 12, suggesting clonal evolution. Moreover, case 9 showed a small clone with 13q14 deletion (16% of cells).

Case 10 was first studied at diagnosis. At this moment, the cytogenetic study was not succeeded, but FISH analysis showed monoallelic deletion of 13q14 (50% of interphase nuclei). The second study two years later, after treatment (Chlorambucil/Prednisone) and during accelerated lymphocytosis, showed a karyotype with t(11;13)(p15;q22) translocation. Simultaneously, a reduction in the del(13)(q14) clone was observed (27.3% of cells).

Case 11 [110] showed a clone with a dup(12)(q13q22) (50% of cells), confirmed by FISH. The remaining cells were normal. The analysis with D13S319 probe showed 20% of cells with monoallelic deletion.

Finally, in case 12 was detected a clone with del(18)(q11) as a sole anomaly. This study was done two years after diagnosis, without previous treatment, when patient was in disease progression. FISH analysis cannot be performed.

Discussion

The identification of the non-random genetic alterations associated with cancer has great importance not only to basic research but also to disease management as diagnostic and prognostic markers. We present the cytogenetic and FISH analyses of 12 CLL/SLL patients with new or infrequent clonal chromosome rearrangements in their karyotypes, all of them showing adverse clinical course with partial or null response to different chemotherapeutic agents.

Nine out of these 16 chromosome rearrangements described in this work had been previously reported by us as new alterations in CLL/SLL [48, 108, 110]. All of them are still novel rearrangements, and except for t(2;17), not recurrent in CLL/SLL. Two out of 9 were pseudo dicentric chromosomes, two deletions and the remaining five were translocations. In addition, two new chromosome rearrangements that have not been reported yet, and five infrequent chromosome abnormalities are shown. Chromosomes 17, 2, 12 and 14 were the most commonly involved.

Among these novel anomalies, a psu dic (17;2)(p11.2;p21) was found as a sole cytogenetic anomaly meanwhile a psu dic (12;21)(q24;q10) was associated to dic (17;18)(p11.2;p11.2) in the same karyotype. Both chromosome 17 abnormalities resulted in 17p loss, with *TP53* monoallelic deletion.

In hematological malignancies, including lymphoid and myeloid disorders rare dicentric anomalies have been described: dic(7;12), dic(9;12) and dic(17;18) in the novo or therapy-related acute myeloid leukemia and myelodysplastic syndrome [111]. Of interest, not all are associated with adverse clinical outcome, such as dic(9;12) in childhood B-ALL, which has favorable prognosis [112, 113].

Telomere dysfunction has been associated with genomic instability and mitotic disturbances in different type of neoplasms such colorectal and intestinal carcinomas and carcinomas of the head and neck [114-116]. Abnormal shortening of terminal TTAGGG repeats causes different mitotic defects with anaphase bridges triggering both numerical and structural chromosome changes, and with mitotic multipolarity leading to more dramatic genomic changes that may prove deleterious to the cell. It is known that one of them or both mechanisms are associated with the formation of dicentric chromosomes in these carcinomas [114-117]. In contrast, the mechanisms that prevail behind dicentric chromosomes generation in CLL are not known.

In reference to our cases with dicentric chromosomes, it is interesting to point out that case 1 with a psu dic (17;2)(p11.2;p21) [108], develop a HD nineteen months after the diagnosis of CLL. It is known that patients with CLL/SLL have approximately an 8-fold increased risk to develop HD [118], with HD representing one of the most frequent secondary neoplasms. In a few cases, clonality analysis of CLL/SLL and RS cells resulted in the detection of the same clonal immunoglobulin heavy chain gene rearrangement, indicating that the RS cells derived from a common precursor B-cell. However, in a high percentage of cases, EBV positive RS cells were not genetically related to the SLL tumour clone, suggesting that EBV might play a role in the development of the HD component [119-121], situation that is suggested for our patient.

Loss of the short arm of chromosome 17, where the *TP53* tumour suppressor gene localizes, has been observed in 4% of cytogenetically evaluable CLL patients. However, FISH studies have shown that monoallelic deletions of this gene occur in 9-15% of cases [19, 122]. Different reports have shown that *TP53* gene deletions have strong implications in the clinical course of the disease, with significantly shorter median survival, advanced clinical stage and resistance to chemotherapy. In addition, *TP53* abnormalities occur in about 60% of cases with Richter's syndrome [19, 122]. Moreover, Döhner et al. [122] have done FISH and molecular analysis of this gene, revealing a biallelic inactivation of the *TP53* in this pathology.

As for psu dic(12;21), to our knowledge this is the first CLL case where this alteration has been reported. It was found as a part of a karyotype that also include a dic(17;18), which represents the most frequent recurrent dicentric chromosome in CLL/SLL [111]. This patient showed disease progression to Richter's transformation. It is important to remark that, in the first study of this case, we found only one cell with a karyotype 45,XY,der(17;18)(q10;q10) (not presented in table 2), suggesting it as the early event leading to CLL that was followed by stepwise clonal evolution in the leukemic cells leading to lymphoma.

In a substantial subset of patients, new, and often more dangerous, mutations accumulate over time. These new alterations of the original clone may be better suited to survive and thrive, and they become the dominant clones. These clones with enhanced survival capabilities are much harder to kill with distinct protocols and disease progression is precipitated by the emergence of new and more aggressive clones [44, 57, 123]. At first, karyotypic clonal evolution seemed to be rare in CLL, but in last ten years it was detected from 16 to 43% of patients by sequential studies. Alterations more frequently involved are deletions at 6q, 13q and 11q [44, 123, 124]. In addition, the introduction of FISH studies was an important step to identify genomic changes, undetectable by conventional cytogenetic, implicated in the development and evolution of the malignant clone. In our experience, patients with clonal evolution showed a poor prognosis with a treatment-free survival significantly shorter than those observed in cases with 13q14 deletion or normal karyotypes ($p<0.001$) (data not shown).

Balanced translocations t(2;17)(p21;q23) (cases 2 and 3) and t(5;12;19) (q15;p11;q13) (case 5) have not been yet found in any other type of neoplasm. Our study also showed that chromosome 2 with breakpoint in p21 was the partner most frequently involved in rearrangements with chromosome 17. Bands 2p21-22 were found frequently rearranged in leukemia patients, region where, interestingly, the double-stranded RNA activated protein kinase (*PKR*) tumor suppressor gene [125], and the mismatch repair gene *MSH2* were mapped [126].

Moreover, t(10;14)(q12;q32) was described in other oncohematological diseases with a very low frequency [111]. Der(17)t(12;17)(q13;q25), der(14)t(11;14)(q13;q24) and del(3)(p11) were observed in the same karyotype. The first two abnormalities have not been reported just as in the literature neither in CLL/SLL nor in other neoplasm [111] but a der(11)t(11;14)(q13;q24) has been recently described in this disorder [102].

We have two more translocations in our series. One of them, t(11;13)(p15;q22), is a new chromosome rearrangements in CLL/SLL. The literature show a similar alteration but with different breakpoints on 13q arm in a CLL patient [44]. The other one, t(2;14)(p11;q32) (case 6), became recurrent in these pathology from our data [127]. At the moment of the present study, this patient was the only one that did not show an unfavorable clinical outcome.

Translocations are primary events useful for lymphoma classification, whereas secondary alterations are mainly chromosomal imbalances, many of which are associated with clinical outcome. By contrast, it is important to note that reciprocal chromosomal translocations are not frequently observed in CLL/SLL [111]. According to the literature, we did not find a CLL/SLL specific translocation but found instead many different translocations, all of them associated with unfavorable clinical course [88, 102].

As concerns to deletions, del(3)(p11), del(5)(q15) and del(18)(p11) are novel rearrangements in this pathology but they are recurrent anomalies in other neoplasms [111]. It has been described a human receptor tyrosine kinase gene on chromosome region 3p11.2, that is expressed by some pre-B and thymic T-cell lines, but is not detectable on normal adult human tissues. Gene cloning studies established that this gene is a new member of the EPH family of receptor tyrosine kinases, and its inappropriate expression may contribute to oncogenesis [128]. Losses of 5q are observed in different types of neoplasms and it has been suggested the possibility of gene products involved in cancer progression [129]. Recently the putative tumor suppressor gene, *DAL1* has been mapped on chromosome band 18p11.3. Future studies, will determinate whether this or another tumor suppressor gene is the target of the high frequency of allelic deletion found in different types of neoplasms [130, 131].

In addition, case 5 and case 9 showed deletions of a part of the long arm of chromosome 7: del(7)(q11) and del(7)(q34), which are infrequent findings in CLL. The former was previously described in only three cases [47, 132, 133]. Deletions 7q have been reported in myeloid diseases like acute myeloid leukemia and myelodisplastic syndrome but seem to be rare in lymphoproliferative diseases. Particularly, loss of 7q22-qter is common in hairy cell leukemia and splenic marginal zone lymphoma but not in CLL/SLL [134].

Among the remaining alterations, we had one case with dup(12)(q13q24) as a sole cytogenetic anomaly and the coexistence of 13q deletion identified by FISH. Duplication 12q is an uncommon chromosome abnormality not only in CLL but also in other hematological neoplasms [111]. Structural aberrations of the long arm of chromosome 12 have been observed in about 5% of CLL cases [88] and in approximately 15% of patients with Richter's syndrome [135]. Our case showed atypical morphology, high proliferation rate and progression of the disease, indicating that the duplication of this region may be similar to the complete trisomy 12 in CLL patients, in which this association is well known. This region contains several genes directly or indirectly associated with cancer but none of them seems affected in CLL. Some authors proposed that *MDM2* or *P2RX7* genes could be involved, but results are controversial [136-139]. More recently, a novel gene named CLLU1 (located at 12q22), was observed as uniquely overexpressed in CLL, suggesting their participation in CLL development [81].

Breakpoint analysis among these novel or infrequent chromosomal anomalies identified eight recurrent breaks. Each of these eight bands has previously been reported to be involved in lymphoid tumour [111] and they cannot be considered specific of CLL.

In reference to chromosome imbalances, most common gains involved chromosome region 12q13-q24, whereas 17p11.2 was the most frequently deleted regions. It is important to point out that one patient of our series (case 4) showed loss of 14q24-qter chromosome region and had an aggressive clinical course. Our finding and data from other reports suggest that the loss of 14q24-qter region is indicative of adverse clinical outcome and may be a new marker for prognostic stratification [140].

In summary, we have identified 16 novel or infrequent chromosome alterations, two of them reciprocal translocations, that might represent new recurrent abnormalities in CLL/SLL, and we described the salient clinical features of our patients. All these rare chromosome abnormalities, except for case 6, were observed in patients with adverse clinical outcome and short event free survival time. Our findings may facilitate the understanding of genetic events associated with development and progression of CLL/SLL and may have implications in the clinical management of patients with this disease.

Acknowledgments

This work was supported by grants from our National Research Council (CONICET) and the National Agency of Scientific and Technical Promotion (ANPCyT).

References

[1] Stilgenbauer S, Lichter P, Döhner H. Genetics of B-cell chronic lymphocytic leukemia. In: Chronic Lymphocytic Leukemia. *Molecular Genetics, Biology, Diagnosis, and Management.* Ed. Faguet GB 2004.

[2] Rozman C, Montserrat E. Chronic lymphocytic leukemia. *N. Engl. J. Med.* 1995;1333:1052-1057.

[3] Zwiebel JA, Cheson BD. Chronic lymphocytic leukemia: staging and prognostic factors. *Semis. Oncol.* 1998;25:42-59.

[4] SEER Cancer Statistics Review, 1975-2000, http://seer.cancer.gov/cgi-bin/csr/search.pl#results.

[5] American Cancer Society Facts and Figures for professionals, http://www.cancer.org/docroot/STT/stt_o.asp.

[6] Boggs DR, Chen SC, Zhang ZN, Zhang A. Chronic lymphocytic leukemia in China. *Am. J. Hematol.* 1987;25:349-354.

[7] Chang YH, Park J, Kim HC, Chun HK, Kim YR, Kim M, Han K, Lee J-H, Lee K-H, Cho HI, Lee, YS, Lee DS. Korean patients with chronic lymphocytic leukemia show

the similar types of chromosomal aberrations as those in Europe and North America. *Leukemia Res.* 2006;30:695-699.

[8] Yamahihara ET, Blaisdell RK, Hayashi T, Lukes RJ. Malignant lymphoma in Hawaii-Japanese: a retrospective morphologic survey. *Hematol. Oncol.* 1989;7:219-232.

[9] Gale RP, Cozen W, Goodman MT, Wang FF, Bernstein L. Decrease chronic lymphocytic leukemia incidence in Asians in Los Angeles County. *Leukemia Res.* 2000;24:665-669.

[10] Tamura K, Sawada H, Izumi Y, Fukuda T, Utsunomiya A, Ikeda S, Uike N, Tsukada J, Kawano F, Shibuya T, Gondo H, Okamura S, Suzumiya J; Kyushu Hematology Organization for Treatment (K-HOT) Study Group. Chronic lymphocytic leukemia (CLL) is rare, but the proportion of T-CLL is high in Japan. *Eur. J. Haematol.* 2001;67:152-157,

[11] Mauro FR, Foa R, Giannarelli D Cordone I, Crescenzi S, Pescarmona E, Sala R, Cerretti R, Mandelli F. Clinical characteristics and outcome of young chronic lymphocytic leukemia patients: a single institution study of 204 cases. *Blood.* 1999;94:448-454.

[12] Dighiero G. CLL Biology and Prognosis. *Hematology Am. Soc. Hematol. Educ. Program.* 2005:278-284.

[13] Byrd JC, Stilgenbauer S, Flinn I. Chronic Lymphocytic Leukemia. Hematology Am. Soc. Hematol. Educ. Program. 2004:163-183.

[14] Rozman C, Montserrat E, Rodriguez-Fernandez JM, Ayats R, Vallespi T, Parody R, Rios A, Prados D, Morey M, Gomis F. Bone marrow histologic pattern--the best single prognostic parameter in chronic lymphocytic leukemia: a multivariate survival analysis of 329 cases. *Blood.* 1984;64:642-648.

[15] Montserrat E, Sanchez-Bisono J, Vinolas N, Rozman C. Lymphocyte doubling time in chronic lymphocytic leukaemia: analysis of its prognostic significance. *Br. J. Haematol.* 1986;62:567-575.

[16] Kröber A, Seiler T, Benner A, Bullinger L, Bruckle E, Lichter P, Dohner H, Stilgenbauer S. V(H) mutation status, CD38 expression level, genomic aberrations, and survival in chronic lymphocytic leukemia. *Blood.* 2002;100:1410-1416.

[17] Di Giovanni S, Valentini G, Carducci P, Giallonardo P. Beta-2-microglobulin is a reliable tumor marker in chronic lymphocytic leukemia. *Acta Haematol.* 1989;81: 181-185.

[18] Hallek M, Langenmayer I, Nerl C, Knauf W, Dietzfelbinger H, Adorf D, Ostwald M, Busch R, Kuhn-Hallek I, Thiel E, Emmerich B. Elevated serum thymidine kinase levels identify a subgroup at high risk of disease progression in early, nonsmoldering chronic lymphocytic leukemia. *Blood.* 1999;93:1732-1737.

[19] Döhner H, Stilgenbauer S, Benner A, Leupolt E, Krober A, Bullinger L, Dohner K, Bentz M, Lichter P. Genomic aberrations and survival in chronic lymphocytic leukemia. *N. Engl. J. Med.* 2000,343:1910-1916.

[20] Lin K, Sherrington PD, Dennis M, Matrai Z, Cawley JC, Pettitt AR. Relationship between p53 dysfunction, CD38 expression, and IgV(H) mutation in chronic lymphocytic leukemia. *Blood.* 2002,100:1404-1409.

[21] Crespo M, Bosch F, Villamor N, Bellosillo B, Colomer D, Rozman M, Marce S, Lopez-Guillermo A, Campo E, Montserrat E. ZAP-70 expression as a surrogate for

immunoglobulin-variable-region mutations in chronic lymphocytic leukemia. *N. Engl. J. Med.* 2003,348:1764-1775.

[22] Dewald GW, Brockman SR, Paternoster SF, Bone ND, O'Fallon JR, Allmer C, James CD, Jelinek DF, Tschumper RC, Hanson CA, Pruthi RK, Witzig TE, Call TG, Kay NE. Chromosome anomalies detected by interphase fluorescence in situ hybridization: correlation with significant biological features of B-cell chronic lymphocytic leukaemia. *Br. J. Haematol.* 2003,121:287-295.

[23] Raval A, Byrd JC, Plass C. Epigenetics in chronic lymphocytic leukemia. *Semin. Oncol.* 2006,33:157-166.

[24] Damle RN, Wasil T, Fais F, Ghiotto F, Valetto A, Allen SL, Buchbinder A, Budman D, Dittmar K, Kolitz J, Lichtman SM, Schulman P, Vinciguerra VP, Rai KR, Ferrarini M, Chiorazzi N. IgV gene mutation status and CD38 expression as novel prognostic indicators in chronic lymphocytic leukemia. *Blood.* 1999;94:1840-1847.

[25] Hamblin TJ, Davis Z, Gardiner A, Oscier DG, Stevenson FK. Unmutated IgV(H) genes are associated with a more aggressive form of chronic lymphocytic leukemia. *Blood.* 1999;94:1848-1854.

[26] Klein U, Tu Y, Stolovitzky GA, Mattioli M, Cattoretti G, Husson H, Freedman A, Inghirami G, Cro L, Baldini L, Neri A, Califano A, Dalla-Favera R. Gene expression profiling of B cell chronic lymphocytic leukemia reveals a homogeneous phenotype related to memory B cells. *J. Exp. Med.* 2001;194:1625-1638.

[27] Rosenwald A, Alizadeh AA, Widhopf G, Simon R, Davis RE, Yu X, Yang L, Pickeral OK, Rassenti LZ, Powell J, Botstein D, Byrd JC, Grever MR, Cheson BD, Chiorazzi N, Wilson WH, Kipps TJ, Brown PO, Staudt LM. Relation of gene expression phenotype to immunoglobulin mutation genotype in B cell chronic lymphocytic leukemia. *J. Exp. Med.* 2001;194:1639-1647.

[28] Wiestner A, Rosenwald A, Barry TS, Wright G, Davis RE, Henrickson SE, Zhao H, Ibbotson RE, Orchard JA, Davis Z, Stetler-Stevenson M, Raffeld M, Arthur DC, Marti GE, Wilson WH, Hamblin TJ, Oscier DG, Staudt LM. ZAP-70 expression identifies a chronic lymphocytic leukemia subtype with unmutated immunoglobulin genes, inferior clinical outcome, and distinct gene expression profile. *Blood.* 2003;101:4944-4951.

[29] Vasconcelos Y, De Vos J, Vallat L, Reme T, Lalanne AI, Wanherdrick K, Michel A, Nguyen-Khac F, Oppezzo P, Magnac C, Maloum K, Ajchenbaum-Cymbalista F, Troussard X, Leporrier M, Klein B, Dighiero G, Davi F; French Cooperative Group on CLL. Gene expression profiling of chronic lymphocytic leukemia can discriminate cases with stable disease and mutated Ig genes from those with progressive disease and unmutated Ig genes. *Leukemia.* 2005,19:2002-2005.

[30] Nolz JC, Tschumper RC, Pittner BT, Darce JR, Kay NE, Jelinek DF. ZAP-70 is expressed by a subset of normal human B-lymphocytes displaying an activated phenotype. *Leukemia.* 2005;6:1018-1024.

[31] Catherwood MA, Matthews C, Niblock R, Dobbin E, Morris TC, Alexander HD. ZAP-70 mRNA quantification in B-cell chronic lymphocytic leukaemia. *Eur. J. Haematol.* 2006;76:294-298.

[32] Corcoran M, Parker A, Orchard J, Davis Z, Wirtz M, Schmitz OJ, Oscier D. ZAP-70 methylation status is associated with ZAP-70 expression status in chronic lymphocytic leukemia. *Haematologica.* 2005;90:1078-1088.

[33] Shanafelt TD, Geyer SM, Kay NE. Prognosis at diagnosis integrating molecular biologic insights into clinical practice for patients with CLL. *Blood.* 2004;103:1202-1210.

[34] Scroers R, Griesinger F, Trümper I, Haase B, Klein-Hitpass L, Sellmann I, Dührsen U, Dürig L. Combined analysis of ZAP-70 and CD38 expression as a predictor of disease progression in B-cell chronic lymphocytic leukemia. *Leukemia.* 2005;19:750-758.

[35] Autio K, Elonen E, Teerenhovi L, Knuutila S. Cytogenetic and immunologic characterization of mitotic cells in chronic lymphocytic leukaemia. *Eur. J. Haematol.* 1987,39:289-298.

[36] Larramendy ML, Siitonen SM, Zhu Y, Hurme M, Vilpo L, Vilpo JA, Knuutila S. Optimized mitogen stimulation induces proliferation of neoplastic B cells in chronic lymphocytic leukemia: significance for cytogenetic analysis.The Tampere Chronic Lympocytic Leukemia group. *Cytogenet. Cell Genet.* 1998;82:215-221.

[37] Ross FM, Stockdill G. Clonal chromosome abnormalities in chronic lymphocytic leukemia patients revealed by TPA stimulation of whole blood cultures. *Cancer Genet. Cytogenet.* 1987, 25: 109-121.

[38] Nayak BN, Sokal J, Ray M. Clonal chromosomal changes in chronic lymphocytic leukemia. *Cancer Lett.* 1990 Feb;49(2):99-105.

[39] Stephenson CF, Desai ZR, Bridges JM. The proliferative activity of B-chronic lymphocytic leukaemia lymphocytes prior to and after stimulation with TPA and PHA. *Leuk. Res.* 1991;15(11):1005-12.

[40] Juliusson G, Gahrton G. Chromosome aberrations in B-cell chronic lymphocytic leukemia. Pathogenetic and clinical implications. *Cancer Genet. Cytogenet.* 1990,45:143-160.

[41] Stockero KJ, Fink SR, Smoley SA, Paternoster SF, Shanafelt TD, Call TG, Zent CS, Van Dyke DL, Kay NE, Dewald GW. Metaphase cells with normal G-bands have cryptic interstitial deletions in 13q14 detectable by fluorescence in situ hybridization in B-cell chronic lymphocytic leukemia. *Cancer Genet. Cytogenet.* 2006,166:152-156.

[42] Fink SR, Smoley SA, Stockero KJ, Paternoster SF, Thorland EC, Van Dyke DL, Shanafelt TD, Zent CS, Call TG, Kay NE, Dewald GW. Loss of TP53 is due to rearrangements involving chromosome region 17p10 approximately p12 in chronic lymphocytic leukemia. *Cancer Genet. Cytogenet.* 2006;167:177-181

[43] Han T, Sadamori N, Ozer H, Gajera R, Gomez GA, Henderson ES, Bhargava A, Fitzpatrick J, Minowada J, Bloom ML, et al. Cytogenetic studies in 77 patients with chronic lymphocytic leukemia: correlations with clinical, immunologic, and phenotypic data. *J. Clin. Oncol.* 1984,2:1121-1132.

[44] Finn WG, Kay NE, Kroft SH, Church S, Peterson LC. Secondary abnormalities of chromosome 6q in B-cell chronic lymphocytic leukemia: a sequential study of karyotypic instability in 51 patients. *Am. J. Hematol.* 1998,59:223-229.

[45] Buhmann R, Kurzeder C, Rehklau J, Westhaus D, Bursch S, Hiddemann W, Haferlach T, Hallek M, Schoch C. CD40L stimulation enhances the ability of conventional metaphase cytogenetics to detect chromosome aberrations in B-cell chronic lymphocytic leukaemia cells. *Br. J. Haematol.* 2002,118:968-975.

[46] Stilgenbauer S, Döhner H. Molecular genetics and its clinical relevance. *Hematol. Oncol. Clin. North. Am.* 2004,18:827-848.

[47] Reddy KS. Chronic lymphocytic leukaemia profiled for prognosis using a fluorescence *in situ* hybridization panel. *Br. J. Haematol.* 2006,132:705-722.

[48] Chena C, Arrossagaray G, Scolnik M, Palacios MF, Slavutsky I. Interphase cytogenetic analysis in Argentinean B-cell chronic lymphocytic leukemia patients: association of trisomy 12 and 13q14 deletions. *Cancer Genet. Cytogenet.* 2003;146: 154-160.

[49] Liu Y, Szekely L, Grander D, Soderhall S, Juliusson G, Gahrton G, Linder S, Einhorn S. Chronic lymphocytic leukemia cells with allelic deletions at 13q14 commonly have one intact RB1 gene: evidence for a role of an adjacent locus. *Proc. Natl. Acad. Sci. USA.* 1993;90:8697-8701.

[50] Stilgenbauer S, Dohner H, Bulgay-Morschel M, Weitz S, Bentz M, Lichter P. High frequency of monoallelic retinoblastoma gene deletion in B-cell chronic lymphoid leukemia shown by interphase cytogenetics. *Blood.* 1993;81:2118-2124.

[51] Liu Y, Hermanson M, Grander D, Merup M, Wu X, Heyman M, Rasool O, Juliusson G, Gahrton G, Detlofsson R, Nikiforova N, Buys C, Söderhäll S, Yankovsky N, Zabarovsky E, Einhorn S. 13q deletions in lymphoid malignancies. *Blood.* 1995;86:1911-1915.

[52] Stilgenbauer S, Nickolenko J, Wilhelm J, Wolf S, Weitz S, Dohner K, Boehm T, Dohner H, Lichter P. Expressed sequences as candidates for a novel tumor suppressor gene at band 13q14 in B-cell chronic lymphocytic leukemia and mantle cell lymphoma. *Oncogene.* 1998,16:1891-1897.

[53] Migliazza A, Bosch F, Komatsu H, Cayanis E, Martinotti S, Toniato E, Guccione E, Qu X, Chien M, Murty VV, Gaidano G, Inghirami G, Zhang P, Fischer S, Kalachikov SM, Russo J, Edelman I, Efstratiadis A, Dalla-Favera R. Nucleotide sequence, transcription map, and mutation analysis of the 13q14 chromosomal region deleted in B-cell chronic lymphocytic leukemia. *Blood.* 2001;97:2098-2104.

[54] Calin GA, Dumitru CD, Shimizu M, Bichi R, Zupo S, Noch E, Aldler H, Rattan S, Keating M, Rai K, Rassenti L, Kipps T, Negrini M, Bullrich F, Croce CM. Frequent deletions and down-regulation of micro-RNA genes miR15 and miR16 at 13q14 in chronic lymphocytic leukemia. *Proc. Natl. Acad. Sci. USA.* 2002;99:15524-15529.

[55] Calin GA, Ferracin M, Cimmino A, Di Leva G, Shimizu M, Wojcik SE, Iorio MV, Visone R, Sever NI, Fabbri M, Iuliano R, Palumbo T, Picchiorri F, Roldo C, Garzon R, Sevignani C, Rassenti L, Alder H, Volinis S, Liu CG, Kipps TJ, Negrini M, Croce CM. A MicroRNA signature associated with prognosis and progression in chronic lymphocytic leukemia. N Engl J Med 2005;353:1793-1801. *Erratum N. Engl. J. Med.* 2006;355:533.

[56] Döhner H, Stilgenbauer S, James MR, Benner A, Weilguni T, Bentz M, Fischer K, Hunstein W, Lichter P. 11q deletions identify a new subset of B-cell chronic lymphocytic leukemia characterized by extensive nodal involvement and inferior prognosis. *Blood.* 1997;89:2516-2522.

[57] Cuneo A, Bigoni R, Rigolin GM, Roberti MG, Bardi A, Cavazzini F, Milani R, Minotto C, Tieghi A, Della Porta M, Agostini P, Tammiso E, Negrini M, Castoldi G. Late appearance of the 11q22.3-23.1 deletion involving the ATM locus in B-cell chronic lymphocytic leukemia and related disorders. Clinico-biological significance. *Haematologica.* 2002,87:44-51.

[58] Stilgenbauer S, Liebisch P, James MR, Schroder M, Schlegelberger B, Fischer K, Bentz M, Lichter P, Dohner H. Molecular cytogenetic delineation of a novel critical genomic

region in chromosome bands 11q22.3-923.1 in lymphoproliferative disorders. *Proc. Natl. Acad. Sci. USA.* 1996,93:11837-11841.

[59] Monni O, Zhu Y, Franssila K, Oinonen R, Hoglund P, Elonen E, Joensuu H, Knuutila S. Molecular characterization of deletion at 11q22.1-23.3 in mantle cell lymphoma. *Br. J. Haematol.* 1999,104:665-671.

[60] Schaffner C, Stilgenbauer S, Rappold GA, Dohner H, Lichter P. Somatic ATM mutations indicate a pathogenic role of ATM in B-cell chronic lymphocytic leukemia. *Blood.* 1999;94:748-753.

[61] Stilgenbauer S, Winkler D, Ott G, Schaffner C, Leupolt E, Bentz M, Moller P, Muller-Hermelink HK, James MR, Lichter P, Dohner H. Molecular characterization of 11q deletions points to a pathogenic role of the ATM gene in mantle cell lymphoma. *Blood.* 1999;94:3262-3264.

[62] Schaffner C, Idler I, Stilgenbauer S, Dohner H, Lichter P. Mantle cell lymphoma is characterized by inactivation of the ATM gene. *Proc. Natl. Acad. Sci. USA.* 2000;97:2773-2278.

[63] Khanna KK, Jackson SP. DNA double-strand breaks: signaling, repair and the cancer connection. *Nat. Genet.* 2001,27:247-254.

[64] Eclache V, Caulet-Maugendre S, Poirel HA, Djemai M, Robert J, Lejeune F, Raphael M. Cryptic deletion involving the ATM locus at 11q22.3 approximately q23.1 in B-cell chronic lymphocytic leukemia and related disorders. *Cancer Genet. Cytogenet.* 2004;152:72-76.

[65] Stankovic T, Stewart GS, Fegan C, Biggs P, Last J, Byrd PJ, Keenan RD, Moss PA, Taylor AM. Ataxia telangiectasia mutated-deficient B-cell chronic lymphocytic leukemia occurs in pregerminal center cells and results in defective damage response and unrepaired chromosome damage. *Blood.* 2002,99:300-309.

[66] Canman CE, Lim DS. The role of ATM in DNA damage responses and cancer. *Oncogene.* 1998;17:3301-3308.

[67] Juliusson G, Robert KH, Ost A, Friberg K, Biberfeld P, Nilsson B, Zech L, Gahrton G. Prognostic information from cytogenetic analysis in chronic B-lymphocytic leukemia and leukemic immunocytoma. *Blood.* 1985;65:134-141.

[68] Bird ML, Ueshima Y, Rowley JD, Haren JM, Vardiman JW. Chromosome abnormalities in B cell chronic lymphocytic leukemia and their clinical correlations. *Leukemia.* 1989;3:182-191.

[69] Anastasi J, Le Beau M, Vardiman J, Fernald AA, Larson RA, Rowley JD. Detection of trisomy 12 in chronic lymphocytic leukemia by fluorescence in situ hybridization to interphase cells: A simple and sensitive method. *Blood.* 1992;79:1796-1801.

[70] Escudier SM, Pereira-Leahy JM, Drach JW, Weier HU, Goodacre AM. Fluorescent in situ hybridization and cytogenetic studies of trisomy 12 in chronic lymphocytic leukemia. *Blood.* 1993;81:2702-2707.

[71] Matutes E, Oscier D, Garcia-Marco J, Ellis J, Copplestone A, Gillingham R, Hamblin T, Lens D, Swansbury GJ, Catovsky D. Trisomy 12 defines a group of CLL with atypical morphology: correlation between cytogenetic, clinical and laboratory features in 544 patients. *Br. J. Haematol.* 1996;92:382-388.

[72] Crossen PE, Horn HL. Origin of trisomy 12 in B-cell chronic lymphocytic leukemia. *Cancer Genet. Cytogenet.* 1987;28:185-186.

[73] Que TH, Garcia-Marco J, Ellis J, Matutes E, Brito-Babapulle V, Boyle S, Catovsky. Trisomy 12 in chronic lymphocytic leukemia detected by fluorescence in situ hybridization: Analysis by stage, immunophenotype, and morphology. *Blood.* 1993;82:571-575.

[74] Athanasiadou A, Stamatopoulos K, Tsompanakou A, Gaitatzi M, Kalogiannidis P, Anagnostopoulos A, Fassas A, Tsezou A. Clinical, immunophenotypic, and molecular profiling of trisomy 12 in chronic lymphocytic leukemia and comparison with other karyotypic subgroups defined by cytogenetic analysis. *Cancer Genet. Cytogenet.* 2006; 168:109-119.

[75] Dierlamm J, Wlodarska I, Michaux L, Vermeesch JR, Meeus P, Stul M, Criel A, Verhoef G, Thomas J, Delannoy A, Louwagie A, Cassiman JJ, Mecucci C, Hagemeijer A, Van den Berghe H. FISH identifies different types of duplications with 12q13-15 as the commonly involved segment in B-cell lymphoproliferative malignancies characterized by partial trisomy 12. *Genes Chrom. Cancer.* 1997;20:155-166.

[76] Merup M, Juliusson G, Wu X, Jansson M, Stellan B, Rasool O, Roijer E, Stenman G, Gahrton G, Einhorn S. Amplification of multiple regions of chromosome 12, including 12q13-15, in chronic lymphocytic leukaemia. *Eur. J. Haematol.* 1997;58:174-180.

[77] Wessendorf S, Schwaenen C, Kohlhammer H, Kienle D, Wrobel G, Barth TF, Nessling M, Moller P, Dohner H, Lichter P, Bentz M. Hidden gene amplifications in aggressive B-cell non-Hodgkin lymphomas detected by microarray-based comparative genomic hybridization. *Oncogene.* 2003;22:1425-1429.

[78] Schwaenen C, Nessling M, Wessendorf S, Salvi T, Wrobel G, Radlwimmer B, Kestler HA, Haslinger C, Stilgenbauer S, Dohner H, Bentz M, Lichter P. Automated array-based genomic profiling in chronic lymphocytic leukemia: development of a clinical tool and discovery of recurrent genomic alterations. *Proc. Natl. Acad. Sci. USA.* 2004,101:1039-1044.

[79] Haslinger C, Schweifer N, Stilgenbauer S, Dohner H, Lichter P, Kraut N, Stratowa C, Abseher R. Microarray gene expression profiling of B-cell chronic lymphocytic leukemia subgroups defined by genomic aberrations and VH mutation status. *J. Clin. Oncol.* 2004;22:3937-3949.

[80] Kienle DL, Korz C, Hosch B, Benner A, Mertens D, Habermann A, Krober A, Jager U, Lichter P, Döhner H, Stilgenbauer S. Evidence for distinct pathomechanisms in genetic subgroups of chronic lymphocytic leukemia revealed by quantitative expression analysis of cell cycle, activation, and apoptosis-associated genes. *J. Clin. Oncol.* 2005; 23:3780-3792.

[81] Buhl AM, Jurlander J, Jorgensen FS, Ottesen AM, Cowland JB, Gjerdrum LM, Hansen BV, Leffers H. Identification of a gene on chromosome 12q22 uniquely overexpressed in chronic lymphocytic leukemia. *Blood.* 2006;107:2904-2911.

[82] Offit K, Louie DC, Parsa NZ, Filippa D, Gangi M, Siebert R, Chaganti R. Clinical and morphologic features of B-cell small lymphocytic lymphoma with del(6)(q21q23). *Blood.* 1994; 83:2611-2618.

[83] Bigoni R, Cuneo A, Roberti MG, Bardi A, Rigolin GM, Piva N, Scapoli G, Spanedda R, Negrini M, Bullrich F, Veronese ML, Croce CM, Castoldi GL. Chromosome aberrations in atypical chronic lymphocytic leukemia: a cytogenetic and interphase cytogenetic study. *Leukemia.* 1997; 11:1933-1940.

[84] Gaidano, RS Hauptschein, NZ Parsa, K Offit, PH Rao, G Lenoir, DM Knowles, RS Chaganti and R Dalla-Favera. Deletions involving two distinct regions of 6q in B-cell non-Hodgkin lymphoma. *Blood.* 1992; 80:1781-1787.

[85] Menasce LP, Orphanos V, Santibanez-Koref M, Boyle JM, Harrison CJ. Common region of deletion on the long arm of chromosome 6 in non-Hodgkin's lymphoma and acute lymphoblastic leukaemia. *Genes Chromosomes Cancer.* 1994;10 :286-288.

[86] Zhang Y, Matthiesen P, Harder S, Siebert R, Castoldi G, Calasanz MJ, Wong KF, Rosenwald A, Ott G, Atkin NB, Schlegelberger B. A 3-cM commonly deleted region in 6q21 in leukemias and lymphomas delineated by fluorescence in situ hybridization. *Genes Chrom. Cancer.* 2000; 27:52-58.

[87] Merup M, Moreno TC, Heyman M, Ronnberg K, Grander D, Detlofsson R, Rasool O, Liu Y, Soderhall S, Juliusson G, Gahrton G, Einhorn. 6q deletions in acute lymphoblastic leukemia and non-Hodgkin's lymphomas. *Blood.* 1998; 91:3397-3400.

[88] Juliusson G, Oscier DG, Fitchett M, Ross FM, Stockdill G, Mackie MJ, Parker AC, Castoldi GL, Guneo A, Knuutila S, Elonen E, Gahrton G. Prognostic subgroups in B-cell chronic lymphocytic leukemia defined by specific chromosomal abnormalities. *N. Engl. J. Med.* 1990;323:720-724.

[89] Barbouti A, Stankiewicz P, Nusbaum C, Cuomo C, Cook A, Hoglund M, Johansson B, Hagemeijer A, Park SS, Mitelman F, Lupski JR, Fioretos T. The breakpoint region of the most common isochromosome, i(17q), in human neoplasia is characterized by a complex genomic architecture with large, palindromic, low-copy repeats. *Am. J. Hum. Genet.* 2004;74:1-10.

[90] Michaux L, Dierlamm J, Wlodarska I, Bours V, Van Den Berbhe H, Hagemeijer A. t(14;19)/BCL3 rearrangements in lymphoproliferative disorders: A review of 23 cases: *Cancer Genet. Cytogenet.* 1997;94:36-43.

[91] Dyer MJ, Zani VJ, Lu WZ, O'Byrne A, Mould S, Chapman R, Heward JM, Kayano H, Jadayel D, Matutes E. BCL2 translocations in leukemias of mature B cells. *Blood.* 1994;83:3682-3688.

[92] Chena C, Cerretini R, Noriega MF, Narbaitz M, Scolnk M, Palacios MF, Neme D, Bruno S, Slavutsky I. Cytogenetic, FISH and molecular studies in a case of B-cell chronic lymphocitic leukemia with karyotypic evolution. *Eur. J. Haematol.* 2002; 69:309-314.

[93] Seiler T, Dohner H, Stilgenbauer S. Risk stratification in chronic lymphocytic leukemia. *Semin. Oncol.* 2006; 33:186-194.

[94] Dohner H, Stilgenbauer S, Dohner K, Bentz M, Lichter P. Chromosome aberrations in B-cell chronic lymphocytic leukemia: reassessment based on molecular cytogenetic analysis. *J. Mol. Med.* 1999;77:266-81.

[95] Pittman S, Catovsky D. Prognostic significance of chromosome abnormalities in chronic lymphocytic leukaemia. *Br. J. Haematol.* 1984;58:649-660.

[96] Oscier DG, Stevens J, Hamblin TJ, Pickering RM, Lambert R, Fitchett M. Correlation of chromosome abnormalities with laboratory features and clinical course in B-cell chronic lymphocytic leukaemia. *Br. J. Haematol.* 1990;76:352-358.

[97] Gahrton G, Juliusson G. Clinical implication of chromosomal aberrations in chronic B-lymphocytic leukaemia cells. *Nouv. Rev. Fr. Hematol.* 1988;30:389-392.

[98] Glassman AB, Harper-Allen EA, Hayes KJ, Hopwood VL, Gutterman EE, Zagryn SP. Chromosome 6 abnormalities associated with prolymphocytic acceleration in chronic lymphocytic leukemia. *Ann. Clin. Lab. Sci.* 1998;28:24-29.

[99] Cuneo A, Rigolin GM, Bigoni R, De Angeli C, Veronese A, Cavazzini F, Bardi A, Roberti MG, Tammiso E, Agostini P, Ciccone M, Della Porta M, Tieghi A, Cavazzini L, Negrini M, Castoldi G. Chronic lymphocytic leukemia with 6q- shows distinct hematological features and intermediate prognosis. *Leukemia.* 2004; 18:476-483.

[100] Chevallier P, Penther D, Avet-Loiseau H, Robillard N, Ifrah N, Mahe B, Hamidou M, Maisonneuve H, Moreau P, Jardel H, Harousseau JL, Bataille R, Garand R. CD38 expression and secondary 17p deletion are important prognostic factors in chronic lymphocytic leukaemia. *Br. J. Haematol.* 2002;116:142-150.

[101] Jahrsdörfer B, Wooldridge JE, Blackwell SE, Taylor CM, Link BK, Weiner GJ. Good prognosis cytogenetics in B-cell chronic lymphocytic leukemia is associated in vitro with low susceptibility to apoptosis and enhanced immunogenicity. *Leucemia.* 2005;19:759-766.

[102] Mayr C, Speicher MR, Kofler DM, Buhmann R, Strehl J, Busch R, Hallek M, Wendtner CM. Chromosomal translocations are associated with poor prognosis in chronic lymphocytic leukemia. *Blood.* 2006;107:742-751.

[103] Harris NL, Jaffe ES, Stein H, Banks PM, Chan JK, Cleary ML, Delsol G, De Wolf-Peeters C, Falini B, Gatter KC, Grogan TM, Isaacson PG, Knowles DM, Manson DY, Muller-Hermelink HK, Pileri SA, Piris MA, Ralfkier E, Warnke RA. A revised European-American classification of lymphoid neoplasms: a proposal from the International Lymphoma Study Group. *Blood.* 1994;84:1361-1392.

[104] Harris NL, Jaffe ES, Diebold J, Flandrin G, Muller-Hermelink HK, Vardiman J, Lister TA, Bloomfield CD. The World Health Organization classification of neoplastic diseases of the hematopoietic and lymphoid tissues. Report of the Clinical Advisory Committee meeting, Airlie House, Virginia, November, 1997. *Ann. Oncol;* 1999; 10: 14191432.

[105] Cheson BD, Bennett JM, Grever M, Kay N, Keating MJ, O'Brien S, Rai KR. National Cancer Institute-sponsored Working Group guidelines for chronic lymphocytic leukemia: revised guidelines for diagnosis and treatment. *Blood.* 1996;87:4990-4997.

[106] Carbone PP, Kaplan HS, Musshoff K, Smithers DW, Tubiana M. Report of the Committee on Hodgkin's Disease Staging Classification. *Cancer Res.* 1971;31: 1860-1861.

[107] Rai KR, Sawitsky A, Cronkite EP, Chanana AD, Levy RN, Pasternack BS. Clinical staging of chronic lymphocytic leukemia. *Blood.* 1975;46:219-234.

[108] Cerretini R, Chena C, Giere I, Sarmiento M, Arrossagaray G, Rodriguez A, Perez Bianco R, de Dios Soler M, Narbaitz M, Slavutsky I. Structural aberrations of chromosomes 17 and 12 in chronic B-cell disorders. *Eur. J. Haematol.* 2003;71: 433-438.

[109] ISCN. An International System for Human Cytogenetic Nomenclature. In Mitelman F, ed. Basel; Karger, 1995.

[110] Chena C, Sarmiento M, Palacios MF, Scolnik M, Slavutsky I. Dup(12)(q13-q22) and 13q14 deletion in a case of B-cell chronic lymphocytic leukemia. *Acta Haematol.* 2000;104:197-201.

[111] Mitelman Database of Chromosome Aberrations in Cancer. Mitelman I, Johansson B, Mertens I; eds. Available at http//cgap.nci.nih.gov/Chromosomes/Mitelman. 2006.

[112] Carroll JA, Raimondi SC, Williams DL, Behm FG, Botowite M, Castelbery RP, Harris MB, Patterson D, Pullen DJ, Crist WM. tdic(9;12): a non random chromosome abnormality in childhood B-cell precursor acute lymphoblastic leukemia: a Pediatric Oncology Study group. *Blood.* 1987;6:1962-1965.

[113] Callet-Bauchu E, Salles G, Gazzo S, Poncet C, Morel D, Pages J, Coiffier B, Coeur P, Felman P. Translocations involving the short arm of chromosome 17 in chronic B-lymphoid disorders: frequent occurrence of dicentric rearrangements and possible association with adverse outcome. *Leukemia.* 1999;13:460-468.

[114] Rudolph KL, Millard M, Bosenberg MW, DePinho RA. Telomere dysfunction and evolution of intestinal carcinoma in mice and humans. *Nat. Genet.* 2001;28:155-159.

[115] Gisselsson D, Jonson T, Yu C, Martins C, Mandahl N, Wiegant J, Jin Y, Mertens F, Jin C. Centrosomal abnormalities, multipolar mitoses, and chromosomal instability in head and neck tumors with dysfunctional telomeres. *Br. J. Cancer.* 2002;87:202-207.

[116] Stewenius Y, Gorunova L, Jonson T, Larsson N, Hoglund M, Mandahl N, Mertens F, Mitelman F, Gisselsson D. Structural and numerical chromosome changes in colon cancer develop through telomere-mediated anaphase bridges, not through mitotic multipolarity. *Proc. Natl. Acad. Sci. USA.* 2005;102:5541-5546.

[117] Gisselsson D, Jonson T, Petersen A, Strombeck B, Dal Cin P, Hoglund M, Mitelman F, Mertens F, Mandahl N. Telomere dysfunction triggers extensive DNA fragmentation and evolution of complex chromosome abnormalities in human malignant tumors. *Proc. Natl. Acad. Sci. USA.* 2001;98:12683-12688.

[118] Travis LB, Curtis RE, Hankey BF, Fraumeni JF. Second cancers in patients with chronic lymphocytic leukemia. *J. Natl. Can. Inst.* 1992;84:1422-1427.

[119] Ohno T, Smir BN, Weisenburger DD, Gascoyne RD, Hinrichs SD, Chan WC. Origin of the Hodgkin/Reed-Sternberg cells in chronic lymphocytic leukemia with "Hodgkin´s transformation". *Blood.* 1998;91:1757-1761.

[120] Kanzler H, Küppers R, Helmes S, Wacker H-H, Chott A, Hansmann M-L, Rajewsky K. Hodgkin and Reed-Sternberg-like cells in B-cell chronic lymphocytic leukemia represent the outgrowth of single germinal–centre B-cell-derived clones: potential precursors of Hodgkin and Reed-Sternberg cells in Hodgkin´s disease. *Blood.* 2000; 95:1023-1031.

[121] Küppers R, Sousa AB, Baur AS, Strickler JG, Rajewsky K, Hansmann ML. Common germinal-centre B-cell origin of the malignant cells in two composite lymphomas, involving classical Hodgkin´s disease and either follicular lymphoma of B-CLL. *Mol. Med.* 2001;7:285-292.

[122] Döhner H, Fisher K, Bentz M, Hansen K, Benner A, Cabot G, Diehl D, Schlenk R, Coy J, Stilgenbauer S, Volkmann M, Galle PR, Poustka A, Hunstein W, Lichter P. p53 gene deletion predicts for poor survival and non-response to therapy with a purine analogs in chronic B-cell leukemias. *Blood.* 1995;85:1580-1589.

[123] Fegan C, Robinson H, Thompson P, Whittaker JA, White D. Karyotypic evolution in CLL: identification of a new sub-group of patients with deletions of 11q and advanced or progressive disease. *Leukemia.* 1995;9:2003-2008.

[124] Oscier D, Fitchett M, Herbert T, Lambert R. Karyotypic evolution in B-cell chronic lymphocytic leukaemia. *Genes Chrom. Cancer.* 1991;3:16-20.

[125] Haus O. The genes of interferons and interferon-related factors: localization and relationships with chromosome aberrations in cancer. *Arch. Immunol. Ther. Exp.* 2000;48:95-100.

[126] Kotoula V, Hytiroglou P, Kaloutsi V, Barbanis S, Kouidou S, Papadimitriou CS. Mismatch repair gene expression in malignant lymphoproliferative disorders of B-cell origin. *Leuk. Lymphoma.* 2002;43:393-399.

[127] Geisler CH, Philip P, Christensen BE, Hou-Jensen K, Pedersen NT, Jensen OM, Thorling K, Andersen E, Birgens HS, Drivsholm A, Ellegaard J, Larsen JK, Plesner T, Brown P, Andersen PK, Hansen MM. In B-cell chronic lymphocytic leukaemia chromosome 17 abnormalities and not trisomy 12 are the single most important cytogenetic abnormalities for the prognosis: a cytogenetic and immunophenotypic study of 480 unselected newly diagnosed patients. *Leuk. Res.* 1997;21:1011-1023.

[128] Wicks IP, Lapsys NM, Baker E, Campbell LJ, Boyd AW, Sutherland GR. Localization of a human receptor tyrosine kinase (ETK1) to chromosome region 3p11.2. *Genomics.* 1994;19:38-41.

[129] Jain AN, Chin K, Borresen-Dale AL, Erikstein BK, Eynstein Lonning P, Kaaresen R, Gray JW. Quantitative analysis of chromosomal CGH in human breast tumors associates copy number abnormalities with p53 status and patient survival. *Proc. Natl. Acad. Sci. USA.* 2001; 98:7952-7957.

[130] Tran YK, Bogler O, Gorse KM, Wieland I, Green MR, Newsham IF. A novel member of the NF2/ERM/4.1 superfamily with growth suppressing properties in lung cancer. *Cancer Res.* 1999;59:35-43.

[131] Karkera JD, Ayache S, Ransome RJ Jr, Jackson MA, Elsayem AF, Sridhar R, Detera-Wadleigh SD, Wadleigh RG. Refinement of regions with allelic loss on chromosome 18p11.2 and 18q12.2 in esophageal squamous cell carcinoma. *Clin. Cancer Res.* 2000; 6:3565-3569.

[132] Offit K, Louie DC, Parsa NZ, Noy A, Chaganti RSK. Del(7)(q32) is associated with a subset of small lymphocytic lymphoma with plasmacytoid features. *Blood.* 1995;86:2365-2370.

[133] Oscier DG, Gardiner A, Mould S. Structural abnormalities of chromosome 7q in chronic lymphoproliferative disorders . *Cancer Genet. Cytogenet.* 1996;92:24-27.

[134] Andersen CL, Gruszka-Westwood A, Ostergaard M, Koch J, Jacobsen E, Kjeldsen E, Nielsen B. A narrow deletion of 7q is common to HCL, and SMZL, but not CLL. *Eur. J. Haematol.* 2004;72:390-402.

[135] Hebert J, Jonveaux P, Dágay M-F, Berger R. Cytogenetic studies in patients with Richter´s syndrome. *Cancer Genet. Cytogenet.* 1994;73:65-68.

[136] Watanabe T, Hotta T, Ichikawa A, Kinoshita T, Nagai H, Uchida T, Murate T, Saito H. The MDM2 oncogene overexpression in chronic lymphocytic leukemia and low-grade lymphoma of B-cell origin. *Blood.* 1994;84:3158-3165.

[137] Huang YQ, Raphael B, Buchbinder A, Li, JJ, Zhang WG, Friedman-Kien AE. Rearrangement and expression of MDM2 oncogene in chronic lymphocytic leukemia. *Am. J. Hematol.* 1994;47:139-141.

[138] Wiley JS, Dao-Ung LP, Gu BJ, Sluyter R, Shemon AN, Li C, Taper J, Gallo J, Manoharan A. A loss-of-function polymorphic mutation in the cytolytic P2X7 receptor

gene and chronic lymphocytic leukaemia: a molecular study. *Lancet.* 2002; 359:1114-1119.

[139] Nuckel H, Frey UH, Durig J, Duhrsen U, Siffert W.1513A/C polymorphism in the P2X7 receptor gene in chronic lymphocytic leukemia: absence of correlation with clinical outcome. *Eur. J. Haematol.* 2004;72:259-263.

[140] La Starza R, Barba G, Matteucci C, Crescenzi B, Romoli S, Pierini V, Beacci D, Cantaffa R, Martelli MF, Mecucci C. Chronic lymphocytic leukaemia Is terminal del(14)(q24) a new marker for prognostic stratification? *Leuk. Res.* 2006; Apr 29. Epub ahead of print.

In: Genetic Translocations and Other Chromosome Aberrations
Editor: Guy T. Leyden, pp. 253-271
ISBN: 978-1-60456-344-3
© 2008 Nova Science Publishers, Inc.

Chapter 10

Constitutional Chromosome Aberrations in Hematologic Disorders: Evaluation of 4884 Patients from a Sole Institution

Roxana Cerretini[1,2*], Susana Acevedo[1], Christian Chena[1], Carolina Belli[1], Irene Larripa[1] and Irma Slavutsky[1**]

[1]Departamento de Genética, Instituto de Investigaciones Hematológicas "Mariano R. Castex," Academia Nacional de Medicina, Buenos Aires, Argentina
[2]Centro Nacional de Genética Médica, Buenos Aires, Argentina

Abstract

We have reviewed 4884 patients with different hematologic disorders, cytogenetically studied in our laboratory during the last 27 years, in order to analyze the frequency of constitutional chromosome aberrations (CCA) and to evaluate their association with hematologic malignancies. Our population included 1235 pediatric patients and 3649 adults. Twenty-six (0.53%) cases showed CCA. They included 4 patients with Robertsonian translocations, one patient with a balanced translocation, 2 patients with sex chromosome abnormalities and 19 cases with Down´s Syndrome (DS). Non significant differences between the frequency of patients with CCA from our hematologic series and those observed in the two largest combined published surveys of liveborns (0.65%-0.78%) were found. The incidence of DS patients in our population (0.38%) was about three times higher than that seen at birth (0.12%-0.14%) ($p<0.001$). The total incidence of constitutional chromosome abnormalities in the non DS hematologic patients was 0.14% (7/4884) lower than that observed in the newborn population (0.53%-0.64%) ($p<0.001$), supporting the idea that DS patients are at greater risk of developing acute leukemia than

* E-mail address: rcerretini@Argentina.com. Correspondence to: Roxana Cerretini, Departamento de Genética, Instituto de Investigaciones Hematológicas "Mariano R.Castex", Academia Nacional de Medicina, Pacheco de Melo 3081, 1425 - Buenos Aires, Argentina, Phone: (5411) 4805-5759 ext.241/291, FAX: (5411) 4803-9475.
** E-mail address: islavutsky@hematologia.anm.edu.ar

normal subjects of the same age. Non-significant differences were found when the incidences of structural aberrations and sex chromosome anomalies were individually compared with the data of the overall population. Moreover, our hematologic population was stratified in adults (over 17 years of age), children (1-17 years of age) and infants (0-1 year of age). Higher frequencies of chromosome aberrations in children and infants compared to adults ($p<0.0001$) and in infants respect to children ($p<0.05$) were found. These differences were related to the DS patients that showed a clear different distribution by age: infants had the highest incidence compared to adults ($p<0.0001$) and children ($p<0.03$). In conclusion, our results did not show significant differences in the incidence of individuals with CCA compared to the overall population, suggesting that the presence of a CCA, other than DS, would not predispose hematologic malignancies.

Introduction

It is known that genes are major determinants for human variations and that the influence of genetically determined characteristics span in a continuum fashion, that may be nil in one extreme to lethal at the other. Different mechanisms seem to be involved in determining a trait. One or two alleles may be important, but more commonly genes interact with others and with one or more environmental factors. This interaction between genetic and environmental factors is now thought to be the origin for numerous diseases that were previously believed to have only environmental etiology. As known, genetic susceptibility for several carcinogens has now been identified (e.g., bladder cancer in aniline dye workers who are slow acetylators) [1].

There are different types of genetic diseases: a) chromosomal disorders, b) single gene disorders, c) multifactorial disorders and d) somatic cell genetic disorders [1, 2]. The determination of frequencies for each type is not easy and requires a large sample study of a well defined population. One of the first attempts to estimate the frequency of genetic disease was the creation, in 1952, of the British Columbia Health Surveillance Registry [3]. This Registry used multiple sources of ascertainment in a consecutive series of more than one million livebirths and constituted the largest combined survey of liveborns. The results of these studies showed that before the age of 25 years, 53 of 1000 liveborns presented a genetic disease, corresponding 3.6 per 1000 to single gene disorders (1.4 autosomal dominant, 1.7 autosomal recessive, and 0.5 X-linked recessive); 1.8 per 1000 to chromosomal disorders; 46.4 per 1000 to multifactorial disorders, and 1.2 per 1000 to genetic type uncertain disorders. Although this analysis included several sources of ascertainment, mild conditions or inherited traits without medical significance, materials of pregnancy loss, stillbirths and diseases of middle and old age were not considered. Therefore, these frequencies represent minimum estimates.

Chromosomal disorders represent a group of genetic diseases with visible alterations in the number or structure of the chromosomes that correspond to constitutional rearrangements that are inherited from a carrier parent or occur *de novo* in the gametes that result in the zygote [1]. Several newborn surveys, some of them including material from pregnancy loss, stillbirths and/or data derived from prenatal diagnosis studies, have been done in order to know the incidence of constitutional chromosome aberrations (CCA) in the overall population. These multiple cytogenetic analysis, using routine light microscopy (including non-banded and banded studies) reveal that 0.60-0.92% (1/167-1/109) of newborns have a major chromosomal abnormality [4-15]. Particularly, the general population incidence reported by Hsu [16] consists of two large series of livebirths in which the relevant incidences of all CCA can be individually analyzed. One of these series, which include only liveborn population with 0.65% of CCA frequency, shows that around 0.44% of all newborn infants have an imbalance chromosomal rearrangement (including 14% structural unbalanced rearrangements, 32% autosomal trisomies and 54% sex chromosome anomalies) and, the remaining 0.21% infants present structural balanced abnormalities [16]. Moreover, the frequency of CCA in different periods of age, before 25 years (mostly livebirth), after 25 years and during the lifetime has been reported [1]. The analysis of these data shows that the frequency of CCA by 25 years per 1000 livebirths is 1.8/1000 as we have previously described, 2/1000 after 25 years, being the lifetime frequency of 3.8/1000 (this study excludes asymptomatic carriers of balanced structural rearrangements) [1].

It is well known that banding methods allow the identification of the chromosomes and the bands involved in rearrangements. However, some chromosomal anomalies affect segments smaller than the band resolution obtained by routine or high resolution karyotyping (2-5Mb). Thus, molecular cytogenetic methods are necessary to detect these submicroscopic abnormalities such microdeletions or microduplications, which would be missed by conventional microscopy [17]. Multiple microdeletion disorders have been defined in recent years and undoubtedly more will be discovered [18, 19]. In this way, the frequency of submicroscopic rearrangements is not known and their incidence in the overall population remains to be determined.

Structural Constitutional Chromosome Aberrations and Malignancy

In Western countries, the most frequent cause of death is related to somatic cell (cumulative) genetic disorders (cancer and cardiovascular diseases), which have a life time frequency of 240/1000 (assumes that all cancers represent cumulative genetic mutations and excludes the inherited single gene cancer syndromes) [1]. As known, all cancers are the result of accumulation of different genetic lesions as activation of proto-oncogenes, inactivation of tumor suppressor genes, or acquired chromosomal abnormalities (ACA). Particularly, ACA arise during development or during the life of a person. A large number of ACA, as reciprocal and balanced exchanges of chromosomal material between two chromosomes, have been described characterizing various human neoplastic diseases and some of them are used as a diagnostic criterion [20-22].

On the other hand, CCA in humans are diverse and unlike what happens with ACA, they are identified through the study of individuals with congenital anomalies, developmental delay or mental retardation, abnormalities of growth, different neoplasias, couples with recurrent miscarriages and prenatal studies.

Some CCA determine an increased risk of malignancy, mainly familial tumors in children and young people. A few congenital structural chromosome alterations are known to predispose to the development of neoplastic disorders: del(11)(p13) to Wilms' tumor (WT), del(13)(q14) to Retinoblastoma and t(3;8)(p14.2;q24) to renal cell carcinoma (RCC). WT is the most common renal neoplasm of childhood and may occur both sporadically (95%) or rarely as a familial, autosomal dominant disease [21, 23]. The most frequent CCA found in WT is a *de novo* deletion variable in size with an SRO (shortest region overlapping) in the midpoint of 11p13. In some cases, complex *de novo* CCA with or without visible deletion can also be observed. Particularly, in familial cases of WT, parental rearrangements are observed, mostly insertions, pericentric inversions and apparently balanced familial translocations with breakpoint in 11p [21, 24]. Different candidate genes for the pathogenesis of WT have been identified, including those at chromosomes 11p13, WT1 gene, linked to WAGR (WT, aniridia, genital anomalies and mental retardation) syndrome; 11p15, WT2 gene, linked to Beckwith-Wiedemann and 16q, WT3 gene [25-27]. Recurrent abnormalities of other loci, including 1p and 7p have indicated that these sites may harbor WT genes [27].

Retinoblastoma is an embryonic retinal tumor that occurs in infants or small children and may develop both sporadically (90%) or as an autosomal dominant disease. Different types of CCA can result in retinoblastoma through involvement of 13q14. The most common anomaly is a *de novo* interstitial deletion of variable length that always includes 13q14.2 and the adjacent part of 13q14.1, which constitute the SRO [28-30]. Its frequency is about 80-90% of all cytogenetic forms. *De novo* apparently balanced translocations with a breakpoint in or near 13q14 have also been reported. As for the familial occurrence of retinoblastoma, insertions are the most frequent CCA. In 1986, Friend et al. [31] were able to clone the RB1 gene, being the first tumor suppressor gene to be cloned, which maps within this SRO.

Different studies showed that inherited RCC segregated with a CCA. The first family with RCC and a constitutional t(3,8)(p21;q24) was described by Cohen et al. [32], as a translocation which was later re-examined by high resolution banding and reinterpreted as t(3;8)(p14.2;q24.1) [33]. Later, other constitutional translocations: t(3;12)(q13;q24), t(3;6)(p13;q25) and t(2;3)(q35;q21) were found [34-36]. Recently, an association of novel constitutional translocations t(1;3) (q32;q13.3) have been reported in inherited RCC patients [37]. Molecular genetic analysis of familial and non-familial cases of RCC revealed a critical role(s) for multiple genes on human chromosome 3, particularly the VHL gene, of which a role has been firmly established [38].

Other congenital chromosome aberrations such as Robertsonian or reciprocal balanced translocations have been found in patients with malignant disease [39-41]. Particularly the most common constitutional reciprocal translocation in man, the t(11;22)(q23;q11) [42, 43], may increase the risk of breast cancer [44] and carriers of this translocation are considered as a group at high risk for the development of this neoplasia. The finding of a significant association between breast cancer and the constitutional translocation t(11;22)(q23;q11) suggests the involvement in tumorigenesis of additional breast cancer gene(s) on 11q23, 22q11 or both and a study of larger series is required to confirm this idea.

Constitutional Sex Chromosome Abnormalities and Cancer Development

In regards to individuals with constitutional sex chromosome anomalies (structural and numerical), it is known that they have an increased risk of developing specific types of tumors. Klinefelter syndrome (KS) has a higher risk of developing germ cell tumors (GCT), the vast majority being extragonadal, particularly located in the mediastinum [45]. Compiled data from the literature demonstrate a frequency of KS among male patients with mediastinal GCT of at least 8%, or 50 times the expected frequency [46, 47]. In contrast to the many reports of extragonadal GCT, there have been only a few reports of testicular tumours [48-50]. Furthermore, an increased risk of breast cancer has also been reported in these patients

[51, 52]. On the other hand, Turner's syndrome (TS) has a strong association with gonadal tumors and an increased risk of developing neuroblatoma and related tumors such as pheochromocytoma, schwanommas, carcinoids and melanoma [53].

As for XYY patients, little is known about the incidence or type of neoplastic diseases associated with this sex chromosome abnormality. Several cases of malignant diseases have been reported in association with this syndrome, most of them concerned myeloid or lymphoid proliferations [20, 54, 55]. In addition, some growth and differentiation genes such as CSF2RA (α subunit of the granulocyte and macrophage colony-stimulating factor receptor), IL3RA (α subunit of the interleukin 3 receptor), and GCY (growth control, Y chromosome influenced) have been mapped on Yp11, suggesting that these genes might be related to the oncogenesis or to a growth promotion effect on the haematopoietic system [54].

Numerical Constitutional Chromosome Abnormalities and Malignant Diseases

The risk of neoplasms in patients with constitutional numeric chromosome anomalies is well known for trisomies of chromosomes 8, 9, 13, 18 and 21 [56]. These individuals have an excess of hematologic and various solid tumors compared to the general population. These tumors are observed during fetal and neonatal life, are incompletely developed and subsequently regress. Only a small proportion of solid tumors of trisomic children are fully developed during the fetal and neonatal periods. Embryonic tumors are the most frequent at this age but mesenchymal neoplasms, benign and malignant epithelial tumors, blood and lymphoid tumors, and germ cell tumors, have also been observed. The decreasing order of frequency of constitutional trisomies associated with these tumors is: trisomy 21, 18, 13, 8 and 9. These tumors more commonly affect particular organs for a given type of trisomy. Particularly, the neoplasms associated with trisomy 18 affect the kidney, liver, heart and gonads while those associated with trisomy 13 affect the kidney and adrenal gland and consist of neuroblastomas and carcinomas. Renal tumors and hematologic disorders are more common in patients with trisomies 8 and 9. In older children or even adults, tumors are less frequent, often found on the same organ as earlier life, more aggressive and do not involute. The mechanisms responsible for the development of a neoplasm in constitutional aneuploidy are still unknown; perhaps, the oncogenic effect could result from the excess of genetic material through a gene dosage effect. On the other hand, the specificity of a given trisomy for certain tissues will help to identify genes possibly involved in a particular neoplasia.

Particularly, Down syndrome is the most common human aneuploidy at birth (1/700-1000 live births) and results from trisomy of at least the distal long arm of chromosome 21. Trisomy of chromosome 21 is found in 95% of DS patients, except in rare instances of chromosome 21 translocation (4%-5%) or mosaicism (2%-4%) [57-59]. Trisomy 21 results from maternal meiosis I errors in 68%, maternal meiosis II errors in 20%, paternal meiosis I errors in 3%, paternal meiosis II errors in 4% and mitotic non disjunction in 5% of cases [2, 60]. DS patients have a 10-18 fold increased risk for leukemia. The types of leukemia vary with age, with a high frequency of a clonal leukemoid reaction in the newborn period, which usually spontaneously remits [61-63]. This transient myeloproliferative disorder (TMD), also known as transient leukemia affects 10% of DS neonates. Although spontaneous recovery, 25-30% of individuals with TMD will develop acute myelogenous leukemia (AML) with the

FAB M7 megakaryoblastic (AMKL) being the most common subtype in the first four years of life, some of them following a myelodysplastic phase [64-66]. DS neonates and children who do not develop TMD and/or AMKL may have other hematologic abnormalities including thrombocytosis, polycythemia, and/or macrocytosis suggesting a different form of disturbed erythro-megakaryocytopoiesis. Since M7 AML is relatively rare in the general population, the specific increase in risk for M7 AML among DS individuals is estimated at up to 400-fold. After the age of 3 years, acute lymphocytic leukemia (ALL) is predominant [63, 67]. Individuals mosaic for Down syndrome who develop leukemia or transient leukemoid reactions invariably have leukemic involvement of the trisomy 21 clones, suggesting that a locus or loci on this chromosome is responsible for it [63]. Moreover, different studies of DS show that they have an increased risk to develop other neoplasias. Non-Hodgkin lymphoma (NHL) is another hematological malignancy that is frequently found in DS [68, 69] and may represent the second most common malignancy after leukemia. On the contrary, the association of Hodgkin´s disease with DS has not been extensively documented, and only a few cases have been reported [70]. Testicular cancer, particularly testicular seminomas are found with an increased risk in DS. [71-74]. Although undescended testes is considered a risk factor for a testicular tumour and is more common than expected in DS, different studies suggest that the chromosome 21 gene expression might predispose to testicular cancer development in an independent fashion. Risk of penile cancer has also been described. Stomach, liver and gallbladder cancers were reported to occur in an increased risk in DS individuals, particularly, of older ages.

Constitutional trisomy 8 is extremely rare, and it occurs in approximately one in 25,000 livebirths. Pure trisomy 8 is usually not compatible with life, and, presumably owing to the lethality of the multiple congenital abnormalities, exists primarily in the mosaic state (CT8M) [75]. There are several case reports of trisomy 8 mosaic individuals developing aplastic anemia, myelodysplasia, and acute and chronic myelogenous leukemia. In all cases, the malignant clone being derived from the trisomy 8 cells [76]. This trisomy itself may be regarded as the first mutation in a multistep carcinogenetic process. The frequency of CT8M in hematological dysplastic and neoplastic disorders is approximately 15-20% according to data from Maserati et al. [77]. Moreover, acquired trisomy 8 is common in lympho- and myeloproliferative disorders [22]. Haas and Seyger [78] postulated that a significant number of trisomic neoplasms is derived from constitutional trisomic cells of meiotic origin, leading to a constitutional mosaic with the trisomic line limited to neoplastic cells. However, Seghezzi et al. [76] differ somewhat with this previous hypothesis in that the origin of trisomy 8 in neoplasms is not necessarily meiotic and may often be mitotic-postzygotic events. This CT8M may arise from a postzygotic mitotic non disjunction in an early embryonic multipotent cell, from which different tissues may originate.

Although the general population with CCA incidence has been entirely settled and several large series have been studied little has been reported about the frequency of CCA in patients with hematologic disorders. There are only a few reports of large series with different results. Alimena et al. [79] identified several CCA from consecutive series of 1400 patients with malignant hematologic disorders, studied in their own laboratories during the period 1975-1985. However, Benítez et al. [80] reviewed 5500 patients with different hematologic diseases, some of them (718) from their own laboratory during a period of 7 years (1978-1985) and the remaining one (4782) from an unselected series in the literature. In addition, the Groupe Francais de Cytogénétique Hématologique [81] studied 56 patients with CCA and malignant hematologic disorders, from 20 different cytogenetic laboratories. This was a cooperative and retrospective work.

We have been especially interested in evaluating the possible relationship between CCA and hematological disorders. For this purpose, in the present article that includes previous data [82], we have reviewed 4884 patients with different hematologic disorders citogenetically studied in our laboratory during the last 27 years (1975-2002) to analyze the frequency of CCA and to evaluate their association with hematologic malignancies. Moreover, it is significant to point out that our work constitutes the largest series of patients with hematological diseases from a sole institution studied so far.

Material and Methods

Bone marrow (BM) cells of 4884 patients with different hematologic diseases: acute leukemias (AL) (2052), myelodysplastic syndromes (MDS) (586), malignant lymphomas (ML) (749), and myeloproliferative disorders (MPD) (1497) were processed for cytogenetic analysis by short-term cultures (24-48 h) in RPMI-1640 medium supplemented with 15% fetal calf serum. GTG-banding technique was used for chromosome identification. Karyotype abnormalities were described using the International System for Human Cytogenetic Nomenclature (ISCN, 1995) [83]. Patients with a single chromosome alteration compatible with life were studied in phytohemagglutinin-stimulated cultures (72h) of peripheral blood lymphocytes (PBL) to establish the existence of a constitutional anomaly at diagnosis or during remission. Fluorescence in situ hybridization (FISH) using centromeric and whole chromosome painting probes was performed according to standard protocols.

Our population included 3649 adults (74.7%) and 1235 (25.3%) pediatric patients (0-17 years of age). DS was observed in 19 (0.38%) individuals. The sex ratio was 1.3:1 (male: female). The statistical analysis to compare both populations (hematologic versus overall population) has been performed with Fisher exact Test. This comparison was performed taking into account the frequencies found in the two largest surveys: They were: the combined study reported by Hsu [16] that included 68,159 livebirths that showed a total of 442 CCA (0.65%) and the investigation performed by Nielsen and Wohlert [14], in which chromosome examination was made in 34,910 newborns (included 34,891 liveborn children and 19 fetuses, who were aborted after prenatal chromosome study). The liveborn incidence of CCA observed in this study was of 0.78% (275/34,891). For statistical analysis, we have also stratified our hematologic population into three groups: adults over 17 years of age (Group 1) (3649; 74.7%), children (1-17 years of age; Group 2) (1114; 22.8%) and infants (0-1 year of age; Group 3) (121; 2.5%).

Results

Twenty-six out of 4884 (0.53%) cases showed CCA. They included four patients with Robertsonian translocations (RT) (Cases 1-4), one patient with a balanced translocation (Case 5), two patients with sex chromosome abnormalities (Cases 6 and 7) and 19 cases with DS (Cases 8-26). Clinical and cytogenetic data are summarized in Table 1.

Table 1. Patients with hematologic diseases and constitutional chromosome aberrations.

Patient	Sex/Age (yrs)	Hematologic disorder	Constitutional anomaly	Additional Karyotypic change(s)
	Robertsonian Translocation			
1	F/ 2.9	MDS	der(13;14)(q10;q10)	None
2	F/5	MDS	der(13;14)(q10;q10)	None
3	F/42	PV	der(14;21)(q10;q10)	None
4	M/60	AML	der(21;22)(q10;q10)	del(6)(q21), +mar
	Reciprocal Translocation			
5	M/54	CLL	t(9;20)(p12;p12)	a) +12 (FISH) b) +12, +22
	Sex Chrom Anomalies			
6	M/46	NHL	XXY*	None
7	F/62	NHL	XYY*	+22
	Down Syndrome			
8	F/17	ALL	+21	None
9	F/10	ALL	+21	None
10	F/14 d	ALL	+21	None
11	M/0.7	ALL	+21*	None
12	M/2	ALL	+21	Bq+,Dq+
13	F/17 d	ALL	+21	None
14	M/17	ALL	+21	t(14;14)(q32;q11), t(9;22)(q34;q11)
15	M/5	AML	+21	del(12)(p12)
16	F/2	AML	inv(9)(p12;q21),+21,	None
17	F/3	MDS	+21*	None
18	F/10	MDS	+21	None
19	M/1.83	MDS	+21	a) del(6)(q26),del(8)(q11q21),del (18)(p11). b) Idem, +3.
20	M/21	MDS	+21	None
21	F/11	MDS	+21	-9
22	M/3	MDS	+21*	del(17)(q21)
23	M/6	MDS	+21	None
24	M/1.5	MDS	+21	None
25	F/0.75	MDS	+21	None
26	F/0.92	MDS	+21	None

* Mosaic karyotype; a) First study; b) Second study; d: day; MDS: myelodysplastic syndrome; CLL: chronic lymphocytic leukemia; NHL: non-Hodgkin lymphoma; PV: polycythaemia vera; ALL: acute lymphoblastic leukemia; AML: acute myeloid leukemia.

Cases 1 and 2 with MDS had der(13;14)(q10;q10) and Case 3 with a diagnosis of Polycythaemia vera (PV), a MPD, showed der(14;21)(q10;q10). These cases did not show additional karyotypic aberrations. Case 4 presented MDS with AML evolution, der(21;22)(q10;q10) and the following acquired changes in the karyotype: del(6)(q21) and a non identified marker chromosome (mar).

Case 5 with a diagnosis of CLL showed a balanced reciprocal translocation t(9;20)(p12;p12), confirmed by FISH, with extra changes in the karyotype. The first study at diagnosis, showed a clone with trisomy 12 by FISH (12% of cells) and the second study, 1 year after treatment, showed a karyotype: 46,XY,t(9;20)(p12;p12)[19]/ 47,XY,t(9;20) (p12;p12), +22[4]. In addition, FISH study showed 5% of cells with trisomy 12.

Cases 6 and 7 with diagnosis of diffuse large B-cell lymphoma had mosaic karyotypes with sex chromosome anomalies. Case 6 showed a karyotype 46,XY[6]/47,XXY[15] in BM cells and 46,XY[2]/47,XXY[8] in PBL. The chromosome examination of Case 7 showed 46,XY[2]/47,XYY[10]/47,XYY,+22[2] in BM cells and 46,XY[4]/ 47,XYY[7] in PBL.

Sixteen of the 19 DS patients (84.2%) had non-mosaic constitutional karyotypes with trisomy 21, a value only slightly lower than the 93%-96% observed in the general DS population (84). Nine out of 19 (47.4%) DS cases presented acute leukemia (7 acute lymphoblastic leukemia and 2 AML) and 10 (52.6%) presented MDS. Additional structural and numerical changes in the karyotypes were observed in 6 patients (Cases 12, 14, 15, 19, 21, 22). Deletions were the most frequent rearrangements: del(6)(q26), del(8)(q11q21), del(12)(p12), del(18)(p11), del(17)(q21), followed by translocations: t(14;14)(q32;q11), t(9;22) (q34;q11) and numerical anomalies: +3, -9.

Table 2 shows the frequencies of patients with CCA in our hematologic population and in the two largest surveys reported [14, 16]. Non significant differences between the incidences of CCA patients (0.53%) of both our hematologic and the overall population (0.65%-0.78%) [16, 14] was found. DS patients represent 0.38% (19 of 4,884) of cases from this series. This incidence was approximately three times higher than of that seen at birth (0.12%-0.14%; $p<0.0001$). Conversely, the total incidence of CCA in the non-DS hematologic patients was 0.14% (7 of 4,884) lower than of that observed in the newborn population (0.53%-0.64%; $p<0.001$) [14, 16]. The incidence of balanced alterations was similar in both populations (hematologic and newborn) although, in hematologic patients, reciprocal translocations presented an incidence about four to seven times lower than those seen at birth; this difference was the only significant respect in the series reported by Nielsen and Wohlert (14) ($p<0.04$). Non-significant differences were found when the incidences of patients with sex chromosome anomalies were compared to the overall population.

Table 2. Frequencies of chromosome alterations in our hematologic population and the two largest combined surveys.

Type of Abnormality	Our Population of 4,884 patients		Combined Surveys of 68,159 liveborns [16]		Survey of 34,891 Newborns [14]	
	N° of Cases (%)	Rate	N° of Cases (%)	Rate	N° of Cases (%)	Rate
Total abnormalities	26 (0.53)	1/187	442 (0.65)	1/154	275 (0.78)	1/126
Robertsonian Translocations	4 (0.08) 1/	1221	62 (0.09)	1/1099	43 (0.12)	1/811
Reciprocal Translocations	1 (0.02)#	1/4884	64 (0.09)	1/1065	50 (0.14)	1/697
47,XXY	1 (0.03)	1/2784	45 (0.10)	1/969	27 (0.15)	1/638
47,XYY	1 (0.03)	1/2784	45 (0.10)	1/969	21 (0.18)	1/851
47,+21	19 (0.38)*	1/257	82 (0.12)	1/831	51 (0.14)	1/684

* Significant differences respect to the overall populations: $p<0.001$

Significant differences respect to the Nielson and Wohlert´s survey (14): $p<0.04$.

Table 3. Frequencies of chromosome alterations in our hematologic population stratified by age.

Type of abnormality	Our hematologic population (n=4884) Group 1(%) (n=3649)	Group 2 (%) (n=1114)	Group 3(%) (n=121)	Groups 2+3 (%) (n=1235)
Total of CCA	6 (0.16)	15 (1.34)*	5 (4.13)* #	20 (1.61)*
Robertsonian Translocations	2 (0.05)	2 (0.17)	0	2 (0.16)
Reciprocal Translocations	1 (0.03)	0	0	0
47,XXY	1 (0.04)	0	0	0
47,XYY	1 (0.04)	0	0	0
47,+21	1 (0.02)	13 (1.16)*	5 (4.13)*##	18 (1.45)*

* Significant differences respect to Group 1: p<0.0001
# Significant differences respect to Group 2: p<0.05.
Significant differences respect to Group 2: p<0.03

In addition, we have compared the incidence of CCA in our three stratified group population. Table 3 shows higher frequencies of chromosome aberrations in Groups 2 and 3 compared to Group 1 (p<0.0001), in Groups 2+3 respect to Group 1 (p<0.0001) and in Group 3 respect to Group 2 (p<0.05). When the distribution of each chromosome abnormality by group was analyzed in DS patients, a clear different distribution by age was found. Group 3 showed the highest incidence compared to Group 1 (p<0.0001) and Group 2 (p<0.03). Particularly, the incidence of CCA found in Group 1 was about 4-4.8 times lower than of that seen at birth (0.16% vs 0.65%-0.78%, p<0.001), whereas, Group 2, Group 3 and Group 2+3 presented incidences about 1.7-2, 5.3-6.4 and 2-2.5 times higher than the frequency of the overall population, respectively (1.34%, 4.13,% and 1.61% vs 0.65%-0.78%). Interestingly, 68.4% (13/19) of DS patients were found in Group 2, 26.3% (5/19) in Group 3 and only one (1/19) in Group 1. Thus, our results showed that the frequency for leukaemia varied substantantially with age, with a very high risk in infants and childhood.

Discussion

In an approach to know the incidence of CCA in the hematologic population, we cytogenetically studied 4,884 unselected patients with different hematologic disorders over the last 27 years. Our data suggest that the presence of CCA other than DS, would not predispose to hematologic malignancies.

The literature reports two large series of unselected patients with hematologic disorders in which the frequency of CCA were evaluated (Table 4). The first one, reported by Alimena et al. [79], including children and adults, found 1.5% cases with CCA (21/1400), 47.6% were DS, 23.8% had balanced translocation, 4 were RT, one patient had a translocation Y-autosome and the remaining 28.5% of the cases presented sex chromosome anomalies (structural and numerical). Particularly, all 10 DS patients (median age 4.5 years) had primary trisomy 21; 7 of them (median age 10 years) had ALL and the remaining ANLL (median age 2 years).

Table 4. Distribution of chromosome alterations in our hematologic population and two other series of the literature.

Type of abnormality	Present study (%)	Alimena et al. [79] (%)	Benitez et al. [80] (%)
Total of CCA	26/4884 (0.53)	21/1400 (1.5)	59/5500 (1.1)
Robertsonian Translocations	4/26 (15)	4/21 (19)	10/59 (16.9)
Reciprocal Translocations	1/26 (3.8)	1/21 (4.8)	7/59 (11.9)
47,XXY	1/26 (3.8)	1/21 (4.8)	3/59 (5.1)
47,XYY	1/28 (3.8)	2/21 (9.5)	6/59 (10.2)
47,+21	19/26 (73)	10/21 (47.6)	28/59 (47.5)
Other anomalies	-	3/21 (14.3)	5/59 (8.5)

Benitez et al. [80] found an incidence of 1.25% (9/718) of patients with CCA (95% adults), from their own laboratory, similar to the frequency observed by Alimena et al. [79]. Only one of their patients presented with Down´s syndrome and ANLL. From these data plus those obtained from an unselected series of the literature, they observed 1.1% frequency of CCA (59/5500), 47.5% were DS, 16.9% had RT, 11.9% with balanced translocations, 1.7% with an inversion, 3.4% with unbalanced chromosome aneuploidy different from DS and 18.6% presented numerical sex chromosome anomalies. Particularly, 82.1% (23/28) of DS patients were observed in children (1-17 years of age) (8 ANLL and 15 ALL) and 5 cases in infants (0-1 year of age) (4 ALL and 1 chronic myeloproliferative disorder). No adult patients with DS were found. The frequencies observed in both studies were approximately double of that observed in the newborn population (1.5%-1.1% vs 0.65%) and these differences were highly significant ($p<0.001$). However, when Benitez et al. [80] excluded the DS patients from the series, this difference disappeared, and the frequency of patient with CCA other than DS and hematological disorders was similar to that observed in the general population (0.57% vs. 0.53%). An analysis of the age stratified population showed that adults and general population had a similar incidence of congenital alterations, whereas children and infants presented highly significant differences. The incidence of balanced alterations and sex chromosome anomalies were similar in both populations (hematological and newborn) and although their frequency was double that seen at birth, these differences were not significant.

Additionally, another study reported in the literature by the Group Français de Cytogénétique Hématologique [81] describes 56 patients with CCA and malignant hematologic disorders. Among them, 44.6% (25/56) were DS patients, 11 cases (44%) with ALL (median age 5 years), eight patients (32%) with ANLL (median age 2 years), and six infants that developed TLR (transient leukemoid reaction) in the neonatal period. Three of them developed AL within 14-19 months following TLR and the other three children remained in remission. Their findings were consistent with the previous reports about DS and leukemia. The remaining patients presented: 16% sex chromosome aberrations, and 39.2% with Robertsonian, reciprocal translocation and inversions. In contrast with their DS results, their data were inconclusive with respect to the presence of CCA other than DS and a greater risk of hematologic malignancy.

Our study presented 26 patients with CCA and hematological disorders with an incidence of CCA similar to that seen in the overall population. Unlike the previous works [79-81], we had a greater incidence of DS patients with hematological disorders (73%; 19/26). Seven out of 19 DS patients (36.8%) developed ALL and the remaining 12 (63%) presented myeloid leukemia of DS (10 MDS and 2 AML), which is defined in these patients as a unique disorder that included MDS and AML with the same biological and clinical features [85]. Most of them (7/12, 58.3%) were under 5 years of age. This distribution of acute leukemias according to age is in concordance to that observed in the literature, which recently reports a rate of about fourfold higher for AML than for ALL in DS children under 5 years of age. Conversely, the rate for AML and ALL was similar in the age group 5-29 years [86]. Earlier studies had found a ratio of ALL to AML among DS and non-DS children under 15 years of age of 6.5 [20, 87-89].

Little is known about the mechanisms leading to the increased risk of leukemia in DS individuals. A higher risk of leukaemia has been reported in patients with DS and unusual karyotypes but there is no solid evidence for this link [90, 91]. Additionally, genetic mapping studies suggested an increase in the frequency of disomic homozygosity, especially in proximal 21q in DS individuals with transient leukemia and ANLL-M7 [92]. Futhermore, several genes located on chromosome 21 have been found disrupted in leukemia. One of them, the AML1 gene on chromosome 21q22 encoding a transcription factor, is involved in 25% of childhood ALL and 15% of childhood AML cases and is a critical regulator of normal haemopoiesis [93]. Moreover, a chromosome-21-encoded tumour invasion and metastasis factor called TIAM1 was found to increase in DS patients with AML [94]. However, further studies are necessary to clarify these data and/or to find other genes on chromosome 21 involved in the crucial step of leukemogenesis in DS. It is of interest to note that trisomy 21, the most frequent trisomy observed as a constitutional chromosome abnormality, is also one of the most common primary, single acquired autosomal trisomies in hematological disorders, more commonly associated to disease progression than diagnostic features [95].

Different reports refer, that AML and MDS in DS often show additional chromosomal abnormalities other than constitutional trisomy 21, those being seen in acute non-lymphoid leukemias (ANLL) higher than in lymphoid forms [96-99]. This data agree with our findings showing acquired chromosome aberrations in 33% of our DS patients with AML or MDS. Chromosome aberrations specific for ANLL and those for ALL were absent in most of our DS subjects.

Thus, when DS patients were excluded from our series, the incidence of CCA was significantly lower than of that observed in non-DS newborn population [14, 16]. An analysis of our age stratified population showed that the three groups presented significant differences respect to general population. Particularly, Groups 2 and 3 showed an increased incidence of CCA due to the presence of DS patients 86.7% of Group 2 and 100% of Group 3. These results agree with data from the literature showing that hematologic malignancies in DS most commonly occur during the first four years of life [85, 86].

Furthermore, the incidences of each sex chromosome anomalies and structural aberrations were individually analyzed. Our cases with sex chromosome abnormalities showed a diagnosis of diffuse large B cell non-Hodgkin lymphoma, an association that was rarely observed. In fact, a number of XYY and XXY males with hematologic malignancies have been reported, most of them show leukemia or MDS [20, 54]. Conversely, only sporadic cases with non-Hodgkin lymphomas were described [100-105]. The relation between the appearance of hematologic malignancies and constitutional sex chromosome pathologies (numerical and structural anomalies) is not yet well defined, whereas a definite increased incidence of epithelial tumors of the gonads has been well documented. When the incidences of sex chromosome abnormalities were compared with the overall population series of Benitez et al. [80] and Alimena et al., [79] they found an increased frequency of XXY and XYY patients compared to those seen at birth whereas our study showed lower incidences of these type of anomalies. However, both our results and previous reports did not find statistical differences with the overall population [14, 16]. These findings are in agreement with the hypothesis of an incidental occurrence of hematological disorders in these subjects [54, 106-107]. Moreover, evidence of leukemic transformation has been found in cytogenetically normal cell lines of patients with mosaicism for sex chromosome anomalies suggesting that an abnormal chromosome complement alone does not necessarily predispose to the neoplastic transformation [95].

Several case reports have described the simultaneous occurrence of carriers with constitutional balanced rearrangements and hematologic diseases [39-41, 108-110]. In most of them, it could not be determined whether a real association existed or if they represented random events. In our series of patients with CCA, four cases with Robertsonian translocations and only one patient with a reciprocal translocation were found: t(9;20)(p12;p12), which to our knowledge, has not been reported yet in the literature. Our findings and previous data of the literature [79-80] did not show differences in the incidence of balanced alterations with the overall population [14, 16], except when reciprocal translocations were compared with the data reported by Nielson and Wohlert [14]. Various reports suggest that CCA could affect genes that would be mutated during meiosis or somatic malignant transformation [108]. Thus, the chromosome rearrangements may not be sufficient for full neoplastic transformation and other acquired gene mutations may be necessary. This possibility could explain the penetrance of the disease in families with constitutional chromosome anomalies [41]. On the other hand, hereditary predisposition is quite uncommon for the malignancies that are associated with specific translocations. It may be that these translocations would be lethal to embryos carrying them in the germline. However, there can be inherited mutations in genes, which predispose to translocations [111].

In conclusion, the association between hematologic disorders and constitutional balanced alterations and sex chromosome anomalies can be random events. Our results confirmed that patients with DS have a high risk of developing leukemia, that ranged from 8.3-9.7 times more during childhood and to 29.5-34.4 during the first period of life. Thus, our data supports the hypothesis of no association between CCA, other than DS, and hematologic malignancies.

Acknowledgments

This work was supported by grants from: our National Research Council (CONICET), from the National Center of Medical Genetics and from the National Agency of Scientific and Technical Promotion (ANPCyT).

References

[1] Rimoin DL, Connor JM, Pyeritz RE. Nature and frequency of genetic disease. In: Rimoin DL, Connor JM, Pyeritz RE, eds. *Principles and practice of medical genetics*. Emery and Rimoin´s, Third Edition, 1996: 31-34.

[2] Mc Kinlay RJ, Sutherland GR. *Chromosome abnormalities and genetic counseling. Oxford monographs on medical genetics* No 29. Second Edition, 1996.

[3] Baird PA, Anderson TW, New Combe HB, Lowry RB. Genetic disorders in children and young adults: A population study. *Am J Hum Genet* 1988; 42: 677-693.

[4] Jacobs P, Meville M, Ratcliffe S, Keay AJ, Syme J. A cytogenetic survey of 11,680 newborn infants. *Ann Hum Genet* 1974; 37: 359-376.

[5] Sergovich F, Valentine GH, Chen ATL, Minch RA, Sment MS. Chromosome aberrations in 2159 consecutive newborn babies. *N Engl J Med* 1969; 280: 852-855.

[6] Hamerton LJ, Ray M, Abbott J, Williamson C, Ducasse GC. Chromosome studies in a neonatal population. *Canad Med Assoc J* 1972; 106: 776-779.

[7] Gerald P, Walzer S. Chromosome studies of normal newborn infants. In: Jacobs PA, ed. *Human population cytogenetics*. The University Press, Edinburgh, 1970: 143-151.

[8] Walzer S, Breau G, Gerald P. A chromosome survey of 2400 normal newborn infants. *J Pediat* 1969; 74: 438-448.

[9] Walzer S, Gerald P. Chromosome abnormalities in 11,154 newborn infants. *Am J Hum Genet* 1972; 24: 38a.

[10] Lubs HA, Rudler FH. Chromosomal abnormalities in the human population: Estimation of rates based on New Haven newborn study. *Science* 1970; 169: 495-497.

[11] Bochnov NP, Nuleshov NP, Chevotareu AN, Alekhin VI, Mudiam SA. Population cytogenetic investigation of newborns in Moscow. *Hum Genet* 1974; 22: 139-152.

[12] Friedrich V, Nielsen J. Chromosomes studies in 5049 consecutive newborn children. *Clin Genet* 1973; 4: 333-343.

[13] Nielsen J, Sillesen I. Incidence of chromosome aberrations among 11,149 newborn children. *Hum Genet* 1975; 30: 1-12.

[14] Nielsen J, Wohlert M. Chromosome abnormalities found among 34,910 newborn children: results from a 13-year incidence study in Århus, Denmark. *Hum Genet* 1991; 87: 81-83.

[15] Jacobs PA, Browne C, Gregson N, Joyse C, White H. Estimates of the frequency of chromosome abnormalities detectable in unselected newborns using moderate levels of banding. *J Med Genet* 1992; 20: 103-108.

[16] Hsu LYF. Prenatal Diagnosis of chromosomal abnormalities through amniocentesis. In: Milunsky A, ed. *Genetic Disorders and the fetus: Diagnosis, Prevention and Treatment.* Baltimore and London, The Johns Hopkins University Press, Fourth Edition, 1998: 179-248.

[17] Heng NHQ, Spyropoulos B, Moens PB. FISH technology in chromosome and genome research. *Bioassays* 1997; 19: 75-84.

[18] Flint J, Wilkie AOM, Buckle VJ, et al. The detection of subtelomeric chromosomal rearrangements in idiopathic mental retardation. *Nature Genet* 1995; 9: 132-139.

[19] Ledbetter DN, Ballabio A. Molecular cytogenetics of contiguous gene syndromes: Mechanisms and consequences of gene dosage imbalance. In: Shriver CR, Beaudet AL, Sly WS, Valle D, eds. *The metabolic and molecular basis of inherited diseases*. Mc Graw-Hill, New York, 1995: 811-839.

[20] Sandberg AA. *The chromosomes in human cancer and leukemia.* Second Edition, Elsevier Science Publishing Co., Inc, 1990.

[21] Heim S, Mitelman F. *Cancer Cytogenetics. Chromosomal and molecular genetic aberrations of tumor cells.* Wiley-Liss, Second Edition, 1995.

[22] Mitelman database of chromosome aberrations in cancer. Mitelman F, Johansson B, Mertens F, eds http://cgap.nci.nih.gov/chromosomes/Mitelman.
[23] Coppes MJ, Williams BRG. The molecular genetics of Wilms′ tumor. *Cancer Invest* 1994; 12: 57-65.
[24] de Grouchy J, Turleau C. Mendelian disorders and autosomal structural rearrangements. In: Daniel ART, ed. *The cytogenetics of mammalian autosomal rearrangements.* Alan Liss, Inc. 1998: 853-854.
[25] Hirose M. The role of Wilms′ tumor genes. *J Med Invest* 1999; 46: 130-140.
[26] Coppes MJ, Egeler RM. Genetics of Wilm′s tumor. *Sem Urol Oncol* 1999; 17: 2-10.
[27] Dome JS, Coppes MJ. Recent advances in Wilm′s tumor. *Curr Opin Pediatr* 2002; 14: 5-11.
[28] Knudson AG, Meadows AT, Nichols WW, Hill R. Chromosomal deletion and Retinoblastoma. *N Engl J Med* 1976; 295: 1120-1123.
[29] Yunis JJ, Ramsey N (1978): Retinoblastoma and suband deletion of chromosome 13. *Am J Dis Child* 132: 161-163.
[30] Galle BL, Squire JA, Goddard A, et al. Biology of disease. Mechanism of oncogenesis in retinoblastoma. *Lab Invest* 1990; 62: 394-407.
[31] Friend SH, Bernards R, Rogelj S, et al. Human DNA segment with properties of the gene that predisposes to retinoblastoma and osteosarcoma. *Nature* 1986; 323: 643-646.
[32] Cohen AJ, Li FP, Berg S, et al. Hereditary renal cell carcinoma associated with chromosomal translocation. *N Engl J Med* 1979; 301: 592-595.
[33] Wang N, Perkins KL. Involvement of band 3p14 in t(3;8) in hereditary renal carcinoma. *Cancer Genet Cytogenet* 1984; 11: 479-481.
[34] Kovacs G, Hoene E. Loss of der(3) in renal carcinoma cells of a patient with constitutional t(3;12). *Hum Genet* 1988; 78: 148-150.
[35] Kovacs G, Brusa P. Clonal chromosome aberrations in normal kidney tissue from patients with renal cell carcinoma. *Cancer Genet Cytogenet* 1989; 37: 289-290.
[36] Bodmer D, Eleveld MJ, Ligtenberg MJ, et al. An alternative route for multistep tumorigenesis in a novel case of hereditary renal cell cancer and a t(2;3)(q35;q21). *Am J Hum Genet* 1998; 62: 1475-1483.
[37] Kanayama H, Lui WO, Takahashi M, et al. Association of a novel constitutional translocation t(1q;3q) with familial renal cell carcinoma. *J Med Genet* 2001; 38: 165-170.
[38] Bodmer D, van den Hurk LU, van Groningen JJ, et al. Understanding familial and non-familial renal cell cancer. *Hum Mol Genet* 2002; 11: 2489-2498.
[39] De Greef I, Geraedts JPM, Leeksma CHW. Malignant hematologic disorders in two Robertsonian 13;14 translocation carriers. *Cancer Genet Cytogenet* 1982; 7: 181-184.
[40] Martín Ramos ML, Barreiro E, López-Pérez J, González-Aguilera, Fernández-Peralta. Acute Megakaryoblastic leukemia in a patient with a familial pericentric inversion of chromosome 8, inv(8)(p23.1q13). *Cancer Genet Cytogenet* 1998; 105: 74-78.
[41] Markkanen A, Ruutu T, Rasi V, Franssila K, Knuutila S, De la Chapelle A Constitutional translocation t(3;6)(p14;p11) in a family with hematologic malignancies. *Cancer Genet Cytogenet* 1987; 25: 87-95.
[42] Fraccaro M, Lindsten J, Ford CE. The 11q; 22q translocation: A European collaborative analysis of 43 cases. *Hum Genet* 1980; 156: 21-51.

[43] Iselius L, Lindsten J, Aurias A, et al. The 11q; 22q translocation: A collaborative study of 20 new cases and analysis of 110 families: *Hum Genet* 1983; 64: 343-355
[44] Lindblom A, Sandelin K, Iselius L, et al. Predisposition for breast cancer in carriers of constitutional translocation 11q; 22q. *Am J Hum Genet* 1994; 54: 871-876.
[45] Hasle H, Mellemgaard A, Nielse L, Hansen J Cancer incidence in men with Klinefelter Syndrome. *Brit J Cancer* 1995; 71: 416-420.
[46] Hasle H, Jacobsen BB, Asschenfeldt P, Andersen K Mediastinal germ cell tumor associated with Klinefelter Syndrome: a report of case and review of the literature. *Eur J Pediatr* 1992; 151: 735-739.
[47] Dere Non Court AN, Castro-Magana M, Janes KL Mediastinal teratoma and precocious puberty in a boy with mosaic Klinefelter Syndrome. *Am J Med Genet* 1995; 55: 38-42.
[48] Carroll PR, Morse MJ, Koduru PPK, Chaganti RSK. Testicular germ cell tumor in patient with Klinefelter syndrome. *Urology* 1988; 31: 72-74.
[49] Deyeus FH, Logothetis CJ, Chong C, Sella A, Ogden S. Genetic abnormalities in men with germ cell tumors. *J Urol* 1988; 140: 80-84.
[50] Reddy SR, Svec F, Richardson P. Seminoma of the testis in a patient with 48,XXYY variant of Klinefelter´s syndrome. *South Med J* 1991; 84: 773-775.
[51] Hultborn R, Hanson C, Kopf I, Verbiene I, Warnhammar E, Weimarck A. Prevalence of Klinefelter´s syndrome in male breast cancer patients. *Anticancer Res* 1997; 17: 4293-4297.
[52] Swerdlow AJ, Hermon C, Jacobs PA, et al. Mortality and cancer incidence in persons with numerical sex chromosome abnormalities: a cohot study. *Ann Hum Genet* 2001; 65: 177-188.
[53] Blatt J, Olshan AF, Lee PA, Ross JL Neuroblastoma and related tumors in Turner's syndrome. *J Pediatr* 1997; 131: 5, 666-670.
[54] Oguma N, Shigeta C, Kamada N. XYY male and hematologic malignancy. *Cancer Genet Cytogenet* 1996; 90:179-181.
[55] Limacher J-M, Girard-Lemaire F, Jeandidier E, et al. Gastrointestinal stromal tumor in an XYY/XY male. *Cancer Genet Cytogenet* 2002; 133: 152-155.
[56] Satge D, van den Berghe H. Aspects of the neoplasms observed in patients with constitutional autosomal trisomy. *Cancer Genet Cytogenet* 1996; 87: 63-70.
[57] Stoll C, Alembik Y, Dott B, Roth MP. Epidemiology of the Down syndrome in 118,265 consecutive births. *Am J Med Genet (Suppl)* 1990; 7: 79-83.
[58] Down syndrome prevalence at birth. United States, 1983-1990. *MMWR Morb Mortal WKLY Rep* 1994; 43: 617-622.
[59] Baird PA, Sadovnick AD. Causes of death to age 30 in Down syndrome. *J Hum Genet* 1988; 43: 239-248.
[60] Antorakis SE. 10 years of genomics, chromosome 21, and Down syndrome. *Genomics* 1998; 52: 1-16.
[61] Evans DIK, Steward JK. Down´s syndrome and leukemia. *Lancet* 1972; 2: 1322.
[62] Iselius I, Jacobs P, Morton N. Leukemia and transient leukemia in Down syndrome. *Hum Genet* 1990; 85: 477-485.
[63] Horwitz M. The genetics of familial leukemia. *Leukemia* 1997; 11: 1347-1359.
[64] Windham GC, Bjerkedal T, Langmark F. A population based study of cancer incidence in twins and in children with congenital malformations or low birth weight, Norway 1967-1980. *Am J Epidemiol* 1985; 121: 49-56.

[65] Zipursky A, Thoner P, de Harven E, et al. Myelodysplasia and acute megakaryoblastic leucemia in Down´s syndrome. *Leuk Res* 1994; 18: 163-171.

[66] Zipursky A, Brown E, Christersen H, Sutherland R, Doyle J. Leukemia and/or myeloproliferative syndrome in neonates with Down syndrome. *Sem Perinatol* 1997; 21: 97-101.

[67] Savasan S, Tave JW. Down syndrome and leukemia: What have we learned? *Pediatr Pathol Mol Med* 2000; 19: 269-281.

[68] Lorenzana AN, Schorin MA. Non-Hodgkin´s lymphoma in a neonate with Down´s syndrome. Case reports and literature review. *Am J Pediatr Hematol Oncol* 1989; 11: 186-190.

[69] Satge D, Le Tourneau A, Verger JP, et al. A case report of Down syndrome and centroblastic lymphoma. *Pathol Res Pract* 1996; 192: 1266-1269.

[70] Kusumakumary P, Jyothitmayi R, Chellam VG, Krishnan NM. Hodgkin´s disease in association with Down syndrome: A case report. *Pediatr Hematol Oncol* 1996; 13: 469-471.

[71] Dieckmann KP, Rube C, Henke RP. Association of Down's Syndrome and testicular cancer. *J Urol* 1997; 157:1701-1704.

[72] Satgé D, Sasco AJ, Curé H, Ledue B, Sommelet D, Vekemans MJ. An Excess of testicular germ cell tumors in Down's syndrome: Three case reports and a review of the literature. *Cancer* 1997: 80: 929-935.

[73] Satge D, Sommelet d, Geneix A, Nishi M, Malet C, Vekemans M. A tumor profile in Down syndrome. *Am J med Genet* 1998; 78: 207-216.

[74] Hill DA, Gridley G, Cnattingius S, et al. Mortality and cancer incidence among individuals with Down syndrome. *Arch Intern Med* 2003; 163: 705-711.

[75] Hasle H, Clausen N, Pedersen B, Bendix-Hansen K. Myelodysplastic syndrome in a child with constitutional trisomy 8 mosaicism and normal phenotype. *Cancer Genet Cytogenet* 1995; 79: 79-81.

[76] Seghezzi L, Maserati E, Minelli A, et al. Constitutional trisomy 8 as first mutation in multistep carcinogenesis: Clinical, cytogenetic, and molecular data on three cases. *Genes Chrom Cancer* 1996; 17: 94-101.

[77] Maserati E, Aprili F, Vinnante F, et al. Trisomy 8 in myelodysplasia and acute leucemia is constitutional in 15-20% of cases. *Genes Chrom Cancer* 2002; 33: 93-97.

[78] Haas OA, Seyger M. Meiotic origin of trisomic neoplasms. *Cancer Genet Cytogenet* 1993; 70: 112-116.

[79] Alimena G, Billstrom R, Casalone R, Gallo E, Mitelman F, Pasquali F. Cytogenetic pattern in leukemic cells of patients with constitutional chromosome anomalies. *Cancer Genet Cytogenet* 1985; 16: 207-218.

[80] Benítez J, Valcarcel E, Ramos C, Ayuso C, Cascos AS. Frequency of constitutional chromosome alterations in patients with hematologic neoplasias. *Cancer Genet Cytogenet* 1987; 24: 345-354.

[81] Groupe Francais de Cytogénétique Hématologique. Cytogenetic findings in leukemic cells of 56 patients with constitutional chromosome abnormalities: a cooperative study. *Cancer Genet Cytogenet* 1988; 35: 243-252.

[82] Cerretini R, Acevedo S, Chena C, Belli C, Larripa I, Slavutsky I. Evaluation of constitutional aberrations in hematologic disorders. *Cancer Genet Cytogenet* 2002; 134: 133-137.

[83] ISCN (1995): *An International System for Human Cytogenetic Nomenclature*, Mitelman F (ed); S. Karger, Basel, 1995.

[84] Hook EB. Down syndrome: frequency in human populations and factors pertinent to variation in rates. In: de la Cruz F, Gerald P. eds. *Trisomy 21 (Down Syndrome): research perspectives.* University Park Press, Baltimore, 1981: 3-67.

[85] Hasle H. Pattern of malignant disorders in individuals with Down´s syndrome. *Lancet Oncol* 2001; 2: 429-436.

[86] Hasle H, Clemmensen IH, Mikkelsen M. Risks of leukaemia and solid tumours in individuals with Down´s syndrome. *Lancet* 2000; 355: 165-169.

[87] Hasle H, Kerndrup G, Jacobsen BB. Chilhood myelodysplastic syndrome in Denmark: Incidence and predispose conditions. *Leukemia* 1995; 9: 1569-1572.

[88] Robison LL, Nesbit ME, Sather HN, et al. Down syndrome and acute leukemia in children: a 10 year retrospective survey from Children's Cancer Study Group. *J Pediatr* 1984; 105: 235-242.

[89] Robison L. Down syndrome and leukemia. *Leukemia* 1992; 6: 5-7.

[90] Iselius L, Jacobs P, Morton N. Leukaemia and transient leukaemia in Down´s syndrome. *Hum Genet* 1990; 85: 477-485.

[91] Taub JW, Huang X, Matherly LH, et al. Expression of chromosome 21-localized genes in acute myeloid leukemia: Differences between Down´s syndrome and non-Down´s syndrome blasts cells and relationship to *in vitro* sensitivity to cytosine arabinoside and daunorubicin. *Blood* 1999; 94: 1393-1400.

[92] Shen JJ, Williams BJ, Zipursky A, et al. Cytogenetic and molecular studies of Down´s syndrome individuals with leukemia: *Am J Hum Genet* 1995; 56: 915-925.

[93] Okuda T, Cai Z, Yang S, et al. Expression of knocked-in AML1-eto leukemia gene inhibits the establishment of normal definitive hematopoiesis and directly generates dysplastic hematopoietic progenitors. *Blood* 1998; 91: 3134-3143.

[94] Chen H, Antonarakis SE. Localization of a human homolog of the mouse TIAM-1 gene to chromosome 21q22.1. *Genomics* 1995; 30: 123-127.

[95] United Kingdom Cancer Cytogenetics Group (UKCCG). Primary, single autosomal trisomies associated with haematological disorders. *Leukemia Res* 1992; 16: 841-851.

[96] Alimena G, De Cuia MR, Diverio D, Mancini M, Montefusco E. Leukemias and constitutional chromosome pathology. *Haematologica* 1987; 72 (Suppl 6): 37-41.

[97] Mitelman F, Heim S. Mandhal N. Trisomy 21 in neoplastic cells. *Am J Med Genet Suppl* 1990; 7: 262-266.

[98] Hayashi Y, Hanada R, Yamamoto K. Chromosome abnormalities and prognosis in childhood acute leukemia. *Acta Paedriatr Jpn* 1991; 33: 497-506.

[99] Suarez CR, Le Beau MM, Silberman S, Fresco R, Rowley JD. Acute megakaryoblastic leukemia in Down's syndrome: report of a case and review of the cytogenetic findings. *Med Pediatr Oncol* 1985; 13: 225-231.

[100] Tsung SH, Heckman MG. Klinefelter syndrome, immunological disorders, and malignant neoplasm. *Arch Pathol* 1974; 98: 351-354.

[101] Becher R. Klinefelter´s syndrome and malignant lymphoma. *Cancer Genet Cytogenet* 1986; 21: 271-274.

[102] Attard-Montalto SP, Schuller I, Lastowska MA, Gibbons B, Kingston JE, Eden OB. Non-Hodgkin´s lymphoma and Klinefelter syndrome. *Pediatr Hematol Oncol* 1994; 11: 197-200.

[103] Humphreys M, Lavery P, Morris C, Nevin N. Klinefelter syndrome and non-Hodgkin lymphoma. *Cancer Genet Cytogenet* 1997; 97: 111-113.

[104] Ohyashiki K, Yoshida MA, Ohyashiki J, Block AMW, Dabski K, Sandberg AA. Chromosome changes in mycosis fungoides in an XYY male. *Cancer Genet Cytogenet* 1985; 18. 295-302.

[105] Sandlund JT, Raimondi SC. Small noncleaved cell lymphoma in an adolescent with the XYY syndrome. *Cancer Genet Cytogenet* 1997; 94: 88-90.

[106] Hasle H, Mellemgaard A, Nielse L, Hansen J. Cancer incidence in men with Klinefelter syndrome. *Br J Cancer* 1995; 71: 416-420.

[107] Okada H, Imai T, Itoh S, Isobe K, Onishi S. XYY male with essential throbocythemia in childhood. *Int J Hematol* 2000; 71: 55-58.

[108] Stanley WS, Burkett SS, Segel B, et al. Constitutional inversion of chromosome 7 and hematologic cancers. *Cancer Genet Cytogenet* 1997; 96: 46-49.

[109] Sreekantaiah CH, Han T, Baer MR, Sandberg AA. Acute nonlymphocytic leukemia in a patient with a constitutional inv(4). *Cancer Genet Cytogenet* 1989; 39: 119-123.

[110] Daibata M, Nemoto Y, Komatsu N, Machida H, Miyoshi I, Taguchi H. Constitutional t(3;11)(p21;q23) in a family, including one member with lymphoma: Establishment of novel cell lines with this translocation. *Cancer Genet Cytogenet* 2000; 117: 28-31.

[111] Knudson AG. Chasing the cancer demon. In: Campbell A, Anderson WW, Jones EW, eds, *Annual Review of Genetics*, 2000; 34: 1-19.

Index

A

B

C

D

E

H

I

J

K

L

M

N

O

P

R

S

T

U

V

W

X

Y

Z